Blue Planet, Red and Green Photosynthesis

SCIENCES

Biology, Field Director – Marie-Christine Maurel

Plantae and Photosynthesis, Subject Head –
Brigitte Gontero

Blue Planet, Red and Green Photosynthesis

Productivity and Carbon Cycling in Aquatic Ecosystems

Coordinated by
Stephen Christopher Maberly
Brigitte Gontero

iSTE

WILEY

First published 2022 in Great Britain and the United States by ISTE Ltd and John Wiley & Sons, Inc.

ISTE Ltd
27-37 St George's Road
London SW19 4EU
UK

www.iste.co.uk

John Wiley & Sons, Inc.
111 River Street
Hoboken, NJ 07030
USA

www.wiley.com

Library of Congress Control Number: 2021951334

British Library Cataloguing-in-Publication Data
A CIP record for this book is available from the British Library
ISBN 978-1-78945-082-8

ERC code:
PE10 Earth System Science
PE10_1 Atmospheric chemistry, atmospheric composition, air pollution
PE10_8 Oceanography (physical, chemical, biological, geological)
PE10_9 Biogeochemistry, biogeochemical cycles, environmental chemistry
PE10_17 Hydrology, water and soil pollution
LS8 Ecology, Evolution and Environmental Biology
LS8_1 Ecosystem and community ecology, macroecology
LS8_4 Evolutionary ecology
LS8_11 Marine biology and ecology

Contents

Preface

Stephen Christopher MABERLY[1] and Brigitte GONTERO[2]
[1] *Lake Ecosystems Group, UK Centre for Ecology & Hydrology, Lancaster, United Kingdom*
[2] *Aix-Marseille Université, CNRS, BIP UMR 7281, Marseille, France*

Over 70% of our blue planet is covered by water, and aquatic photosynthesis contributes about 50% of the total global net primary production. There are many differences between primary production in water and on land. Aquatic productivity is largely produced by a diverse array of eukaryotic algae from the red-lineage, as well as cyanobacteria from which the chloroplasts in the eukaryotes originated, while terrestrial photosynthesis is produced by green plants. Aquatic productivity is largely carried out by single cells or microscopic colonies, while terrestrial productivity is performed by multicellular, macroscopic plants. Another distinction between land and water is that on land about three-quarters of primary production results from passive diffusion of CO_2 into a leaf. In contrast, most aquatic productivity relies on active transfer of carbon to the site of CO_2 fixation by the primary carboxylation enzyme (ribulose-1,5-bisphosphate carboxylase-oxygenase, Rubisco) by the so-called CO_2 concentrating mechanisms (CCMs). This helps to overcome carbon limitation caused by lower rates of CO_2 diffusion in water compared to air and depends primarily on the active exploitation of bicarbonate, a source of inorganic carbon that is present in water but absent in air. The diversity of mechanisms employed by different phylogenetic groups of aquatic photoautotrophs to take up bicarbonate and concentrate CO_2 inside their cells, thereby supporting productivity, is a particular focus of this book.

The geochemistry of planet Earth has been fundamentally altered by the evolution of life, perhaps about four billion years ago, and especially by the evolution of aquatic oxygenic photosynthetic organisms, over 2.4 billion years ago.

These organisms are able to split water in order to extract hydrogen that is needed to reduce carbon dioxide (CO_2) to organic compounds, releasing oxygen as a by-product. This transformed the Earth from an anoxic planet with high CO_2 concentrations to an oxic planet with low CO_2 concentrations. Atmospheric O_2 allowed the production of ozone in the atmosphere that absorbs damaging ultraviolet radiation, so facilitating the evolution of life on land. Recently, human activity has increased the release of CO_2 to the atmosphere, of which about a quarter to a third has been absorbed in the oceans. Nevertheless, the global carbon cycle has been disrupted, triggering critical climate change problems for life on Earth. While the oceans play an important role in regulating atmospheric gas composition, particularly that of CO_2, inland waters, despite their small global area, form key connections for greenhouse gas fluxes between the land, atmosphere and ocean. Therefore, understanding the processes involved in aquatic photosynthesis and productivity is fundamental to the ability to forecast future changes in the global carbon cycle.

This book has been written by 20 authors from eight countries who present the most up to date information on their topic in a way that is accessible to non-specialists. The chapters cover timelines from the evolution of photosynthesis to current issues of environmental change, spatial scales from sub-cellular to global, and disciplines encompassing physics, chemistry, geology, biology and ecology. For each chapter to be read independently, some material has necessarily been repeated, but in order to produce a unified and structured book, we have cross-referenced links between the different chapters.

The first three chapters set the scene by presenting information about photosynthesis, carbon cycling and the evolution and diversity of aquatic photoautotrophs. In Chapter 1, Gontero, Lenton and Maberly outline the key processes involved in photosynthesis: the capture of light energy and fixation of inorganic carbon into organic carbon by Rubisco. They also give an overview of the global carbon cycle, the changes that have occurred over geological time and its response to the recent increase in atmospheric CO_2. Raven (Chapter 2) traces the evolution of aquatic photoautotrophs and the inheritance and development of different aspects of the photosynthetic machinery from prokaryotes to eukaryotes including freshwater plants and seagrasses. In Chapter 3, Pierella Karlusich, Nef, Bowler and Dorrell present the very recent insights on the evolution, phylogenetic relationships and biogeography of marine phytoplankton provided by next generation sequencing and extensive environmental sampling.

The next four chapters describe different aspects of the aquatic CCM. In Chapter 4, Rokitta, Kranz and Rost describe how problems of inorganic carbon supply are reduced by actively acquiring bicarbonate and facilitating CO_2 uptake in order to concentrate CO_2 around the site of Rubisco. Discussion of this biophysical CCM is complemented in Chapter 5 by Gontero and Maberly who describe the mechanisms of biochemical CCMs, C_4 photosynthesis and crassulacean acid metabolism, in freshwater and marine plants and algae. Both types of CCM rely on the metalloenzyme carbonic anhydrase that rapidly interconverts CO_2 and bicarbonate, and Matsuda, Nawaly and Yoneda (Chapter 6) present the properties, types and functions of this ubiquitous but structurally variable enzyme. Meyer (Chapter 7) describes how, in order to concentrate CO_2 around Rubisco efficiently, the enzyme can be concentrated in small micro-compartments, the carboxysome in cyanobacteria and the pyrenoid in some eukaryotic algae.

The final two chapters scale up to the ecological aspects of aquatic productivity. In Chapter 8, Norici, Gerotto, Beardall and Raven outline the main resources required by aquatic photoautotrophs, the mechanisms involved in their acquisition and assimilation and how variation in these factors controls phytoplankton productivity. In Chapter 9, Gao, Zhao and Beardall describe how marine primary producers have responded to environmental change, including temperature, ocean acidification and ultraviolet radiation, as well as interactions among these factors, and assess the possible effect of future changes as humanity alters the global environment.

We thank all the authors for their expert contribution, diligence and commitment in the writing of this book.

We dedicate this book to the late Mario Giordano, a colleague and friend to many of us, who made a major contribution to aquatic CCMs in his tragically short career.

December 2021

1

An Introduction to Productivity and Carbon Cycling in Aquatic Ecosystems

Brigitte GONTERO[1], Timothy M. LENTON[2] and Stephen Christopher MABERLY[3]

[1] *Aix-Marseille Université, CNRS BIP UMR 7281, Marseille, France*

[2] *Global Systems Institute, University of Exeter, United Kingdom*

[3] *Lake Ecosystems Group, UK Centre for Ecology & Hydrology, Lancaster, United Kingdom*

1.1. Overview

Seen from space, the Earth, despite its name being synonymous with soil, is a blue planet. The French surrealist poet Paul Eluard wrote, "La terre est bleue comme une orange" and the idea of blueness has been reinforced by photographs taken from space and the 2001 BBC documentary *The Blue Planet*. The blue color derives from light from our nearest star, the Sun, that is scattered back from the oceans that cover 71% of the Earth's area and by molecules of nitrogen and oxygen (O_2) in the atmosphere, the latter gas produced by photosynthesis. The Earth is the only planet in our solar system currently known to contain large amounts of liquid water at the surface. The water is believed to derive from impact by extra-terrestrial bodies such as carbonaceous chondritic meteorites (Alexander *et al.* 2012) and possibly also from the moon-forming impact with the large proto-planet called Theia about 4.5 billion years ago (Budde *et al.* 2019). The presence of liquid surface water has driven the evolution of life on Earth that began around four billion years ago (Tashiro *et al.* 2017) while the oxygen has been produced by splitting of water in the primary phase of

Blue Planet, Red and Green Photosynthesis,
coordinated by Stephen Christopher MABERLY and Brigitte GONTERO.

photosynthesis (Barber 2017). This has not only produced the oxygen in the atmosphere but also allowed ozone to form, reducing the amount of damaging ultraviolet radiation reaching the surface of the Earth. The hydrogen released from water is used to convert carbon dioxide (CO_2) into organic compounds.

1.2. Light and productivity on Earth

Productivity on planet Earth is largely dependent on the energy produced by hydrogen-fusion reactions in the Sun. The so-called solar constant, the average amount of electromagnetic solar radiation impinging on the top of the Earth's atmosphere, is 1370 Jm^{-2} s^{-1} (Kleidon *et al.* 2016) but varies by about 0.1% over the 11-year solar cycle, linked to the amount of sunspots, and was markedly lower in the Earth's early history (see Chapter 2). Taking the geometry and rotation of the Earth into account, on average 343 J m^{-2} s^{-1} is received at the top of atmosphere, of which, as a result of reflection and absorption, ~70% reaches the Earth's surface, that is, globally ~79 PJ s^{-1} (Table 1.1). The photosynthetically available radiation (PAR) in the spectral band 400–700 nm represents about 48.5% of the total radiation (Kleidon *et al.* 2016), and there are on average 4.56 µmol photon J^{-1} (Kirk 1994). This equates to about 1.75 10^{-4} Pmol photon s^{-1} (175 Ga mol photon s^{-1}). Because the oceans cover about 70% of the surface of the Earth, they receive roughly 70% of the total photon flux but slightly more than land on an area basis (10,950 vs. 10,462 Pmol m^{-2} y^{-1}) because of less atmospheric losses over the ocean.

In recent decades, satellite observations of chlorophyll combined with models have been used to estimate net primary productivity (NPP) at global scales. These indicate that ocean phytoplankton are responsible for 58 Pg C y^{-1} and terrestrial vegetation a similar amount, 56.4 Pg C y^{-1} (Behrenfeld *et al.* 2001). In the oceans, NPP is estimated to be an additional 1 Pg C y^{-1} for marine macroalgae (Smith 1981) and 0.4 Pg C y^{-1} for seagrasses using estimates of areal coverage (Duarte *et al.* 2005) and average productivity (van der Heijden and Kamenos 2015). Lakes and reservoirs cover an area of about 2.67 10^{-3} Pm^2 (Messager *et al.* 2016), and rivers and streams are estimated to cover an additional 0.77 10^{-3} Pm^2 (Allen and Pavelsky 2018) giving a total area of 3.44 10^{-3} Pm^2, about 2.3% of land area. There are no modern estimates of NPP in inland waters, but (Likens 1973) estimated a mean areal NPP of 250 gC m^{-2} y^{-1}, for lakes and rivers. Multiplying this by their global area gives a rough estimate of 0.9 Pg C y^{-1}. These revised estimates give a global total of nearly 117 Pg C y^{-1} (Table 1.1) of which 51% is aquatic. On land, the eukaryotic green-lineage of embryophytes, which evolved from green algae, is the major contributor to primary productivity, but in aquatic environments a wide range of phylogenetic groups comprising red-lineage, and green-lineage eukaryotes as well as prokaryotic cyanobacteria, are involved.

Component	Unit	Ocean	Land	Total	Ref.
Area	Pm^2	0.361	0.149	0.510	1
Total energy flux	$PJ\ s^{-1}$	56.7	22.4	79.1	1
Total PAR photon flux	$Pmol\ photon\ s^{-1}$	$1.25\ 10^{-4}$	$4.95\ 10^{-5}$	$1.75\ 10^{-4}$	2, 3
Average PAR irradiance	$Pmol\ photon\ m^{-2}\ y^{-1}$	10,950	10,462	10,817	
Total NPP	$Pg\ C\ y^{-1}$	58.0 (59.4)[a]	56.4 (57.3)[b]	114.4 (116.7)[c]	4
Average NPP per unit area	$mol\ C\ m^{-2}\ y^{-1}$	13.4 (13.7)[a]	31.5 (32.1)[b]	18.7 (19.1)[c]	
Total biomass	Pg C	6	470	476	5
Photoautotrophic biomass	Pg C	1.25	20	21.25	6
Photoautotrophic turnover	day	7.9	129	68	7

Table 1.1. *Global average for oceans and land (including ice-covered areas) of total radiation and photosynthetically active radiation (PAR, 400 to 700 nm) at the surface, net primary productivity (NPP), living biomass stock and production turnover. a, includes macroalgae and seagrasses; b, includes rivers and streams; c, overall total (see text). (1) (Kleidon et al. 2016); (2) Assumes PAR 48.5% of total; (3) Assumes 4.56 μmol photon J^{-1}; (4) (Behrenfeld et al. 2001); (5) (Bar-On et al. 2018); (6) based on plankton for oceans and leaves for land (Bar-On et al. 2018); (7) based on values in Behrenfeld et al. (2001) without additional estimates in parentheses. N.B. P = 10^{15}*

The lower areal NPP in the ocean compared to land (Table 1.1) is caused by competition for photons between water and phytoplankton. In addition, the low nutrient levels result in only about 35% of the absorbed photons being used in photosynthesis, the remainder being lost as heat in non-photochemical quenching or re-emitted as fluorescence (Lin *et al.* 2016). Recent estimates (Bar-On *et al.* 2018) indicate a much lower total and photoautotrophic biomass in the oceans than on land. Consequently, the average turnover time for photoautotrophic biomass is only about eight days in the ocean but about 130 days on land.

1.3. Converting light energy into chemical energy

1.3.1. *Underwater light*

On average, about 5% of the light hitting a water surface is reflected and so does not enter the water. The magnitude of reflection is greater at a low, compared to a high zenith angle, and when the water is calm (Kirk 1994). Within the water, light is attenuated with depth. This occurs by absorbance and scattering by water molecules, colored dissolved organic carbon, phytoplankton and non-living particles such as soil or sediment (see Chapter 8). Thus, while on average 30% of the light is attenuated by the Earth's atmosphere of about 100 km, the same attenuation can occur in less than 5 m in the clearest water and in less than 0.1 m in turbid water. Since wavelengths of light are attenuated differentially, depending on the cause of attenuation, the spectral properties of light vary with depth: red light is lost proportionately more rapidly than blue light in clear coastal waters while colored dissolved organic carbon absorbs strongly in the blue wavebands. Light often limits overall productivity of turbid systems and controls the euphotic zone, the depth of water where positive net photosynthesis can occur. In the clear ocean, living coralline red algae have been found at a depth of 268 m off the Bahamas where light was about 0.0005% of the surface (Littler *et al.* 1985). The maximum PAR photon irradiance these macroalgae would have experienced was estimated to be 0.009 μmol photon m^{-2} s^{-1}, just at the theoretical absolute minimum needed to support photoautotrophic photosynthesis of 0.01 μmol photon m^{-2} s^{-1} (Raven and Cockell 2006). Very clear lakes, such as Lake Tahoe, have freshwater mosses growing to about 120 m, less than half the depth range of the lake (Frantz and Cordone 1967). In sub-surface water in contrast, sun-flecks caused by wave focusing can produce very short bursts of light with an intensity that is five times greater than at the surface (Schubert *et al.* 2001) with the potential to cause photoinhibition (Raven 2011) and photodamage.

1.3.2. *The primary phase of photosynthesis*

The photochemical phase is the first step in photosynthesis (see Chapter 8). It converts the energy from photons of light into chemical energy (adenosine triphosphate, ATP) that is used in carbon-fixation and other metabolic processes, and also produces reducing power (nicotinamide adenine dinucleotide phosphate, NADPH) that is used to reduce CO_2. PAR is absorbed by a range of pigments in pigment-protein, light-harvesting complexes associated with thylakoids (Raven 2021). Chlorophyll (Chl) *a* is involved in all photoautotrophs, but additional accessory pigments also absorb light and pass the energy to Chl *a*. These include other forms of chlorophyll molecules, xanthophylls, carotenes and phycobilins that absorb different wavelengths of light and are distributed differently among algal groups (see Chapters 2, 3 and 8). The Chla/b binding protein is the most abundant membrane protein on Earth (Dittami *et al.* 2010).

Linear flow of electrons derived from splitting water to reducing NADP to NADPH involves two linked photosystems (Figure 1.1). The co-operation of two photosystems is required to provide the energy needed to transfer electrons from water (Em = + 0.8 V) to the terminal electron acceptor NADP+ (-0.3 V). At the reaction center of photosystem II (PSII), water is photolyzed by an oxygen-evolving complex. On average, the energy from five photons is used to split two molecules of water producing one molecule of oxygen (as a by-product) and four reducing equivalents (Barber 2017). For that purpose, the four oxidizing equivalents needed to oxidize the two water molecules are stored by four Mn ions in the catalytic center of the oxygen evolving complex, the oxygen-generating catalytic center of PSII consisting of a $Mn_3Ca^{2+}O_4$ cubane (Ferreira *et al.* 2004). The four electrons released are passed down an electron transfer chain within the thylakoid membrane through cytochrome b_6f (Kurisu *et al.* 2003, Malone *et al.* 2019) that is distributed in grana (40%) and in lamellae (60%) to photosystem I (PSI). Here, the energy from an additional four photons is used to reduce the electrons sufficiently to reduce ferredoxin so that ferredoxin NADP reductase (FNR) can reduce NADP to produce NADPH. Therefore, on average, a minimum of nine photons are absorbed to produce one molecule of oxygen and to fix one molecule of CO_2 (Raven *et al.* 2014).

The thylakoids are membrane bound sacs (Raven 2021), and linear electron transport produces protons in the lumen of the thylakoid firstly in the splitting of water and, secondly, in the passage of electrons through the membrane via cytochrome b6f (Figure 1.1). The potential energy in the electrochemical gradient of protons is converted into chemical energy, by a transmembrane enzyme, ATP synthase, that is mostly found in the lamella grana and creates ATP from adenosine diphosphate (ADP). This enzyme is a "dynamo" made up of two complexes: a water-soluble complex, coupling factor CF_1, that is bound to a membrane embedded complex CF_0 (Hahn *et al.* 2018).

In addition to linear electron transport, PSI can be involved in cyclic electron transfer (Arnon *et al.* 1954) that produces ATP as a result of the vectorial transport of protons by cytochrome b_6f but does not lead to the production of oxygen. Many pathways for the photochemical quenching of electrons have been identified, reducing O_2 to H_2O in "water to water" cycles. These reactions include, for instance, photorespiration and the Mehler reaction (Mehler 1951) that provides an electron sink pathway avoiding the production of reactive oxygen species. In the latter, a superoxide anion is transformed into H_2O_2 by a superoxide dismutase that is then converted into H_2O by a catalase. Electron transfer proceeds from H_2O at the PSII donor side to H_2O generated at the PSI acceptor side, thereby called a "water to water" cycle. Part of H_2O_2 can also be used by an ascorbate peroxidase that produces monodehydroascorbate that serves in turn to regenerate ascorbate with glutathione or NADPH (Asada 1999). Cyclic electron transport and the Mehler reaction are important mechanisms to match the ratio of ATP to NADPH produced with cellular demand (Joliot and Johnson 2011).

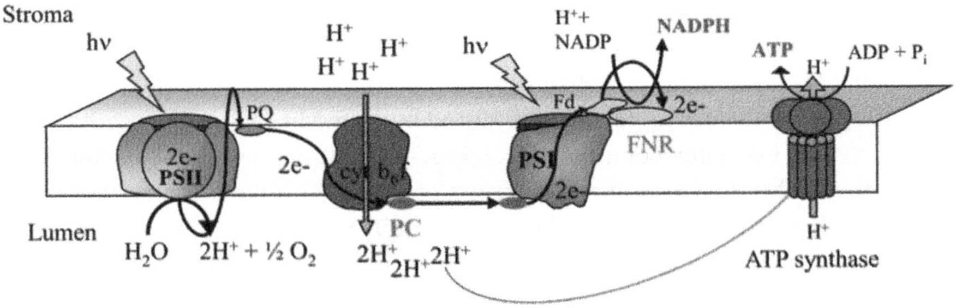

Figure 1.1. *The primary phase of photosynthesis. FNR stands for ferredoxin NADP reductase, PC for plastocyanin and cyt for cytochrome. For a color version of this figure, see www.iste.co.uk/maberly/photosynthesis.zip*

1.4. Carbon fixation

1.4.1. *Inorganic carbon in air and water*

Carbon dioxide is the available form of inorganic carbon for photosynthesis by land plants. The atmospheric concentration has varied considerably since plants started to colonize the land about 500 Ma (Morris *et al.* 2018) and has increased by about 43% in the last two centuries, from a pre-industrial concentration of around 280 ppm to a global annual average for 2020 of about 415 ppm (www.esrl.noaa.gov/gmd/ccgg/trends/). This increase has resulted primarily from burning fossil fuels that release organic carbon that was produced by photosynthesis in previous eras.

The concentration of CO_2 in water in equilibrium with an atmosphere at 400 ppm CO_2 at 15°C and standard pressures is about 18 mmol m^{-3} in fresh water and 15 mmol m^{-3} in seawater (Figure 1.2; Maberly and Gontero 2017). However, the surface concentrations are highly variable with time and location in aquatic systems. In fresh waters, the concentration of CO_2 can be over 20 times higher than equilibrium when respiration in the lake is high or when water with high CO_2 concentration enters from the catchment (Cole *et al.* 1994; Maberly *et al.* 2015; Maberly and Gontero 2017). In contrast, when rates of inorganic carbon uptake exceed rates of input from the catchment or air–water exchange, the concentrations of CO_2 in lakes can be 5,000 times lower than air-equilibrium (Maberly 1996; Maberly and Gontero 2017; Maberly *et al.* 2021). The CO_2 concentration in marine systems can also be out of air-equilibrium, but to a lesser extent than in some fresh waters.

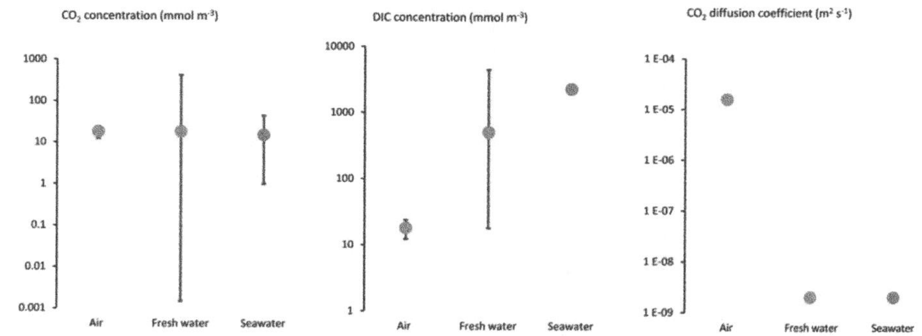

Figure 1.2. *Comparison of inorganic carbon concentrations, and the CO_2 diffusion coefficient, in air, fresh water and seawater. The circles represent the mean and the error bars represent the range. Note that a logarithmic scale is used on each panel. Diffusion coefficients derived from (Zeebe 2011) and concentrations from (Maberly and Gontero 2017). For a color version of this figure, see www.iste.co.uk/maberly/photosynthesis.zip*

$$\mathbf{CO_2 + H_2O \Leftrightarrow H_2CO_3^- \leftrightharpoons HCO_3^- + H^+ \leftrightharpoons CO_3^{2-} + 2H^+}$$

$$CO_2 + H_2O \Leftrightarrow H_2CO_3$$

$$CO_2 + H_2O \Leftrightarrow HCO_3^- + H^+$$

$$CO_2 + OH^- \Leftrightarrow HCO_3^-$$

Figure 1.3. *The carbonate equilibria (in bold) and the hydration and dehydration reactions of CO_2 below (in italics). Rapid reactions are shown with two arrows in opposite directions; slow reactions are shown with a double-headed arrow*

Four forms of inorganic carbon are present in water (see Chapter 4). When CO_2 dissolves in water, a small proportion reacts with water to form carbonic acid (H_2CO_3) that dissociates to form bicarbonate (HCO_3^-) or CO_2 and H_2O can react to form HCO_3^- and H^+ (Figure 1.3). The CO_2 can also react with water or OH^- to form HCO_3^- and this in turn can dissociate, forming carbonate (CO_3^{2-}). Where volumetric rates of transformation between the different forms of inorganic carbon are rapid, CO_2 and HCO_3^- can be out of chemical equilibrium because the rates of CO_2 hydration and dehydration are relatively slow (Johnson 1982). The widespread enzyme carbonic anhydrase increases the rate of interconversion of CO_2 and HCO_3^- (Jensen *et al.* 2019; see Chapter 6).

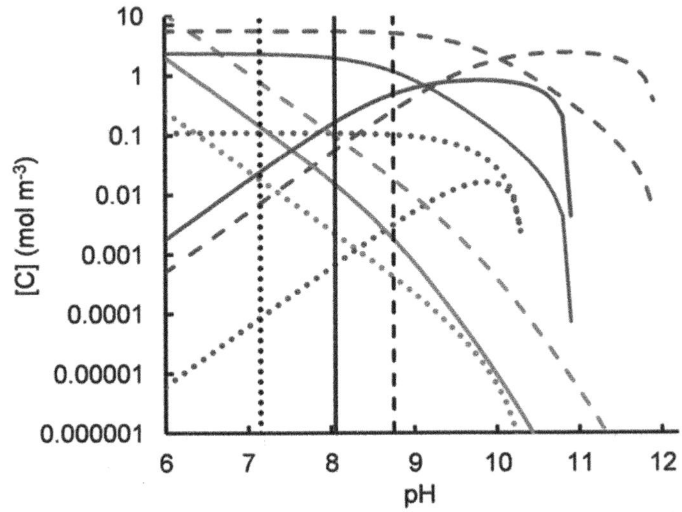

Figure 1.4. *Bjerrum plot of CO_2 (orange), HCO_3^- (green) and CO_3^{2-} (blue) for seawater (solid line) and fresh water with an alkalinity of 0.11 (dotted line) or 5.6 (dashed line) equiv m^{-3} representing the 5th and 95th percentile of global river alkalinities (Hartmann* et al. *2014) calculated for 15°C. The vertical black lines show the air-equilibrium concentrations for the three systems. For a color version of this figure, see www.iste.co.uk/maberly/photosynthesis.zip*

Dissolved CO_2 and carbonic acid are usually treated together as "free CO_2". The equilibrium between the forms of dissolved inorganic carbon (DIC) is controlled primarily by pH and also by temperature and ionic strength (Stumm and Morgan 1996). The first and second dissociation constants, pK_1' and pK_2' represent the pH where the concentrations of CO_2 and HCO_3^- and HCO_3^- and CO_3^{2-}, respectively, are equal. In fresh water, these dissociation constants typically have values around 6.41 and 10.38 at 15°C and an ionic strength of 1.1 mol m^{-3} while at the same temperature in seawater, they are 0.46 and 1.27 pH units lower. Consequently, CO_3^{2-} ions constitute a larger proportion of DIC in seawater than in fresh water (Figure 1.4; Maberly and Gontero 2017). The pH at equilibrium with 400 ppm in the atmosphere is 7.15 and 8.75 in the low and high alkalinity fresh water example and 8.05 in seawater at 15°C.

The concentration of DIC can be highly different among the three realms. It only comprises CO_2 in air, but can be much higher in aquatic environments where HCO_3^- and CO_3^{2-}, produced from the reaction of CO_2 with H_2O and the dissociation of the resulting carbonic acid (as well as geology and biogeochemical processes on land), produce dissolved inorganic carbon (DIC). The concentration of DIC in the sea is high and fairly constant, but extremely variable in fresh waters, where different geological settings result in different pH values and thus different concentration of HCO_3^- and CO_3^{2}. These ionic forms of DIC can be essentially zero in very soft acid water, and up to concentrations similar to those of seawater in hard inland waters draining carbonate-rich catchments. Concentrations in soda-lakes can be even higher (Talling 1985) although these extreme values, arguably not representing fresh water, are excluded from Figure 1.2. The presence of HCO_3^- in aquatic systems provides an additional potential source of exogenous carbon for photosynthesis that many, but not all, aquatic photoautotrophs can make use of (Maberly and Gontero 2017; see Chapter 4). There is currently no evidence for direct use of carbonate ions (Maberly 1992).

All cells or leaves in air and water are surrounded by an unstirred boundary layer through which transport of material occurs largely by diffusion. Because water has a greater density and viscosity than air, diffusion coefficients for gases such as CO_2 and O_2 are about 10,000 times lower (Raven 1970; Figure 1.2). Consequently, when not limited by water-stress, photosynthesis by leaves of terrestrial plants are close to being saturated at atmospheric concentrations of CO_2 (Lloyd and Farquhar 1996), while leaves of freshwater plants have half-saturation CO_2 concentrations that are six to eleven times that of air-equilibrium (Maberly and Madsen 1998). Because of their generally small size, phytoplankton has thinner boundary layers around them and transport limitation of dissolved substances by diffusion is correspondingly lower (Wolf-Gladrow and Riebesell 1997). Transport limitation will not only limit influx of CO_2, but also efflux of O_2, potentially enhancing photorespiration and limiting productivity.

Low rates of diffusion, the generation of low CO_2 concentrations at times of high photoautotroph biomass and the kinetic properties of the primary carboxylation enzyme can limit aquatic productivity if this was based solely on inward passive diffusion of CO_2 "pulled" in by carbon fixation. However, many photoautotrophs counter this potential carbon limitation with different types of CO_2-concentrating mechanisms (CCMs) that actively "push" carbon into the cell (see Chapters 4, 5, 6 and 7).

1.4.2. *Mechanisms of carbon fixation*

Five autotrophic carbon fixation processes are currently known in addition to the Calvin–Benson–Bassham cycle described below (Berg 2011). These are the reductive citric acid (Arnon-Buchanan) cycle, the reductive acetyl-CoA (Wood-Ljungdahl) pathway, the hydroxypropionate (Fuchs–Holo) cycle, the 3-hydroxypropionate/ 4-hydroxybutyrate cycle and the dicarboxylate/4-hydroxybuturate cycle. They are largely present in bacteria or Archaea, often in anaerobic environments and for some pathways, thermophilic environments. Raven (2009) compared the energy cost, reductants, enzyme kinetics and organisms that utilize these different carbon-fixation pathways. These chemolithotrophic and anoxygenic autotrophs contribute about 0.4 Pg C y^{-1}, less than 1% of the production by photoautotrophic phytoplankton (Raven 2009).

The Calvin–Benson–Bassham cycle, also called the C_3 or the reductive pentose phosphate pathway, was elucidated at the end of the 1950s in green freshwater microalgae, and currently is ultimately responsible for more than 99.5% of the CO_2 fixation on Earth (Raven 2009). The stoichiometry of carbon fixation is shown below with the source of oxygen produced shown in bold:

$$6CO_2 + 12H_2\mathbf{O} + h\nu \rightarrow C_6H_{12}O_6 + 6O_2 + 6H_2O \qquad [1.1]$$

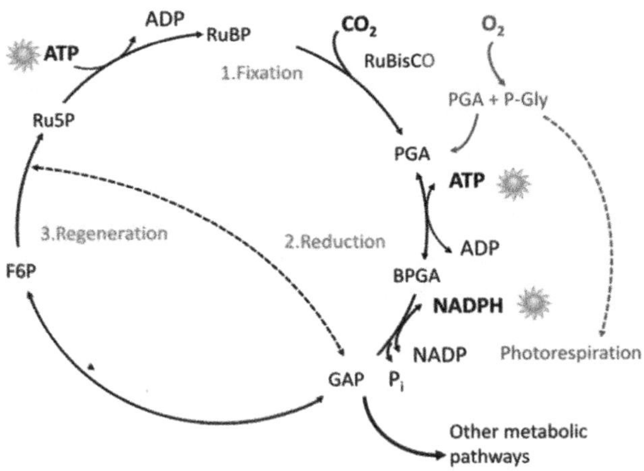

Figure 1.5. *Scheme of the Calvin–Benson–Bassham cycle and photorespiration. BPGA, 1,3-bisphosphoglycerate; F6P, fructose-6-phosphate; GAP, glyceraldehyde-3-phosphate; PGA, 3-phosphoglycerate; P-gly, phosphoglycolate; RuBP, ribulose-1,5-bisphosphate; Ru5P, ribulose-5-phosphate. The "sun" symbols represent products from the primary phase of photosynthesis*

The carboxylation enzyme in this pathway, ribulose-1,5-bisphosphate carboxylase-oxygenase (Rubisco), is the most abundant protein on Earth (Ellis 1979; Raven 2009, 2013; Losh *et al.* 2013; Bar-On and Milo 2019). Rubisco catalyses the incorporation of CO_2 into a five-carbon compound, ribulose-1,5-bisphosphate (RuBP), forming a six-carbon intermediate (Gontero *et al.* 2006; Gontero and Salvucci 2014) (Figure 1.5). This intermediate splits immediately, producing two molecules of 3-phosphoglycerate (PGA), a C_3 compound, that gives the cycle one of its names. The C_3 cycle can be thought of as comprising three steps: first, fixation described above, second, reduction involving ATP and NADPH and third, regeneration of the acceptor RuBP involving ATP (Figure 1.5). For each triose phosphate produced, three molecules of CO_2 are fixed and the primary phase of photosynthesis provides the six molecules of NADPH and nine molecules of ATP that are required.

Although Rubisco is ultimately responsible for almost all carbon-fixation on the planet, it has a low specific activity and can fix O_2 as well as CO_2, depending on their relative concentrations at the active site and the precise properties of the enzyme. The oxygenase reaction produces one molecule of PGA but also one molecule of phosphoglycolate (Figure 1.5). This toxic compound is processed in three different organelles: it is dephosphorylated to glycolate in the chloroplasts, the glycolate is converted into glycine in the peroxisomes and glycine is metabolized in the mitochondria (Sage *et al.* 2012). This whole process, called photorespiration, results in a loss of about 30% of the photosynthetic output of terrestrial C_3 plants (Erb and Zarzycki 2016) and incurs an energy cost. Overall, Rubisco has probably evolved to maximize productivity (Bathellier *et al.* 2018) but in some ecological circumstances, productivity and fitness are enhanced by the possession of a CCM (see Chapters 4 and 5).

Rubisco activity is regulated by carbamylation, by interaction with its substrate RuBP and by interaction with carboxyarabinitol-1-phosphate (CA1P) (Figure 1.6). Carbamylation involves the binding of CO_2 and Mg^{2+} ions, and results in the full catalytic activity of Rubisco (Lorimer and Miziorko 1980). CA1P occurs naturally in some species and inhibits the carbamylated enzyme while RuBP inhibits the un-carbamylated form (Gontero and Salvucci 2014, Mueller-Cajar 2017, Salvucci and Anderson 1987). Rubisco activase is involved in the removal of CA1P and RuBP from Rubisco. The Rubisco activase enzymes belong to the AAA+ family (ATPases Associated with diverse cellular Activities) and hydrolyze ATP to remove the inhibitory sugars bound to Rubisco. There are three types of Rubisco activase enzymes (Mueller-Cajar 2017, Figure 1.7). In the green-lineage, the activase oligomeric state can be monomeric or multimeric; however, the hexamer is believed to be the functional form. Two isoforms of activase have been found (α and β) and both have a highly flexible C-terminus. The C-terminus of the α-isoform has two cysteine residues allowing redox regulation by thioredoxin (Gontero and Salvucci 2014; Carmo-Silva *et al.* 2015; Thieulin-Pardo *et al.* 2015). Both isoforms can be phosphorylated *in vivo* at low light and in the dark, but the importance of this process is unclear (Kim *et al.* 2016). In the red-lineage, the activase function is performed by cbbX (Mueller-Cajar *et al.* 2011; Jensen *et al.* 2017; Launay *et al.* 2020), a hetero-oligomer with two subunits, one encoded by the nucleus and the other by the plastid genome (Loganathan *et al.* 2016). Recently, a detailed study of the mechanism of Rubisco carbamylation in the red alga *Galdieria sulphuraria*, showed that conformation changes triggered by CO_2 and Mg^{2+} binding resulted in the active site becoming open, thereby activating Rubisco (Stec 2012). In chemoautotrophic bacteria and cyanobacteria, the activase function depends on CbbO and CbbQ that form a hetero-oligomeric complex in a $CbbQ_6$: $CbbO_1$ stoichiometry (Tsai *et al.* 2015). In addition, Rubisco kinetics can be altered by interaction with other Calvin–Benson–Bassham enzymes (Gontero *et al.* 2006).

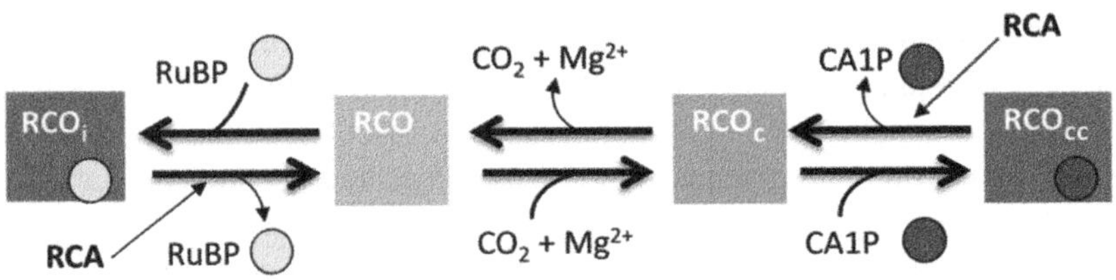

Figure 1.6. *"Traffic-light" regulation of Rubisco. Rubisco (RCO, inactive, in orange) needs to bind CO_2 and Mg^{2+} to form the active carbamylated Rubisco (RCO_c in green). RCO can be inhibited by binding RuBP (yellow circle, producing RCO_i in red) and RCO_c can be inhibited by binding CA1P (blue circle, producing RCO_{cc} in red). Rubisco activase (RCA) relieves the inhibition of the RCO_i and RCO_{cc}) by removing CABP and RuBP, respectively, in the presence of ATP producing RCO and RCO_c and ADP and Pi. The hydrolysis of ATP by RCA is not represented for the sake of clarity. For a color version of this figure, see www.iste.co.uk/maberly/photosynthesis.zip*

1.5. The global carbon cycle

Life is a key part of the carbon cycle (Figure 1.8). A hypothetical world devoid of life but blessed with liquid water would have a partial carbon cycle, but it would be a shadow of the one we participate in on Earth today. The abiotic input of CO_2 degassing from volcanoes and metamorphism is only about one-thousandth of the biogenic exchange fluxes of CO_2 through contemporary net primary production (Lenton 1998; Lenton *et al.* 2020). Thus, one can view today's biosphere as a phenomenal carbon recycling system. Each carbon atom entering the surface Earth system from volcanism is cycled about a thousand times through photosynthesis before it is deposited back in the solid Earth.

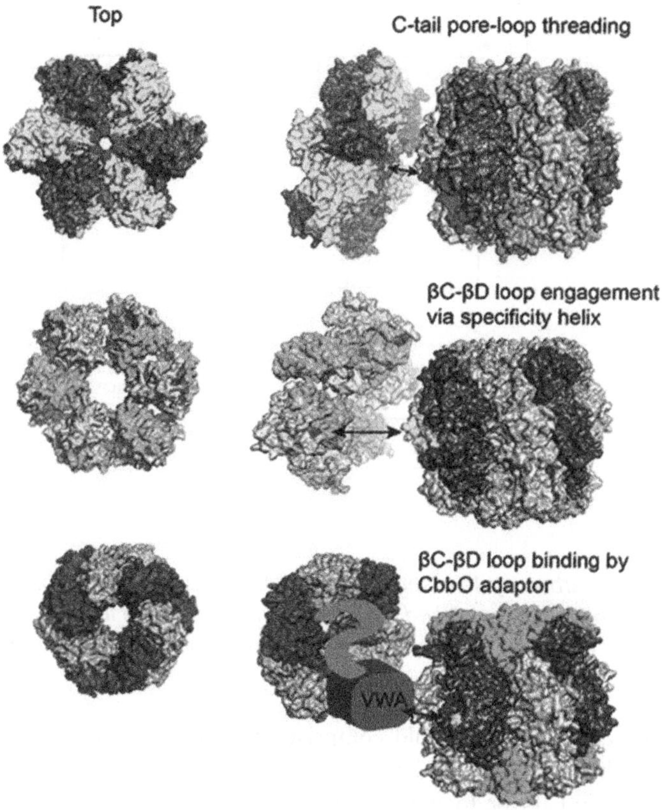

Figure 1.7. *Structures of Rubisco and its activation by Rubisco activases (RCA). The left-hand column shows the activase and the right-hand column shows the activase interacting with Rubisco. The top row is the red-lineage CbbX activase (PDB:3ZUH/1BXN), the middle row is the green-lineage Rubisco activase (PDB:3ZW6/8RUC) and the bottom row is the cyanobacterial activase (PDB:5C3C/1SVD). Residues known to be involved in protein–protein interactions with Rubisco are colored in magenta for red- and green-type activases. Known RCA interacting segments on Rubisco are shown in red (RbcL C-tail) and yellow (interacting βC-βD loop residues). VWA stands for the von Willebrand factor A domain. From (Mueller-Cajar 2017). For a color version of this figure, see www.iste.co.uk/maberly/photosynthesis.zip*

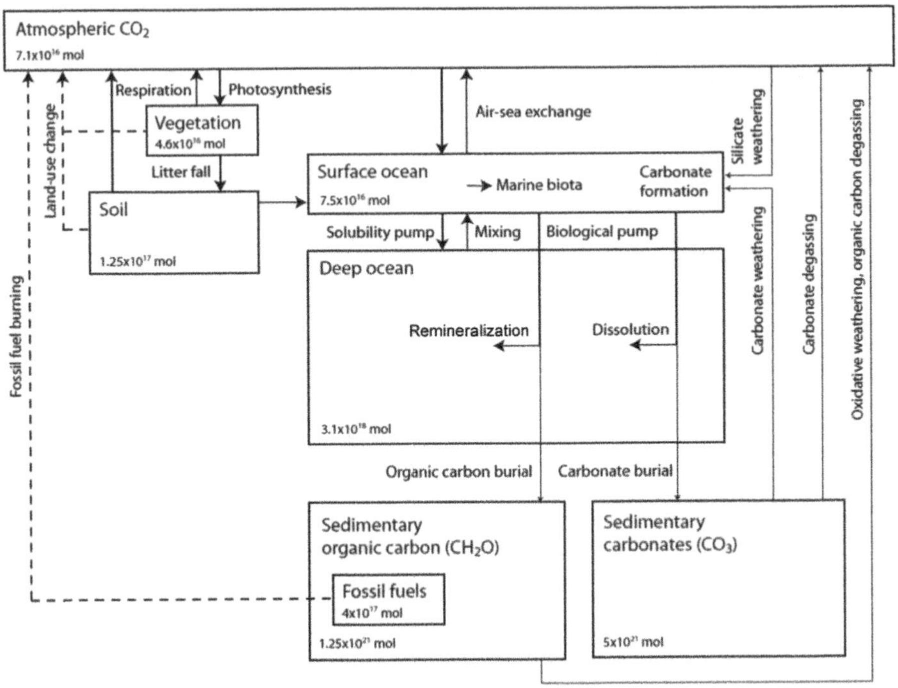

Figure 1.8. *The carbon cycle – showing both short-term cycling within the surface Earth system (bigger arrows) and long-term cycling with the inner Earth (smaller arrows), together with major anthropogenic fluxes (dashed arrows)*

On an abiotic Earth, CO_2 would comprise much of the atmosphere – just as it does on our neighboring planets Mars and Venus – whereas it is a trace gas on Earth today (Lenton 1998). Prior to the industrial revolution, CO_2 was roughly one in every 3,500 molecules in the atmosphere; now it is around one in every 2,400 molecules. Still, CO_2 in the atmosphere is dwarfed by other surface reservoirs of carbon. The amount of carbon in global vegetation is comparable to that in the atmosphere and the amount of carbon in soils (including permafrost) is roughly four times that in the atmosphere (Zimov *et al.* 2006). Even these reservoirs are dwarfed by the ocean, which stores 45 times as much carbon as the atmosphere, thanks to the fact that CO_2 reacts with water. A key consequence is that exchange fluxes of CO_2 with the land and ocean provide pivotal short-term controls on atmospheric CO_2 (Figure 1.8). This brings us to an important distinction: between the short-term and long-term carbon cycles. The short-term carbon cycle refers to exchanges of CO_2 between the atmosphere, ocean, land, and life – that is, within the surface Earth system. The long-term carbon cycle refers to exchanges between the surface Earth system and the solid Earth – the sediments, crust and mantle. Both are influenced by life.

Starting with the short-term carbon cycle (Lenton 2000): The exchange of carbon between the atmosphere and the land is largely biological – involving photosynthetic uptake and release by aerobic respiration (and to a lesser extent, fires). The exchange of carbon between the atmosphere and ocean involves a mixture of biology, chemistry and physics. CO_2 is continually being exchanged between the surface ocean and the atmosphere. As ocean surface waters are physically circulated from the low to the high latitudes they cool down, causing them to take up more CO_2. In some high-latitude regions – the North Atlantic and the Southern Ocean today – these surface waters sink to great depth, taking excess CO_2 with them, producing a "solubility pump" of carbon to the deep ocean. There is also a "biological pump" of carbon whereby organisms in the surface ocean take up CO_2 and their dead bodies sink (or the excrement of organisms that ate them, sinks) transferring carbon to depth. Both carbon pumps produce an excess of carbon in the deep ocean and a deficit in the surface ocean, which tends to lower the concentration of CO_2 in the atmosphere.

In addition to being comprised of organic carbon, some marine organisms also precipitate calcium and magnesium carbonates. These include some of the phytoplankton floating in the surface ocean, some benthic organisms (that live on the seafloor) and corals. Contrary to popular misrepresentation, the corresponding reaction actually releases CO_2, for example

$$2HCO_3^- + Ca^{2+} \rightarrow CaCO_3 + CO_2 + H_2O \qquad [1.2]$$

Then as dead carbonate-producing phytoplankton sink into the deep ocean, the carbonate ($CaCO_3$) tends to dissolve as the pressure increases. This produces a characteristic "snowline" – called the "carbonate compensation depth" (CCD) – below which no carbonate is preserved in ocean sediments, but above which the sediments are white with carbonate. Above the CCD, as the sediments are compressed, new carbonate rocks can form, removing carbonate from the surface Earth system, to enter the long-term carbon cycle (Figure 1.8) (Berner 2004).

Much of the burial of carbonate rocks is balanced by the weathering of carbonate rocks on land, which takes up atmospheric carbon dioxide, for example

$$CaCO_3 + CO_2 + H_2O \rightarrow 2HCO_3^- + Ca^{2+} \qquad [1.3]$$

The result is a null cycle with no net effect on atmospheric CO_2, so long as the rates of weathering and deposition of carbonates are in balance. However, there is another important process at play: The weathering of silicate rocks converts CO_2 from the atmosphere into bicarbonate anions in solution and liberates corresponding cations, for example

$$CaSiO_3 + 2CO_2 + H_2O \rightarrow Ca^{2+} + 2HCO_3^- + SiO_2 \qquad [1.4]$$

When this is followed by the formation of carbonate in the ocean (reaction 1.2), the net result is the removal of CO_2 from the atmosphere to the Earth's crust, for example

$$CaSiO_3 + CO_2 \rightarrow CaCO_3 + SiO_2 \qquad [1.5]$$

This is the dominant long-term removal process for CO_2 injected into the atmosphere by volcanoes. Some of that CO_2 comes direct from the mantle, but much of it today comes from buried carbonate sediments that are subducted by plate tectonics, heated up and the CO_2 they contain "degassed" back to the atmosphere – thus closing the long-term inorganic carbon cycle.

There are roughly $3.4x10^{18}$ moles of carbon in the oceans, land and atmosphere together, and silicate weathering removes around $7x10^{12}$ moles of carbon per year (Lenton *et al.* 2018). Hence, the residence time of CO_2 in the surface Earth system, with respect to removal by silicate weathering, is about 500,000 years. Importantly, the rate of silicate weathering is sensitive to both climate and biology – it happens faster at higher CO_2, at higher temperature, and in the presence of a more productive land biosphere (Lenton *et al.* 2018). This provides a crucial long-term climate regulator: if something tends to increase temperature (such as the steady brightening of the Sun over Earth history), the rate of silicate weathering and associated CO_2 removal increases, tending to lower atmospheric CO_2 and cool the planet, counteracting the initial change. This "negative feedback loop" has been critical to maintaining an equable climate and liquid water on Earth (Walker *et al.* 1981). It has also been radically "amped up" by terrestrial biology, which accelerates silicate weathering by a factor of at least ten (Schwartzman and Volk 1989). The overall result has been to lower atmospheric CO_2 concentration from the dominant gas in the atmosphere (as on Mars and Venus) to a trace gas (Lenton 1998).

There is a further key contributor to the long-term carbon cycle: the burial of organic carbon (dead organic matter) also removes carbon to the crust, put simply:

$$CO_2 + H_2O \rightarrow CH_2O \text{ (buried)} + O_2 \qquad [1.6]$$

Biogeochemists describe this process as the long-term source of atmospheric oxygen (O_2) (Berner 2004; Lenton *et al.* 2018). The small flux of organic carbon burial of around $1x10^{13}$ mol C yr^{-1} represents the net imbalance of all net primary production and aerobic respiration/combustion on the planet. The corresponding net release of oxygen represents about a thousandth of the flux liberated by net primary production. But when integrated over geologic time, the accumulation of organic carbon in sedimentary rocks is responsible for the corresponding accumulation of $3.8x10^{19}$ mol O_2 comprising 21% of the atmosphere (or around 750 times the preindustrial CO_2 level).

Not all of the buried, reduced organic carbon stays in the lithosphere – instead it is returned to the Earth's surface either by tectonic processes of subduction and degassing, or uplift of sediments and subsequent weathering – and in both cases, it reacts with oxygen in the atmosphere, closing the long-term organic carbon cycle:

$$CH_2O + O_2 \rightarrow CO_2 + H_2O \qquad [1.7]$$

The resulting residence time of oxygen molecules in the atmosphere with respect to exchange with crust is long – nearly four million years (Lenton and Watson 2000). Happily, the huge oxygen reservoir means that it cannot get used up by the respiration of organic matter (which totals around $2.4x10^{17}$ mol C), or even by the combustion of all known fossil fuel reserves (which total roughly $4x10^{17}$mol, or about 1% of the O_2 reservoir) (Broecker 1970). Instead, our principal concern in the Anthropocene is how the anthropogenic perturbation to the carbon cycle is affecting atmospheric CO_2 and hence climate.

Prior to the industrial revolution, the massive exchange fluxes of CO_2 between the atmosphere and ocean, and between the atmosphere and land, were approximately in balance. Since then, the emissions of CO_2, primarily from burning fossil fuels and secondarily from land-use change (converting organic carbon from the land biosphere into CO_2) have increased the concentration of CO_2 in the atmosphere from 280 parts per million (ppm) to currently (2021) around 415 ppm. Yet CO_2 is not rising as fast as it is being added to the atmosphere. The reason is that about half of the CO_2 added annually is being taken up by ocean and land "carbon sinks".

The ocean carbon sink exists because adding excess CO_2 to the atmosphere forces some of it to dissolve in and react with the water in the surface ocean. The exchange of gas across the sea surface is relatively rapid, but dissolved CO_2 then reacts more slowly with seawater. Adding a reactant to one side of a chemical reaction always drives the reaction to the other side, until a new equilibrium is achieved – thus CO_2 is transformed into dissolved inorganic carbon (DIC). Some

DIC is transferred to depth by the solubility pump, but the rate of uptake is limited by the relatively slow exchange of surface waters with the bulk of the deep ocean. Some DIC is taken up by phytoplankton and transferred to depth by the biological carbon pump.

The land carbon sink exists because those ecosystems – especially forests – that are not being cleared for agriculture are accumulating carbon in live vegetation and soils. One key reason is that increasing atmospheric CO_2 fertilizes photosynthesis – making the uptake of carbon by plants more efficient. This is because there is a competition between CO_2 and O_2 molecules for the active sites of the carbon-fixing enzyme "Rubisco", so increasing the CO_2/O_2 ratio increases the amount of CO_2 that gets fixed. In addition to this "CO_2 fertilization effect" some land ecosystems are being fertilized by human inputs of nutrients, often carried in gaseous form by the atmosphere. Also, where agricultural land is abandoned, natural ecosystems tend to re-grow and accumulate carbon from the atmosphere.

Measured fluctuations in atmospheric CO_2 are particularly revealing about the land carbon sink, which is more variable from year to year than the ocean carbon sink (Bousquet *et al.* 2000). For example, after the eruption of Mount Pinatubo in 1991, CO_2 rose more slowly because the resulting cooling strengthened the land carbon sink. Whereas after the strong El Niño of 1998, CO_2 rose faster because warming that year and associated fires eliminated the land carbon sink. Subsequently, the switch to La Niña increased marine net primary production (Behrenfeld *et al.* 2001).

As the rise in atmospheric CO_2 also causes temperatures to rise, the land and ocean become less-effective carbon sinks (Lenton 2000). Warming causes respiration by plants and microorganisms in soils to increase more than it increases photosynthesis. Furthermore, increased mortality from droughts, fires and insect infestations increases the loss of land carbon. Warming also makes CO_2 less soluble in water. Furthermore, the uptake of CO_2 is acidifying the ocean and this makes it a less effective chemical store of carbon (by shifting the equilibrium of the reactions with seawater towards gaseous CO_2). How the biological pump will change in a high CO_2 world is unclear (Passow and Carlson 2012). It may get more efficient as increased CO_2 increases photosynthetic carbon fixation by some algal groups, suppresses calcification and increases the release of carbohydrates that stick particles together leading to more efficient sinking (Riebesell 2004). However, warming of surface waters tends to stratify the ocean (see Chapter 9) and reduce the supply of nutrients, weakening the biological pump.

Overall, negative feedback in the carbon cycle wins out at present and is slowing the rise of atmospheric CO_2. Were it not for the land and ocean carbon sinks, atmospheric CO_2 would already be above 500 parts per million and there would have been considerably more global warming.

1.6. Perspectives

Life has changed the biogeochemistry of planet Earth. Incident light from the sun is captured by photoautotrophic organisms and used to remove hydrogen from water, releasing oxygen as a by-product, and converting CO_2 into organic forms of carbon. This process has produced the oxygen in the atmosphere and in the surface of the oceans, and led to the deposition of calcareous rocks and organic carbon. The latter are fossil fuels, produced by primary productivity in the past, that are currently being over-exploited by Man, disrupting the global carbon cycle and triggering rapid climate change.

Plants, eukaryotic algae and cyanobacteria, with very different evolutionary histories, are responsible today for the planet's primary productivity on land and in water. There are marked differences between photosynthesis in aquatic and terrestrial environments. Firstly, photosynthesis on land is almost exclusively performed by large plants that are embryophytes derived from the "green-lineage" of algae. In contrast, photosynthesis in water is mainly the result of microscopic organisms from a large diversity of photoautotrophs, including cyanobacteria (Flombaum *et al.* 2013). However, eukaryotic algae from the "red-lineage" are the most important (Falkowski *et al.* 2004). Second, of the primary productivity on land, about 80% results from passive diffusion of CO_2 down a gradient created by carbon-fixation in the chloroplasts by Rubisco. In aquatic environments, in contrast, especially in the oceans, most organisms possess a CCM to counter the low rates of CO_2 diffusion in water and CO_2 depletion when rapid CO_2 uptake exceeds resupply. The diverse structures and mechanisms involved in CCMs in ancient cyanobacteria and relatively more recently evolved eukaryotic green and red algal lineages drive the global carbon cycle and support over 50% of global net primary productivity on the blue planet.

1.7. Acknowledgments

SCM was supported by a visiting scholarship funded by Aix-Marseille Université. BG thanks CNRS and Aix-Marseille Université for supporting her research.

1.8 References

Alexander, C.M.O., Bowden, R., Fogel, M.L., Howard, K.T., Herd, C.D.K., Nittler, L.R. (2012). The provenances of asteroids, and their contributions to the volatile inventories of the terrestrial planets. *Science*, 337(6095), 721–723.

Allen, G.H. and Pavelsky, T.M. (2018). Global extent of rivers and streams. *Science*, 361(6402), 585–587.

Arnon, D.I., Whatley, F.R., Allen, M.B. (1954). Photosynthesis by isolated chloroplasts II. Photosynthetic phosphorylation, the conversion of light into phosphate bond energy. *Journal of the American Chemical Society*, 76(24), 6324–6329.

Asada, K. (1999). The water–water cycle in chloroplasts: Scavenging of active oxygens and dissipation of excess photons. *Annual Review of Plant Physiology and Plant Molecular Biology*, 50, 601–639.

Bar-On, Y.M. and Milo, R. (2019). The global mass and average rate of Rubisco. *Proceedings of the National Academy of Sciences of the United States of America*, 116(10), 4738–4743.

Bar-On, Y.M., Phillips, R., Milo, R. (2018). The biomass distribution on Earth. *Proceedings of the National Academy of Sciences of the United States of America*, 115(25), 6506–6511.

Barber, J. (2017). A mechanism for water splitting and oxygen production in photosynthesis. *Nature Plants*, 3(4), 17041.

Bathellier, C., Tcherkez, G., Lorimer, G.H., Farquhar, G.D. (2018). Rubisco is not really so bad. *Plant Cell and Environment*, 41(4), 705–716.

Behrenfeld, M.J., Randerson, J.T., Mcclain, C.R., Feldman, G.C., Los, S.O., Tucker, C.J., Falkowski, P.G., Field, C.B., Frouin, R., Esaias, W.E., Kolber, D.D., Pollack, N.H. (2001). Biospheric primary production during an ENSO transition. *Science*, 291(5513), 2594–2597.

Berg, I.A. (2011). Ecological aspects of the distribution of different autotrophic CO_2 fixation pathways. *Applied and Environmental Microbiology*, 77(6), 1925–1936.

Berner, R.A. (2004). *The Phanerozoic Carbon Cycle: CO_2 and O_2*. Oxford University Press.

Bousquet, P., Peylin, P., Ciais, P., Le Quéré, C., Friedlingstein, P., Tans, P.P. (2000). Regional changes in carbon dioxide fluxes of land and oceans since 1980. *Science*, 290(5495), 1342–1346.

Broecker, W.S. (1970). Man's oxygen reserves. *Science*, 168, 1537–1538.

Budde, G., Burkhardt, C., Kleine, T. (2019). Molybdenum isotopic evidence for the late accretion of outer Solar System material to Earth. *Nature Astronomy*, 3(8), 736–741.

Carmo-Silva, E., Scales, J.C., Madgwick, P.J., Parry, M.A.J. (2015). Optimizing Rubisco and its regulation for greater resource use efficiency. *Plant Cell and Environment*, 38(9), 1817–1832.

Cole, J.J., Caraco, N.F., Kling, G.W., Kratz, T.K. (1994). Carbon dioxide supersaturation in the surface waters of lakes. *Science*, 265(5178), 1568–1570.

Dittami, S.M., Michel, G., Collen, J., Boyen, C., Tonon, T. (2010). Chlorophyll-binding proteins revisited – A multigenic family of light-harvesting and stress proteins from a brown algal perspective. *BMC Evolutionary Biology*, 10(365) [Online]. Available at: https://doi.org/10.1186/1471-2148-10-365.

Duarte, C.M., Middelburg, J.J., Caraco, N. (2005). Major role of marine vegetation on the oceanic carbon cycle. *Biogeosciences*, 2(1), 1–8.

Ellis, R.J. (1979). Most abundant protein in the world. *Trends in Biochemical Sciences*, 4(11), 241–244.

Erb, T.J. and Zarzycki, J. (2016). Biochemical and synthetic biology approaches to improve photosynthetic CO_2-fixation. *Current Opinion in Chemical Biology*, 34, 72–79.

Falkowski, P.G., Katz, M.E., Knoll, A.H., Quigg, A., Raven, J.A., Schofield, O., Taylor, F.J.R. (2004). The evolution of modern eukaryotic phytoplankton. *Science*, 305(5682), 354–360.

Ferreira, K.N., Iverson, T.M., Maghlaoui, K., Barber, J., Iwata, S. (2004). Architecture of the photosynthetic oxygen-evolving center. *Science*, 303(5665), 1831–1838.

Flombaum, P., Gallegos, J.L., Gordillo, R.A., Rincon, J., Zabala, L.L., Jiao, N.A.Z., Karl, D.M., Li, W.K.W., Lomas, M.W., Veneziano, D., Vera, C.S., Vrugt, J.A., Martiny, A.C. (2013). Present and future global distributions of the marine Cyanobacteria *Prochlorococcus and Synechococcus*. *Proceedings of the National Academy of Sciences of the United States of America*, 110(24), 9824–9829.

Frantz, T.C. and Cordone, A.J. (1967). Observations on deepwater plants in Lake Tahoe, California and Nevada. *Ecology*, 48(5), 710–714.

Gontero, B. and Salvucci, M. (2014). Regulation of photosynthetic carbon metabolism in aquatic and terrestrial organisms by Rubisco activase, redox-modulation and CP12. *Aquatic Botany*, 118, 14–23.

Gontero, B., Avilan, L., Lebreton, S. (2006). Control of carbon fixation in chloroplasts. *Annual Plant Reviews*, 22, 187–218.

Hahn, A., Vonck, J., Mills, D.J., Meier, T., Kuhlbrandt, W. (2018). Structure, mechanism, and regulation of the chloroplast ATP synthase. *Science*, 360(6389), eaat4318.

Hartmann, J., Lauerwald, R., Moosdorf, N. (2014). A brief overview of the GLObal RIver CHemistry Database, GLORICH. In *Geochemistry of the Earth's Surface Ges-10*, Gaillardet, J. (eds). Elsevier Procedia, Paris.

Jensen, E., Clement, R., Maberly, S.C., Gontero, B. (2017). Regulation of the Calvin–Benson–Bassham cycle in the enigmatic diatoms: Biochemical and evolutionary variations on an original theme. *Philosophical Transactions of the Royal Society of London B: Biological Sciences*, 372(1728), article no. 29220, DOI: 10.1098/rstb.2016.0401.

Jensen, E., Clement, R., Kosta, A., Maberly, S.C., Gontero, B. (2019). A new widespread subclass of carbonic anhydrase in marine phytoplankton. *ISME Journal*, 13(8), 2094–2106.

Johnson, K.S. (1982). Carbon-dioxide hydration and dehydration kinetics in sea-water. *Limnology and Oceanography*, 27(5), 849–855.

Joliot, P. and Johnson, G.N. (2011). Regulation of cyclic and linear electron flow in higher plants. *Proceedings of the National Academy of Sciences of the United States of America*, 108(32), 13317–13322.

Kim, S.Y., Bender, K.W., Walker, B.J., Zielinski, R.E., Spalding, M.H., Ort, D.R., Huber, S.C. (2016). The plastid casein kinase 2 phosphorylates Rubisco activase at the Thr-78 site but is not essential for regulation of Rubisco activation state. *Frontiers in Plant Science*, 7, 404.

Kirk, J.T.O. (1994). *Light and Photosynthesis in Aquatic Ecosystems*, 2nd edition. Cambridge University Press.

Kleidon, A., Miller, L., Gans, F. (2016). Physical limits of solar energy conversion in the Earth system. In *Solar Energy for Fuels*, Tuysuz, H. and Chan, C.K. (eds). Springer, Cham.

Kurisu, G., Zhang, H., Smith, J.L., Cramer, W.A. (2003). Structure of the cytochrome b6f complex of oxygenic photosynthesis: Tuning the cavity. *Science*, 302(5647), 1009–1014.

Launay, H., Huang, W., Maberly, S.C., Gontero, B. (2020). Regulation of carbon metabolism by environmental conditions: A perspective from diatoms and other chromalveolates. *Frontiers in Plant Science*, 11, 1033.

Lenton, T.M. (1998). Gaia and natural selection. *Nature*, 394, 439–447.

Lenton, T.M. (2000). Land and ocean carbon cycle feedback effects on global warming in a simple Earth system model. *Tellus*, 52B(5), 1159–1188.

Lenton, T.M. and Watson, A.J. (2000). Redfield revisited 2: What regulates the oxygen content of the atmosphere? *Global Biogeochemical Cycles*, 14(1), 249–268.

Lenton, T.M., Daines, S.J., Mills, B.J.W. (2018). COPSE reloaded: An improved model of biogeochemical cycling over Phanerozoic time. *Earth Science Reviews*, 178(3), 1–28.

Lenton, T.M., Dutreuil, S., Latour, B. (2020). Life on Earth is hard to spot. *The Anthropocene Review*, 7(3), 248–272.

Likens, G.E. (1973). Primary production – Freshwater ecosystems. *Human Ecology*, 1(4), 347–356.

Lin, H.Z., Kuzminov, F.I., Park, J., Lee, S., Falkowski, P.G., Gorbunov, M.Y. (2016). The fate of photons absorbed by phytoplankton in the global ocean. *Science*, 351(6270), 264–267.

Littler, M.M., Littler, D.S., Blair, S.M., Norris, J.N. (1985). Deepest known plant life discovered on an uncharted seamount. *Science*, 227(4682), 57–59.

Lloyd, J. and Farquhar, G.D. (1996). The CO_2 dependence of photosynthesis, plant growth responses to elevated atmospheric CO_2 concentrations and their interaction with soil nutrient status 1: General principles and forest ecosystems. *Functional Ecology*, 10(1), 4–32.

Loganathan, N., Tsai, Y.C.C., Mueller-Cajar, O. (2016). Characterization of the heterooligomeric red-type Rubisco activase from red algae. *Proceedings of the National Academy of Sciences of the United States of America*, 113(49), 14019–14024.

Lorimer, G.H. and Miziorko, H.M. (1980). Carbamate formation on the epsilon-amino group of a lysyl residue as the basis for the activation of ribulose bisphosphate carboxylase by CO_2 and Mg^{2+}. *Biochemistry*, 19(23), 5321–5328.

Losh, J.L., Young, J.N., Morel, F.M.M. (2013). Rubisco is a small fraction of total protein in marine phytoplankton. *New Phytologist*, 198(1), 52–58.

Maberly, S.C. (1992). Carbonate ions appear to neither inhibit nor stimulate use of bicarbonate ions in photosynthesis by *Ulva lactuca*. *Plant Cell and Environment*, 15(2), 255–260.

Maberly, S.C. (1996). Diel, episodic and seasonal changes in pH and concentrations of inorganic carbon in a productive lake. *Freshwater Biology*, 35(3), 579–598.

Maberly, S.C. and Gontero, B. (2017). Ecological imperatives for aquatic CO_2-concentrating mechanisms. *Journal of Experimental Botany*, 68(14), 3797–3814.

Maberly, S.C. and Madsen, T.V. (1998). Affinity for CO_2 in relation to the ability of freshwater macrophytes to use HCO_3^-. *Functional Ecology*, 12(1), 99–106.

Maberly, S.C., Berthelot, S.A., Stott, A.W., Gontero, B. (2015). Adaptation by macrophytes to inorganic carbon down a river with naturally variable concentrations of CO_2. *Journal of Plant Physiology*, 172, 120–127.

Maberly, S.C., Gontero, B., Puppo, C., Villain, A., Severi, I., Giordano, M. (2021). Inorganic carbon uptake in a freshwater diatom, *Asterionella formosa* (Bacillariophyceae): From ecology to genomics. *Phycologia*, 1–12.

Malone, L.A., Qian, P., Mayneord, G.E., Hitchcock, A., Farmer, D.A., Thompson, R.F., Swainsbury, D.J.K., Ranson, N.A., Hunter, C.N., Johnson, M.P. (2019). Cryo-EM structure of the spinach cytochrome b(6) f complex at 3.6 angstrom resolution. *Nature*, 575(7783), 535–539.

Mehler, A.H. (1951). Studies on reactions of illuminated chloroplasts 1: Mechanism of the reduction of oxygen and other hill reagents. *Archives of Biochemistry and Biophysics*, 33(1), 65–77.

Messager, M.L., Lehner, B., Grill, G., Nedeva, I., Schmitt, O. (2016). Estimating the volume and age of water stored in global lakes using a geo-statistical approach. *Nature Communications*, 7(13603) [Online]. Available at: https://doi.org/10.1038/ncomms13603.

Morris, J.L., Puttick, M.N., Clark, J.W., Edwards, D., Kenrick, P., Pressel, S., Wellman, C.H., Yang, Z.H., Schneider, H., Donoghue, P.C.J. (2018). The timescale of early land plant evolution. *Proceedings of the National Academy of Sciences of the United States of America*, 115(10), E2274–E2283.

Mueller-Cajar, O. (2017). The diverse AAA plus machines that repair Inhibited Rubisco active sites. *Frontiers in Molecular Biosciences*, 4(00031) [Online]. Available at: https://doi.org/10.3389/fmolb.2017.00031.

Mueller-Cajar, O., Stotz, M., Wendler, P., Hartl, F.U., Bracher, A., Hayer-Hartl, M. (2011). Structure and function of the AAA(+) protein CbbX, a red-type Rubisco activase. *Nature*, 479(7372), 194–199.

Passow, U. and Carlson, C.A. (2012). The biological pump in a high CO2 world. *Marine Ecology Progress Series*, 470, 249–272.

Raven, J.A. (1970). Exogenous inorganic carbon sources in plant photosynthesis. *Biological Reviews of the Cambridge Philosophical Society*, 45(2), 167–220.

Raven, J.A. (2009). Contributions of anoxygenic and oxygenic phototrophy and chemolithotrophy to carbon and oxygen fluxes in aquatic environments. *Aquatic Microbial Ecology*, 56(2–3), 177–192.

Raven, J.A. (2011). The cost of photoinhibition. *Physiologia Plantarum*, 142(1), 87–104.

Raven, J.A. (2013). Rubisco: Still the most abundant protein of Earth? *New Phytologist*, 198(3), 970–970.

Raven, J.A. (2021). Determinants, and implications, of the shape and size of thylakoids and cristae. *Journal of Plant Physiology*, 257(153342), DOI:10.1016/j.jplph.2020.153342.

Raven, J.A. and Cockell, C.S. (2006). Influence on photosynthesis of starlight, moonlight, planetlight, and light pollution (reflections on photosynthetically active radiation in the universe). *Astrobiology*, 6(4), 668–675.

Raven, J.A., Beardall, J., Giordano, M. (2014). Energy costs of carbon dioxide concentrating mechanisms in aquatic organisms. *Photosynthesis Research*, 121(2–3), 111–124.

Riebesell, U. (2004). Effects of CO_2 enrichment on marine phytoplankton. *Journal of Oceanography*, 60(4), 719–729.

Sage, R.F., Sage, T.L., Kocacinar, F. (2012). Photorespiration and the evolution of C_4 photosynthesis. In *Annual Review of Plant Biology, Vol. 63*, Merchant, S.S. (eds). San Mateo, CA.

Salvucci, M.E. and Anderson, J.C. (1987). Factors affecting the activation state and the level of total activity of ribulose bisphosphate carboxylase in tobacco protoplasts. *Plant Physiology*, 85(1), 66–71.

Schubert, H., Sagert, S., Forster, R.M. (2001). Evaluation of the different levels of variability in the underwater light field of a shallow estuary. *Helgoland Marine Research*, 55(1), 12–22.

Schwartzman, D.W. and Volk, T. (1989). Biotic enhancement of weathering and the habitability of earth. *Nature*, 340(457–460) [Online]. Available at: https://doi.org/10.1038/340457a0.

Smith, S.V. (1981). Marine macrophytes as a global carbon sink. *Science*, 211(4484), 838–840.

Stec, B. (2012). Structural mechanism of RuBisCO activation by carbamylation of the active site lysine. *Proceedings of the National Academy of Sciences of the United States of America*, 109(46), 18785–18790.

Stumm, W. and Morgan, J.J. (1996). *Aquatic Chemistry: Chemical Equilibria and Rates in Natural Waters*. Wiley-Intersciences, New York.

Talling, J.F. (1985). Inorganic carbon reserves of natural waters and ecophysiological consequences of their photosynthetic depletion: Microalgae. In *Inorganic Carbon Uptake by Aquatic Photosynthetic Organisms*, Lucas, W.J. and Berry, J.A. (eds). American Society of Plant Physiologists, Rockville, MA.

Tashiro, T., Ishida, A., Hori, M., Igisu, M., Koike, M., Mejean, P., Takahata, N., Sano, Y., Komiya, T. (2017). Early trace of life from 3.95 Ga sedimentary rocks in Labrador, Canada. *Nature*, 549(7673), 516–518.

Thieulin-Pardo, G., Avilan, L., Kojadinovic, M., Gontero, B. (2015). Fairy "tails": Flexibility and function of intrinsically disordered extensions in the photosynthetic world. *Frontiers in Molecular Biosciences*, 2, 23.

Tsai, Y.C.C., Lapina, M.C., Bhushan, S., Mueller-Cajar, O. (2015). Identification and characterization of multiple Rubisco activases in chemoautotrophic bacteria. *Nature Communications*, 6, 8883.

Van Der Heijden, L.H. and Kamenos, N.A. (2015). Reviews and syntheses: Calculating the global contribution of coralline algae to total carbon burial. *Biogeosciences*, 12(21), 6429–6441.

Walker, J.C.G., Hays, P.B., Kasting, J.F. (1981). A negative feedback mechanism for the long-term stabilisation of Earth's surface temperature. *Journal of Geophysical Research*, 86(C10), 9776–9782.

Wolf-Gladrow, D. and Riebesell, U. (1997). Diffusion and reactions in the vicinity of plankton: A refined model for inorganic carbon transport. *Marine Chemistry*, 59(1–2), 17–34.

Zeebe, R.E. (2011). On the molecular diffusion coefficients of dissolved CO_2, HCO_3^-, and CO_3^{2-} and their dependence on isotopic mass. *Geochimica Et Cosmochimica Acta*, 75(9), 2483–2498.

Zimov, S.A., Schuur, E.A., Chapin, F.S. (2006). Permafrost and the global carbon budget. *Science*, 312, 1612–1613.

2

Evolution of Aquatic Photoautotrophs

John A. RAVEN[1,2,3]

[1] *Division of Plant Science, University of Dundee at the James Hutton Institute, Invergowrie, United Kingdom*

[2] *University of Technology, Climate Change Cluster, Faculty of Science, Sydney, Australia*

[3] *School of Biological Sciences, University of Western Australia, Crawley, Australia*

2.1. Background

The Earth Science background to the origin of photoautotrophy (Box 2.1) includes the Faint Young Sun paradox, that is, the presence of liquid H_2O in Earth in the Archean (3.8–2.5 Ga ago) despite the 20–25% lower radiant energy output from the Sun (Charnay *et al.* 2020). Among the probable solutions to this problem is a high CO_2 partial pressure in the atmosphere contributing to a larger greenhouse effect (Charnay *et al.* 2020), with implications for photoautotrophic CO_2 assimilation. The occurrence of life in the Archean, the associated trophic modes and the taxonomy of the organism, depends on fossils *sensu lato*, largely the occurrence of organic matter and its ^{13}C:^{12}C ratio relative to inorganic C, and the chemical nature of the organic matter. The syngenicity of the fossils must be considered; they were formed at the same time as the encapsulating material and must be established (French *et al.* 2015). The same question of syngenicity applies to some palaeoenvironmental indicators. In the Proterozoic (2.4–0.542 Ga ago, and, especially, the Phanerozoic (0.541 Ga ago to today), there is increasing availability of environmental and biological data from the rock record.

Until the global oxidation event (GOE) 2.4–2.3 Ga ago, there were only “whiffs” of O_2 on Earth (Luo *et al.* 2016). The earliest photoautotrophs in the anoxic Archean (possibly late Hadean) era were obligately anaerobic anoxygenic photosynthetic bacteria. They used (and continue to use) electron donors with lower (more reducing) redox potential than the H_2O used in oxygenic photosynthetic organisms, that is, cyanobacteria and photosynthetic eukaryotes (Hamilton 2019). Primary productivity by anoxygenic photosynthesis in the Archean was apparently limited by the supply of reductants such as H_2, S^{2-} and Fe^{2+} (Ward *et al.* 2019b).

Photoautotrophy (= photolithotrophy) involves the photoproduction of ATP, and of NAD(P)H with an inorganic electron source, and ATP and NAD(P)H consumption in CO_2 conversion to reduced organic C using an autotrophic CO_2 fixation pathway.

Photoorganotrophs use energy from exogenous organic matter as well as from photons, and comprise photo-osmo-mixotrophs and photo-phago-mixotrophs.

Photo-osmo-mixotrophs depend on dissolved organic C as C source, but have energetic supplements from photons absorbed by bacteriochlorophyll, chlorophyll or proteorhodopsin.

Photo-phago-mixotrophs are eukaryotic photosynthetic organisms that can also ingest, digest and metabolize living or dead organic particles, providing not just organic carbon but also other nutrients.

Box 2.1. *Terms*

2.2. Anoxygenic photosynthetic bacteria

There are six known clades of anoxygenic photosynthetic bacteria: Acidobacteria, Chlorobi, Chloroflexi, Firmicutes (Heliobacteria), Gemmatimonadetes and Proteobacteria, plus *Candidatus* Eremiobacterota (Hamilton 2019; Ward *et al.* 2019a; Rogers 2021; Ward and Shih 2021). Some of the clades acquired phototrophy by horizontal gene transfer, possibly after the global oxygenation event related to the occurrence of oxygenic photosynthesis in cyanobacteria (Hamilton 2019; Ward *et al.* 2019a; Oliver *et al.* 2021; Ward and Shih 2021). The anoxygenic photosynthetic bacteria have a range of bacteriochlorophyll-protein complexes used in photochemistry in either a Type I photochemical reaction center (evolutionarily related to PSI of oxygenic photoautotrophs) in Acidobacteria, Chlorobi and Heliobacteria or a Type II photochemical reaction center (evolutionarily related to PSII of oxygenic photoautotrophs) in Chloroflexi, Gemmatimonadetes and Proteobacteria (Hamilton 2019; Ward and Shih 2021). The six clades all have membrane-integral light harvesting bacteriochlorophyll-protein complexes that pass excitation energy to the reaction centers. In two cases (Chlorobi and Chloroflexi), the membrane-associated light harvesting complexes are supplemented with chlorosomes, single polar lipid layer-bounded bacteriochlorophyll- bacteriochlorophyll complexes on the cytosol side of the plasma membrane (Orf and Blankenship 2013). Bacteriochlorophyll- bacteriochlorophyll light harvesting complexes are unique to chlorosomes (Orf and Blankenship 2013).

Photosynthesis in Acidobacteria, Firmicutes and Gemmatimonadetes (and *Candidatus* Eremiobacterota), involves only photo-osmo-mixotrophy (less precisely known as photoorganotrophic) organisms (Box 2.1). Photo-osmo-mixotrophs also occur in the Chlorobi and, with a great variety of other trophic modes, in Proteobacteria. Anoxygenic photolithotrophic bacteria have three modes of autotrophic CO_2 assimilation: the Chlorobi use the reverse tricarboxylic acid cycle, Chloroflexi use the 3-hydroxypropionate bi-cycle, and Proteobacteria use the Benson–Calvin–Bassham cycle with Form I or Form II Rubisco (ribulose-1,5-bisphosphate carboxylase/oxygenase) (Hamilton 2019; Ward and Shih 2021). However, Ward and Shih (2021) suggest the occurrence of the Benson–Calvin–Bassham cycle in the Chloroflexi. After the origin and spread of oxygenic photoautotrophy, the anoxygenic autotrophic photosynthetic bacteria were (and are) limited to anoxic habitats where sedimented organic carbon from oxygenic photosynthesis has not been biologically oxidized by the O_2 produced in parallel with the organic carbon (Ward and Shih 2021).

The aerobic anoxygenic photosynthetic Proteobacteria in the oxygenated planktonic habitats of oceans and inland waters are photo-osmo-mixotrophs, lacking inorganic electron donor oxidation and autotrophic CO_2 fixation pathways (Stoecker *et al.* 2017; Lehours *et al.* 2018; Gomez-Consarnau *et al.* 2019; Ward and Shih 2021). This is also the case for planktonic photo-osmo-mixotrophic bacteria that use ion-pumping rhodopsins in light-harvesting and generation of ion electrochemical potential differences (Stoecker *et al.* 2017; Gomez-Consarnau *et al.* 2019; Raven and Beardall 2020). Ion-pumping rhodopsins occur in otherwise chemo-organotrophic Archaea, Bacteria and Eukarya, as well as some oxygenic photoautotrophic cyanobacteria and algae (Stoecker *et al.* 2017; Raven and Beardall 2020). In otherwise chemoorganotrophic organisms the occurrence of aerobic anoxygenic photosynthetic reactions, or ion-pumping rhodopsins, transduction of light energy can spare respiratory use of organic carbon in growth and maintenance in the light, or at least prolong survival during starvation (Stoecker *et al.* 2017; Gomez-Consarnau *et al.* 2019). An early origin (Archean?) has been suggested for ion-pumping rhodopsins (Raven and Smith 1981; DasSarma and Schwieterman 2021).

2.3. Cyanobacteria

Oxygenic photoautotrophs originated more than 2.4 Ga ago, and they supplied the photosynthetic O_2 that, with geological processes that limit access of the photosynthetically generated organic carbon to O_2-consuming processes, allowed the Global Oxidation Event 2.4–2.3 Ga ago (Oliver *et al.* 2021). Oxygenic photosynthesis evolved in cyanobacteria (or their unknown ancestors) using chlorophyll a as the photochemically active pigment (Larkum *et al.* 2018). Light harvesting involves chlorophyll a in integral membrane pigment–protein complexes, and phycobilins in phycobilisomes on the cytosol side of the photosynthetic membrane (Larkum *et al.* 2018). This ancestral state is modified in some open ocean cyanobacteria, and a few other cyanobacteria, by supplementing or replacing phycobilisomes by chlorophyll a + b complexes, sometimes with a chlorophyll c-like minor component, in *Prochlorococcus*, *Prochloron* and *Prochlorothrix*, or by chlorophyll d in *Akaryochloris* where chlorophyll d substitutes for chlorophyll a as the photochemically active pigment in photosystem II (PSII) (Larkum *et al.* 2018; Green 2019).

Photochemistry in cyanobacteria (and eukaryotic oxygen photolithotrophs) involves PSII covering a redox range that allows H_2O oxidation to O_2 and reduction of pheophytin. The other photosystem, PSI, oxidizes cytochrome c_6 or plastocyanin and reduces the low-potential reductant (chlorophyll a). Subsequently, ferredoxin and then $NADP^+$ are reduced, and finally CO_2 using the Benson–Calvin–Bassham cycle. The reductant from PSII is linked to the oxidant of PSI by PQ (plastoquinone) and the cytochrome b_6-cytochrome f-non-haem iron (cytochrome b_6-f-Fe_{nh}) complex (Malone *et al.* 2021) in linear electron flow. Cyanobacteria also have cyclic electron flow around PSI involving ferredoxin oxidation by PQ via a H^+ pumping NAD(P)H dehydrogenase, then the cytochrome b_6-f-Fe_{nh} complex and cytochrome c_6 or plastocyanin; oxygenic eukaryotic photolithotrophs have variants on PSI cyclic electron flow (Larkum *et al.* 2018).

Linear electron flow pumps 3 H^+ from the cytosol into the thylakoid lumen per electron moved from H_2O to CO_2, and cyclic electron flow moves 4 H^+ per electron. In the case of the basal cyanobacterium *Gloeobacter*, which lacks thylakoids and has the membrane-associated components of photosynthesis in the plasmalemma, H^+ is pumped from the cytosol to the medium (Larkum *et al.* 2018). The H^+:3 ATP ratio for the ATP synthase in cyanobacteria is 13–15, higher than the ratio of 10 for other photosynthetic bacteria, indicating that the H^+ electrochemical potential across the coupling membrane in cyanobacteria could, for a given *in vivo* free energy of ATP hydrolysis, be lower than that in other photosynthetic bacteria (Larkum *et al.* 2018). As for autotrophic CO_2 assimilation, the cyanobacteria (like all oxygenic eukaryotes) employ the Benson–Calvin–Bassham cycle, and have, ancestrally, Form IB Rubisco in β-cyanobacteria, which has been replaced by Form IA Rubisco by horizontal gene transfer from a proteobacterium in the oceanic α-cyanobacterial species *Prochlorococcus* and *Synechococcus*. These Rubiscos have low CO_2 affinities and low CO_2:O_2 selectivities, and require a CCM (CO_2 concentrating mechanism) for photosynthesis in today's atmosphere, and for many of the atmospheric compositions since the GOE (Raven *et al.* 2017; Iniguez *et al.* 2020). O_2 not only competes with CO_2 as a substrate for Rubisco, but is also involved in photodamage to PSII, and, in diazotrophic cyanobacteria, damaging nitrogenasie, an enzyme central to nitrogen fixation (Falkowski and Godfrey 2008).

Cyanobacteria have by far their greatest global biomass, productivity and diversity in aquatic environments, mainly free-living but with a few in symbioses. However, there are some terrestrial free-living and symbiotic terrestrial cyanobacteria, and it is likely that the cyanobacteria originated on wet rock surfaces on land (Blank and Sánchez-Baracaldo 2010; Nakov *et al.* 2017; Raven and Sánchez-Baracaldo 2021; Sánchez-Baracaldo *et al.* 2017a, b; Raven and Sánchez-Baracaldo 2021). Aquatic cyanobacteria range from picoplanktonic unicells with a volume of 0.1 μm^3 (marine planktonic *Prochlorococcus*) to colonies of over 100 mm^3 volume of the freshwater planktonic *Microcystis* (Xiao *et al.* 2018). This size variation has a very significant influence on photon absorption per unit photosynthetic pigment in a given light field and on the diffusion of inorganic carbon through the diffusion boundary layer, as modulated by sinking or rising (involving gas vesicles) relative to vertical water movement for phytoplankton cells or colonies (Xiao *et al.* 2018).

N_2 fixation, a trait limited to prokaryotes, is not an ancestral trait of cyanobacteria (Boyd and Peters 2013), and occurs in several derived cyanobacteria (Sánchez-Baracaldo *et al.* 2014). While most cyanobacteria are free-living, some cyanobacteria are the photosynthetic and/or diazotrophic components of symbioses (Usher *et al.* 2007; Douglas 2010; Nelsen *et al.* 2020). On land and, less commonly, in the marine intertidal there are cyanolichens (Usher *et al.* 2007; Douglas 2010; Nelsen *et al.* 2020) where the cyanobiont supplies organic carbon and, frequently, combined nitrogen. In marine diatoms such as species of *Hemiaulus*, *Rhizosolenia* and the marine prymnesiophycean *Braarudosphaera*, and the freshwater diatoms *Epithemia* and *Rhopalodia*, there are diazotrophic cyanobionts (Usher *et al.* 2007; Hagino *et al.* 2013; Nakayama *et al.* 2014; Cornejo-Castillo *et al.* 2016; Sánchez-Baracaldo *et al.* 2021). There are also diazotrophic cyanobionts in hornworts and some liverworts, in the water fern *Salvinia*, in cycads and the flowering plant *Gunnera* (Usher *et al.* 2007).

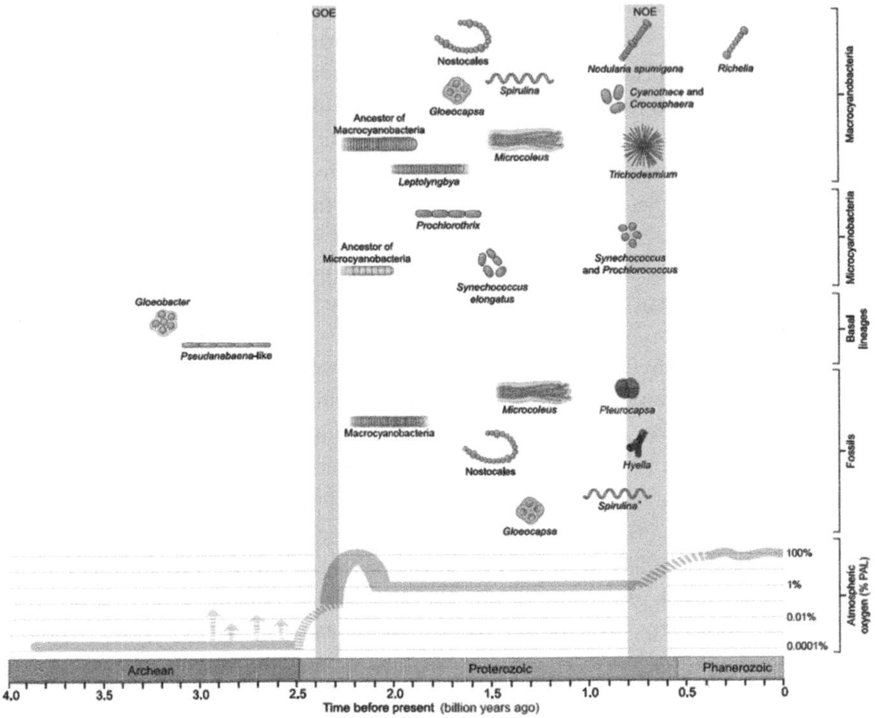

Figure 2.1. *Changes in oxygen concentration in the atmosphere and evolution of cyanobacterial lineages over geological time. (Sánchez-Baracaldo* et al. *2021). For a color version of this figure, see www.iste.co.uk/maberly/photosynthesis.zip*

2.4. Photosynthetic eukaryotes

As will be seen below, the diversity of photosynthetic traits found among present day cyanobacteria can be found among photosynthetic eukaryotes (see also Chapter 3). These include the pigments chlorophylls a, b, c and c-like, and phycobilins, photochemistry and light harvesting, the diversity of redox and H^+ pumping components of thylakoids, the ATP synthase, Form IB Rubisco and the Benson–Calvin–Bassham cycle. The associated genes are retained in the plastid, or have been transferred to the nucleus. All of the extant diversity of photosynthetic eukaryotes (except *Paulinella*: see below) derived their chloroplasts, and the related genes in the plastid and nuclear genome, by linear descent from a primary endosymbiosis of a *Gloeomargarita*-like β-cyanobacterium in a chemoorganotrophic eukaryote (Blank 2013; Nakov *et al.* 2017; Ponce-Toledo *et al.* 2017; Sánchez-Baracaldo *et al.* 2017, 2017). This endosymbiont provided core reactions of the photochemical, redox, H^+ pump and ATP synthase, and Benson–Calvin–Bassham cycle with a Form IB Rubisco.

The timing of this endosymbiosis is not well defined. The origin of major clades of eukaryotes using molecular phylogeny date from at least 1600 Ma ago, with early branching archaeplastida (Glaucophyta) dated 1670–1628 Ma ago (Parfrey *et al.* 2011). The earliest established fossil of a photosynthetic eukaryote, the bangiophycean red alga *Bangiomorpha pubescens*, is suggested to date to 1047 Ma ago (Gibson *et al.* 2018), and a similar age has been suggested for the origin of the later-branching Archaeplastida, the latest-branching, green algae, based on dated molecular phylogeny (Del Cortona *et al.* 2020). In any case, the initial diversification of the Archaeplastida preceded the Neoproterozoic Oxidation Event (NOE) ~635–541 Ma ago, so aquatic photosynthetic eukaryotes might have contributed to, but not necessarily caused, the NOE (Williams *et al.* 2019).

The fossil record of marine eukaryotic microalgae is sparse until about 280 Ma ago when organic cysts of coastal dinoflagellates occur, then coccolithophores (calcified members of the Haptophyta) at 220 Ma ago, and the silicified diatoms (Bacillariophyceae) at 130 Ma ago (Falkowski *et al.* 2004; Westacott *et al.* 2021). Molecular phylogenetic evidence suggests that the origin of these three algal taxa substantially predates the fossil record (Falkowski *et al.* 2004; Brown and Sorhannus 2010; Medlin 2016). The evolution of marine phytoplankton in the Phanerozoic has been related to the availability of essential trace elements (Falkowski *et al.* 2004; Quigg *et al.* 2011). The evolution of algae in the Phanerozoic has been related to the co-evolution of Rubisco and CCMs in response to changes in environmental CO_2 and O_2 (Rickaby and Hubbard 2019). The crown group macroalgal Phaeophyceae, in the red line of secondary photosynthetic algae, are estimated to have evolved 200 Ma ago (Raven 2018).

The euglyphid *Paulinella* is the only known exception to a linear descent from a *Gloeomargarita*-like β-cyanobacterium. This second example of primary photosynthetic endosymbiosis involved an α-cyanobacterium with a form IA Rubisco in their Benson–Calvin–Bassham cycle. This endosymbiotic event occurred about 90–140 Ma ago (Gabr *et al.* 2020; Oberleitner *et al.* 2020), that is, an order of magnitude less than the time of origin of Archaeplastida.

Light harvesting, other than through chlorophyll a-protein complexes, in *Gloeomargarita* involves phycobilins in phycobilisomes that are only retained among some eukaryotes in the Archaeplastida. This is the case in glaucophytes (with no chlorophyll a-protein light harvesting complexes), rhodophytes (with chlorophyll a-protein light harvesting complexes) (Larkum *et al.* 2018; Green 2019), and in the secondarily endosymbiotic Cryptophyta (with phycobilins in the thylakoid lumen associated with chlorophyll a-protein light harvesting complexes) (Larkum *et al.* 2018; Burki *et al.* 2020). While phycobilins provide photon absorption between the blue and red absorption maxima of chlorophylls, phycobilin synthesis has greater nitrogen and energy costs than that of chlorophyll-protein complexes normalized to photon absorptance at the absorption maximum (Raven 1984; Ting *et al.* 2002). Furthermore, the significance of multiple light-harvesting pigments with a range of absorption maxima is less significant for larger organisms (Raven 1984, 1996). Light harvesting in the archaeoplastidial Viridiplantae, and by organisms with plastids derived from algal Viridiplantae by secondary endosymbiosis, the chlorarachniophytes (Rhizaria) and euglenophytes (Excavata), employs chlorophyll b, like the α-cyanobacterium *Prochlorococcus* and the β-cyanobacteria *Prochloron* and *Prochlorothrix* (Larkum *et al.* 2018; Green 2019; Burki *et al.* 2020).

The Form IB Rubisco of β-cyanobacteria is retained by glaucophytes and Viridiplantae (Chlorophyta and streptophytes, i.e. Charophyta and embryophytes), but is replaced in rhodophytes, through horizontal gene transfer from a proteobacterium, by a Form ID Rubisco. The Form ID Rubisco is retained in secondary plastids in the red line of algal evolution (Ochrista, Cryptophyta, Haptophyta), except for the alveolate Chromerids and basal Dinophyta, with a Form II Rubisco derived by horizontal gene transfer from a proteobacterium (Rickaby and Hubbard 2019; Burki *et al.* 2020; Iniguez *et al.* 2020). Light harvesting in the secondary plastids of the red line of algal evolution involves chlorophyll(ide) c forms (lacking in the ochristan Eustigmatophyceae) and carotenoids, predominantly fucoxanthin or its esters, lacking in the ochristan Eustigmatophyceae and Xanthophyceae, and replaced by peridinin in basal dinophytes and chromerids (Falkowski and Raven 2007; Larkum *et al.* 2018). The plastids of some derived dinoflagellates have been replaced in tertiary endosymbiosis by those from algae of the green or red line, kleptoplastids from a variety of sources, or cyanobacterial endo- or ecto-symbionts (Jackson *et al.* 2018). The occurrence of tertiary plastid endosymbioses in dinoflagellates might reflect what has probably happened in other "red line" secondarily endosymbiotic algae, that is, a prior green endosymbiont (e.g. Morozov and Galachyants 2019), and also a prior red endosymbiont for secondarily endosymbiotic "green line" algae (Yang *et al.* 2014; Jackson *et al.* 2018; Ponce-Toledo *et al.* 2018).

Eukaryotic algae span a very large size range, from the picoplanktonic cells with a volume of 0.5 μm^3 (*Ostreococcus*) to phaeophycean macroalgae with a volume of

8 dm^3 (*Macrocystis*) (Vantussenbroek 1993; Raven *et al.* 2019a, 2019b). Even granted the very markedly non-spherical shape of the largest algal individuals, the much lower surface per unit volume of large algae means a much larger package effect (lower absorption per mol chromophore in a given radiation field) and diffusive limitation of the uptake of nutrient such as inorganic C uptake (thicker diffusive boundary layer) (Raven *et al.* 2019a). Microalgae are benthic or planktonic in aquatic habitats, and a few occur on land, for example, in biological crusts (Pushkareva *et al.* 2016; Cheng *et al.* 2019; Kedem *et al.* 2020) and on snow (Gray *et al.* 2020) and ice (Williamson *et al.* 2021). Some microalgae are photosymbionts in symbioses, for example, fungi in lichens on land and, much less commonly submerged, and with a range of non-photosynthetic protists and Metazoa (Yellowlees *et al.* 2008; Douglas 2010; Nelsen *et al.* 2020).

An attribute of some flagellate planktonic microalgae is photo-phago-mixotrophy (Box 2.1), that is, photosynthetic organisms that can also ingest and metabolize living or dead organic particles, providing not just organic carbon but also other nutrients (Stoecker *et al.* 2017). This capacity may be a retention of the mechanism by which cyanobacteria was ingested in producing plastids in archaeplastida and *Paulinella*, and the various endosymbioses producing secondary (and tertiary) plastids. There is also kleptoplasty, that is, plastids from food particles maintained in a photosynthetically active form in the phagotroph for a limited period (Raven 2015; Stoecker *et al.* 2017).

Macroalgae are typically benthic; the Mesoproterozoic *Bangiomorpha* had a holdfast typical of algae growing on rocky substrates (haptophytes) (Raven 2018). Molecular phylogenetic evidence dates the origin of seaweeds related to extant macroalgae with attachment organs in fine-grained material (rhizophytes) to 209 Ma ago (Raven 2018). Some detached benthic algae can grow free-floating (pleustophytes), especially if they have intercellular gas spaces that permit flotation, for example, the *Enteromorpha* morphology of the ulvophycean chlorophyte alga *Ulva* and the phaeophycean *Sargassum* (Bach *et al.* 2021).

Finally, the class Zygnematophyceae, algal streptophytes of the Phylum Charophyta, are the nearest living relative to the ancestors of the terrestrial "higher plants" (embryophytes) (Puttick *et al.* 2018; Cheng *et al.* 2019). Numerous evolutionary returns to floating on, or submerged in, freshwater among plants at the bryophyte, pteridophyte and flowering plant grade have occurred (Maberly 2014). However, only submerged freshwater flowering plants in the Alismatales have made the further transition to life submerged in seawater (Maberly 2014; Larkum *et al.* 2017). Aquatic flowering plants do not express functional stomata, and have no wax layer on the cuticle, of submerged photosynthetic structures (Maberly and Gontero

2018). They retain intercellular gas spaces that increase gas distribution and light scattering and hence light absorptance (Maberly and Gontero 2018). The photo-phago-mixotrophic flowering plant *Utricularia* has aquatic species. Unlike photo-phago-mixotrophic algae, digestion is extracellular in bladders into which prey is sucked by maintaining negative pressure within the bladder (Raven 1997).

2.5. References

Bach, L.T., Tamsitt, V., Gower, J., Hurd, C.L., Raven, J.A., Boyd, P.W. (2021). Testing the climate intervention potential of ocean afforestation using the Great Atlantic Sargassum Belt. *Nature Communications*, 12(1), 2556.

Blank, C.E. (2013). Origin and early evolution of photosynthetic eukaryotes in freshwater environments: Reinterpreting proterozoic paleobiology and biogeochemical processes in light of trait evolution. *Journal of Phycology*, 49(6), 1040–1055.

Blank, C.E. and Sánchez-Baracaldo, P. (2010). Timing of morphological and ecological innovations in the cyanobacteria – A key to understanding the rise in atmospheric oxygen. *Geobiology*, 8(1), 1–23.

Boyd, E.S. and Peters, J.W. (2013). New insights into the evolutionary history of biological nitrogen fixation. *Frontiers in Microbiology*, 4, 201.

Brown, J.W. and Sorhannus, U. (2010). A molecular genetic timescale for the diversification of autotrophic stramenopiles (Ochrophyta): Substantive underestimation of putative fossil ages. *PLOS ONE*, 5(9), e12759.

Burki, F., Roger, A.J., Brown, M.W., Simpson, A.G.B. (2020). The new tree of eukaryotes. *Trends in Ecology & Evolution*, 35(1), 43–55.

Charnay, B., Wolf, E.T., Marty, B., Forget, F. (2020). Is the faint young sun problem for Earth solved? *Space Science Reviews*, 216(5), 90.

Cheng, S., Xian, W., Fu, Y., Marin, B., Keller, J., Wu, T., Sun, W., Li, X., Xu, Y., Zhang, Y., Wittek, S., Reder, T., Guenther, G., Gontcharov, A., Wang, S., Li, L., Liu, X., Wang, J., Yang, H., Xu, X., Delaux, P.-M., Melkonian, B., Wong, G.K.-S., Melkonian, M. (2019). Genomes of subaerial Zygnematophyceae provide insights into land plant evolution. *Cell*, 179(5), 1057–1067.

Cornejo-Castillo, F.M., Cabello, A.M., Salazar, G., Sánchez-Baracaldo, P., Lima-Mendez, G., Hingamp, P., Alberti, A., Sunagawa, S., Bork, P., De Vargas, C., Raes, J., Bowler, C., Wincker, P., Zehr, J.P., Gasol, J.M., Massana, R., Acinas, S.G. (2016). Cyanobacterial symbionts diverged in the late Cretaceous towards lineage-specific nitrogen fixation factories in single-celled phytoplankton. *Nature Communications*, 7, 11071.

Dassarma, S. and Schwieterman, E.W. (2021). Early evolution of purple retinal pigments on Earth and implications for exoplanet biosignatures. *International Journal of Astrobiology*, 20(3), 241–250.

Del Cortona, A., Jackson, C.J., Bucchini, F., Van Bel, M., D'hondt, S., Skaloud, P., Delwiche, C.F., Knoll, A.H., Raven, J.A., Verbruggen, H., Vandepoele, K., De Clerck, O., Leliaert, F. (2020). Neoproterozoic origin and multiple transitions to macroscopic growth in green seaweeds. *Proceedings of the National Academy of Sciences of the United States of America*, 117(5), 2551–2559.

Douglas, A.E. (2010). *The Symbiotic Habit*. Princeton University Press.

Falkowski, P.G. and Godfrey, L.V. (2008). Electrons, life and the evolution of Earth's oxygen cycle. *Philosophical Transactions of the Royal Society B-Biological Sciences*, 363(1504), 2705–2716.

Falkowski, P.G. and Raven, J.A. (2007). *Aquatic Photosynthesis*, 2nd edition. Princeton University Press.

Falkowski, P.G., Katz, M.E., Knoll, A.H., Quigg, A., Raven, J.A., Schofield, O., Taylor, F.J.R. (2004). The evolution of modern eukaryotic phytoplankton. *Science*, 305(5682), 354–360.

French, K.L., Hallmann, C., Hope, J.M., Schoon, P.L., Zumberge, J.A., Hoshino, Y., Peters, C.A., George, S.C., Love, G.D., Brocks, J.J., Buick, R., Summons, R.E. (2015). Reappraisal of hydrocarbon biomarkers in archean rocks. *Proceedings of the National Academy of Sciences of the United States of America*, 112(19), 5915–5920.

Gabr, A., Grossman, A.R., Bhattacharya, D. (2020). Paulinella, a model for understanding plastid primary endosymbiosis. *Journal of Phycology*, 56(4), 837–843.

Gibson, T.M., Shih, P.M., Cumming, V.M., Fischer, W.W., Crockford, P.W., Hodgskiss, M.S.W., Worndle, S., Creaser, R.A., Rainbird, R.H., Skulski, T.M., Halverson, G.P. (2018). Precise age of *Bangiomorpha pubescens* dates the origin of eukaryotic photosynthesis. *Geology*, 46(2), 135–138.

Gomez-Consarnau, L., Raven, J.A., Levine, N.M., Cutter, L.S., Wang, D., Seegers, B., Aristegui, J., Fuhrman, J.A., Gasol, J.M., Sanudo-Wilhelmy, S.A. (2019). Microbial rhodopsins are major contributors to the solar energy captured in the sea. *Science Advances*, 5(8), eaaw8855.

Gray, A., Krolikowski, M., Fretwell, P., Convey, P., Peck, L.S., Mendelova, M., Smith, A.G., Davey, M.P. (2020). Remote sensing reveals antarctic green snow algae as important terrestrial carbon sink. *Nature Communications*, 11(1), 2527.

Green, B.R. (2019). What happened to the phycobilisome? *Biomolecules*, 9(11), 748.

Hagino, K., Onuma, R., Kawachi, M., Horiguchi, T. (2013). Discovery of an endosymbiotic nitrogen-fixing cyanobacterium UCYN-A in *Braarudosphaera bigelowii* (Prymnesiophyceae). *PLOS ONE*, 8(12), e81749.

Hamilton, T.L. (2019). The trouble with oxygen: The ecophysiology of extant phototrophs and implications for the evolution of oxygenic photosynthesis. *Free Radical Biology and Medicine*, 140, 233–249.

Iniguez, C., Capo-Bauca, S., Niinemets, U., Stoll, H., Aguilo-Nicolau, P., Galmes, J. (2020). Evolutionary trends in RuBisCO kinetics and their co-evolution with CO_2 concentrating mechanisms. *Plant Journal*, 101(4), 897–918.

Jackson, C., Knoll, A.H., Chan, C.X., Verbruggen, H. (2018). Plastid phylogenomics with broad taxon sampling further elucidates the distinct evolutionary origins and timing of secondary green plastids. *Scientific Reports*, 8(1), 1523.

Kedem, I., Treves, H., Noble, G., Hagemann, M., Murik, O., Raanan, H., Oren, N., Giordano, M., Kaplan, A. (2020). Keep your friends close and your competitors closer: Novel interspecies interaction in desert biological sand crusts. *Phycologia*, 60, 419–426. DOI: https://doi.org/10.1080/00318884.2020.1843349.

Larkum, A.W.D., Davey, P.A., Kuo, J., Ralph, P.J., Raven, J.A. (2017). Carbon-concentrating mechanisms in seagrasses. *Journal of Experimental Botany*, 68(14), 3773–3784.

Larkum, A.W.D., Ritchie, R.J., Raven, J.A. (2018). Living off the sun: Chlorophylls, bacteriochlorophylls and rhodopsins. *Photosynthetica*, 56(1), 11–43.

Lehours, A.C., Enault, F., Boeuf, D., Jeanthon, C. (2018). Biogeographic patterns of aerobic anoxygenic phototrophic bacteria reveal an ecological consistency of phylogenetic clades in different oceanic biomes. *Scientific Reports*, 8(1), 4105.

Luo, G., Ono, S., Beukes, N.J., Wang, D.T., Xie, S., Summons, R.E. (2016). Rapid oxygenation of Earth's atmosphere 2.33 billion years ago. *Science Advances*, 2(5), e1600134.

Maberly, S.C. (2014). The fitness of the environments of air and water for photosynthesis, growth, reproduction and dispersal of photoautotrophs: An evolutionary and biogeochemical perspective. *Aquatic Botany*, 118, 4–13.

Maberly, S.C. and Gontero, B. (2018). Trade-offs and synergies in the structural and functional characteristics of leaves photosynthesizing in aquatic environments. In *The Leaf: A Platform for Performing Photosynthesis*, Adams III, W.W. and Terashima, I. (eds). Springer International Publishing AG, Cham.

Malone, L.A., Proctor, M.S., Hitchcock, A., Hunter, C.N., Johnson, M.P. (2021). Cytochrome b(6)f – Orchestrator of photosynthetic electron transfer. *Biochimica and Biophysica Acta-Bioenergetics,* 1862(5), 148380.

Medlin, L.K. (2016). Evolution of the diatoms: Major steps in their evolution and a review of the supporting molecular and morphological evidence. *Phycologia*, 55(1), 79–103.

Morozov, A.A. and Galachyants, Y.P. (2019). Diatom genes originating from red and green algae: Implications for the secondary endosymbiosis models. *Marine Genomics*, 45, 72–78.

Nakayama, T., Kamikawa, R., Tanifuji, G., Kashiyama, Y., Ohkouchi, N., Archibald, J.M., Inagaki, Y. (2014). Complete genome of a nonphotosynthetic cyanobacterium in a diatom reveals recent adaptations to an intracellular lifestyle. *Proceedings of the National Academy of Sciences of the United States of America*, 111(31), 11407–11412.

Nakov, T., Boyko, J.D., Alverson, A.J., Beaulieu, J.M. (2017). Models with unequal transition rates favor marine origins of cyanobacteria and photosynthetic eukaryotes. *Proceedings of the National Academy of Sciences of the United States of America*, 114(50), E10606–E10607.

Nelsen, M.P., Luecking, R., Boyce, C.K., Lumbsch, H.T., Ree, R.H. (2020). The macroevolutionary dynamics of symbiotic and phenotypic diversification in lichens. *Proceedings of the National Academy of Sciences of the United States of America*, 117(35), 21495–21503.

Oberleitner, L., Poschmann, G., Macorano, L., Schott-Verdugo, S., Gohlke, H., Stuehler, K., Nowack, E.C.M. (2020). The puzzle of metabolite exchange and identification of putative octotrico peptide repeat expression regulators in the nascent photosynthetic organelles of *Paulinella chromatophora*. *Frontiers in Microbiology*, 11, 607182.

Oliver, T., Sánchez-Baracaldo, P., Larkum, A.W., Rutherford, A.W., Cardona, T. (2021). Time-resolved comparative molecular evolution of oxygenic photosynthesis. *Biochimica and Biophysica Acta (BBA) – Bioenergetics*, 1862(6), 148400.

Orf, G.S. and Blankenship, R.E. (2013). Chlorosome antenna complexes from green photosynthetic bacteria. *Photosynthesis Research*, 116(2), 315–331.

Parfrey, L.W., Lahr, D.J.G., Knoll, A.H., Katz, L.A. (2011). Estimating the timing of early eukaryotic diversification with multigene molecular clocks. *Proceedings of the National Academy of Sciences*, 108(33), 13624–13629.

Ponce-Toledo, R.I., Deschamps, P., Lopez-Garcia, P., Zivanovic, Y., Benzerara, K., Moreira, D. (2017). An early-branching freshwater cyanobacterium at the origin of plastids. *Current Biology*, 27(3), 386–391.

Ponce-Toledo, R.I., Moreira, D., Lopez-Garcia, P., Deschamps, P. (2018). Secondary plastids of euglenids and chlorarachniophytes function with a mix of genes of red and green algal ancestry. *Molecular Biology and Evolution*, 35(9), 2198–2204.

Pushkareva, E., Johansen, J.R., Elster, J. (2016). A review of the ecology, ecophysiology and biodiversity of microalgae in Arctic soil crusts. *Polar Biology*, 39(12), 2227–2240.

Puttick, M.N., Morris, J.L., Williams, T.A., Cox, C.J., Edwards, D., Kenrick, P., Pressel, S., Wellman, C.H., Schneider, H., Pisani, D., Donoghue, P.C.J. (2018). The interrelationships of land plants and the nature of the ancestral embryophyte. *Current Biology*, 28(5), 733–745.

Quigg, A., Irwin, A.J., Finkel, Z.V. (2011). Evolutionary inheritance of elemental stoichiometry in phytoplankton. *Proceedings of the Royal Society B – Biological Sciences*, 278(1705), 526–534.

Raven, J.A. (1984). A cost-benefit-analysis of photon-absorption by photosynthetic unicells. *New Phytologist*, 98(4), 593–625.

Raven, J.A. (1996). The bigger the fewer: Size, taxonomic diversity and the range of chlorophyll(ide)pigments in oxygen-evolving marine photolithotrophs. *Journal of the Marine Biological Association of the United Kingdom*, 76(1), 211–217.

Raven, J.A. (1997). Phagotrophy in phototrophs. *Limnology and Oceanography*, 42(1), 198–205.

Raven, J.A. (2015). Implications of mutation of organelle genomes for organelle function and evolution. *Journal of Experimental Botany*, 66(19), 5639–5650.

Raven, J.A. (2018). Blue carbon: Past, present and future, with emphasis on macroalgae. *Biology Letters*, 14(10), 20180336.

Raven, J.A. and Beardall, J. (2020). Energizing the plasmalemma of marine photosynthetic organisms: The role of primary active transport. *Journal of the Marine Biological Association of the United Kingdom*, 100(3), 333–346.

Raven, J.A. and Sánchez-Baracaldo, P. (2021). *Gloeobacter* and the implications of a freshwater origin of Cyanobacteria. *Phycologia*, 60, 402–418 [Online]. Available at: https://doi.org/10.1080/00318884.2021.1881729.

Raven, J.A. and Smith, F.A. (1981). H^+ transport in the evolution of photosynthesis. *Biosystems*, 14(1), 95–111.

Raven, J.A., Beardall, J., Sánchez-Baracaldo, P. (2017). The possible evolution and future of CO_2-concentrating mechanisms. *Journal of Experimental Botany*, 68(14), 3701–3716.

Raven, J.A., Knight, C.A., Beardall, J. (2019a). Cell size has gene expression and biophysical consequences for cellular function. *Perspectives of Phycology*, 6, 81–94.

Raven, J.A., Knight, C.A., Beardall, J. (2019b). Genome and cell size variation across algal taxa. *Perspectives of Phycology*, 6, 59–80.

Rickaby, R.E.M. and Hubbard, M.R.E. (2019). Upper ocean oxygenation, evolution of RuBisCO and the Phanerozoic succession of phytoplankton. *Free Radical Biology and Medicine*, 140, 295–304.

Rogers, S.O. (2021). Photosynthetic systems suggest an evolutionary pathway to Diderms. *Acta Biotheoretica*, 69, 343–358.

Sánchez-Baracaldo, P., Ridgwell, A., Raven, J.A. (2014). A neoproterozoic transition in the marine nitrogen cycle. *Current Biology*, 24(6), 652–657.

Sánchez-Baracaldo, P., Bianchini, G., Huelsenbeck, J.P., Raven, J.A., Pisani, D., Knoll, A.H. (2017a). Reply to Nakov *et al.*: Model choice requires biological insight when studying the ancestral habitat of photosynthetic eukaryotes. *Proceedings of the National Academy of Sciences*, 114(50), E10608.

Sánchez-Baracaldo, P., Raven, J.A., Pisani, D., Knoll, A.H. (2017b). Early photosynthetic eukaryotes inhabited low-salinity habitats. *Proceedings of the National Academy of Sciences of the United States of America*, 114(37), E7737–E7745.

Sánchez-Baracaldo, P., Bianchini, G., Wilson, J.D., Knoll, A.H. (2021). Cyanobacteria and biogeochemical cycles through Earth history. *Trends in Microbiology*. DOI: doi:10.1016/j.tim.2021.05.008.

Stoecker, D.K., Hansen, P.J., Caron, D.A., Mitra, A., Annual, R. (2017). Mixotrophy in the marine plankton. *Annual Review of Marine Science*, 9, 311–335.

Ting, C.S., Rocap, G., King, J., Chisholm, S.W. (2002). Cyanobacterial photosynthesis in the oceans: The origins and significance of divergent light-harvesting strategies. *Trends in Microbiology*, 10(3), 134–142.

Usher, K.M., Bergman, B., Raven, J.A. (2007). Exploring cyanobacterial mutualisms. *Annual Review of Ecology Evolution and Systematics*, 38, 255–273.

Vantussenbroek, B.I. (1993). Plant and frond dynamics of the giant-kelp, *Macrocystis Pyrifera*, forming a fringing zone in the Falkland Islands. *European Journal of Phycology*, 28(3), 161–165.

Ward, L.M. and Shih, P.M. (2021). Granick revisited: Synthesizing evolutionary and ecological evidence for the late origin of bacteriochlorophyll via ghost lineages and horizontal gene transfer. *PLOS ONE*, 16(1), e0239248.

Ward, L.M., Cardona, T., Holland-Moritz, H. (2019a). Evolutionary implications of anoxygenic phototrophy in the bacterial phylum *Candidatus* Eremiobacterota (WPS-2). *Frontiers in Microbiology*, 10, 1658.

Ward, L.M., Rasmussen, B., Fischer, W.W. (2019b). Primary productivity was limited by electron donors prior to the advent of oxygenic photosynthesis. *Journal of Geophysical Research-Biogeosciences*, 124(2), 211–226.

Westacott, S., Planavsky, N.J., Zhao, M.-Y., Hull, P.M. (2021). Revisiting the sedimentary record of the rise of diatoms. *Proceedings of the National Academy of Sciences of the United States of America*, 118(27), e2103517118. DOI: doi:10.1073/pnas.2103517118.

Williams, J.J., Mills, B.J.W., Lenton, T.M. (2019). A tectonically driven Ediacaran oxygenation event. *Nature Communications*, 10, 2690.

Williamson, C.J., Turpin-Jelfs, T., Nicholes, M.J., Yallop, M.L., Anesio, A.M., Tranter, M. (2021). Macro-nutrient stoichiometry of glacier algaefrom the southwestern margin of the Greenland ice sheet. *Frontiers in Plant Science*, 12, 673614.

Xiao, M., Li, M., Reynolds, C.S. (2018). Colony formation in the cyanobacterium Microcystis. *Biological Reviews*, 93(3), 1399–1420.

Yang, Y., Matsuzaki, M., Takahashi, F., Qu, L., Nozaki, H. (2014). Phylogenomic analysis of "red" genes from two divergent species of the "green" secondary phototrophs, the chlorarachniophytes, suggests multiple horizontal gene transfers from the red lineage before the divergence of extant chlorarachniophytes. *PLOS ONE*, 9(6), e101158.

Yellowlees, D., Rees, T.A.V., Leggat, W. (2008). Metabolic interactions between algal symbionts and invertebrate hosts. *Plant Cell and Environment*, 31(5), 679–694.

3

Biogeographical Patterns and Genomes of Aquatic Photoautotrophs

Juan José PIERELLA KARLUSICH, Charlotte NEF, Chris BOWLER and Richard G. DORRELL
Institut de Biologie de l'École Normale Supérieure (IBENS), École Normale Supérieure, CNRS, INSERM, Université PSL, Paris, France

3.1. Introduction – the changing face of algal genomes

Almost half of photosynthesis on Earth occurs in marine and freshwater environments; hence, aquatic photosynthetic organisms have an enormous impact on food webs and on the biogeochemistry of the entire Earth System. Photosynthesis in marine and freshwater habitats is supported by cyanobacteria and a diverse range of eukaryotic algae that acquired photosynthesis through the endosymbiotic uptake of chloroplasts (see Chapter 2). These eukaryotic algae are highly important for global environmental processes, accounting for over one-third of planetary primary production (Dorrell and Smith 2011), and may be particularly sensitive to environmental change (Li *et al.* 2009; Hinder *et al.* 2012). Eukaryotic algae, however, comprise huge functional diversity, ranging from tiny single-celled organisms of less than one micrometer diameter to giant seaweeds a hundred meters in length, and including obligate phototrophs, predators and symbionts of other organisms (Figure 3.1). The phylogenetic relationships and evolutionary histories of these groups are also extremely complex, with different algal groups separated by one another, and from plants, by hundreds of millions of years of evolution, and multiple horizontal and endosymbiotic gene exchanges (Figure 3.1) (Dorrell and

Blue Planet, Red and Green Photosynthesis,
coordinated by Stephen Christopher MABERLY and Brigitte GONTERO.

Smith 2011; Sibbald and Archibald 2017; Novák Vanclová *et al.* 2020). The diversity of eukaryotic algae renders understanding their biology extremely challenging; dependent on resolving the genomic content of multiple, distantly related lineages, even before this information can be placed into a functional or environmental context.

The advent of next-generation sequencing has revolutionized our understanding of microbial evolution, physiology and ecology. Within the context of microalgae, this commenced with the completion of chloroplast (Kowallik *et al.* 1995; Reith and Munholland 1995) and mitochondrial genomes (Wolff *et al.* 1994; Ohta *et al.* 1998) in the 1990s, followed by the complete nuclear genomes of the red alga *Cyanidioschyzon merolae* and the diatom *Thalassiosira pseudonana*, and the green algae *Chlamydomonas reinhardtii* and *Ostreococcus tauri* in the early 2000s (Armbrust *et al.* 2004; Matsuzaki *et al.* 2004; Derelle *et al.* 2006; Merchant *et al.* 2007). At the date of writing this publication (February 2021), well over a hundred eukaryotic algal genomes have been completed (Grigoriev *et al.* 2020; Hanschen and Starkenburg 2020; Nelson *et al.* 2021). Alongside these, hundreds of transcriptome libraries, that is, surveys of expressed genomic sequences through high-throughput sequencing of cellular RNA, have been completed for eukaryotic algae (Figure 3.2), providing unprecedented insight into the macro- and micro-diversity, and physiological genetics of marine photoautotrophs.

Despite the impressive scale of sequence information assembled in the last three decades, the genomes and transcriptomes of cultured algae may provide somewhat limited insights into the functional biology of algae in their native environment. Many of the algal species in the world's ocean rest uncultured, and may remain unculturable (Mangot *et al.* 2017; Absolon *et al.* 2019). Here, environmental data, exemplified by ambitious projects such as the BioMarKs, Malaspina and *Tara* Oceans expeditions (Karsenti *et al.* 2011; Dunthorn *et al.* 2014; Fernández-Castro *et al.* 2014; Bork *et al.* 2015) can provide valuable insights into this microbial "dark matter", which can then be correlated to spatial information and environmental parameters. In this chapter, we provide an overview of the main groups of aquatic photoautotrophs, exploring their complex evolutionary histories, and the diversity and limitations of genome and transcriptome data from cultured species for understanding algal biology. Subsequently, we consider how metagenomic and metatranscriptomic data generated by the *Tara* Oceans expeditions (Carradec *et al.* 2018; Pierella Karlusich *et al.* 2020a) has unveiled patterns of algal biogeography and mechanisms of adaptation and acclimation to the marine environment, changing our understanding of how the dominant groups of algae function in the open ocean.

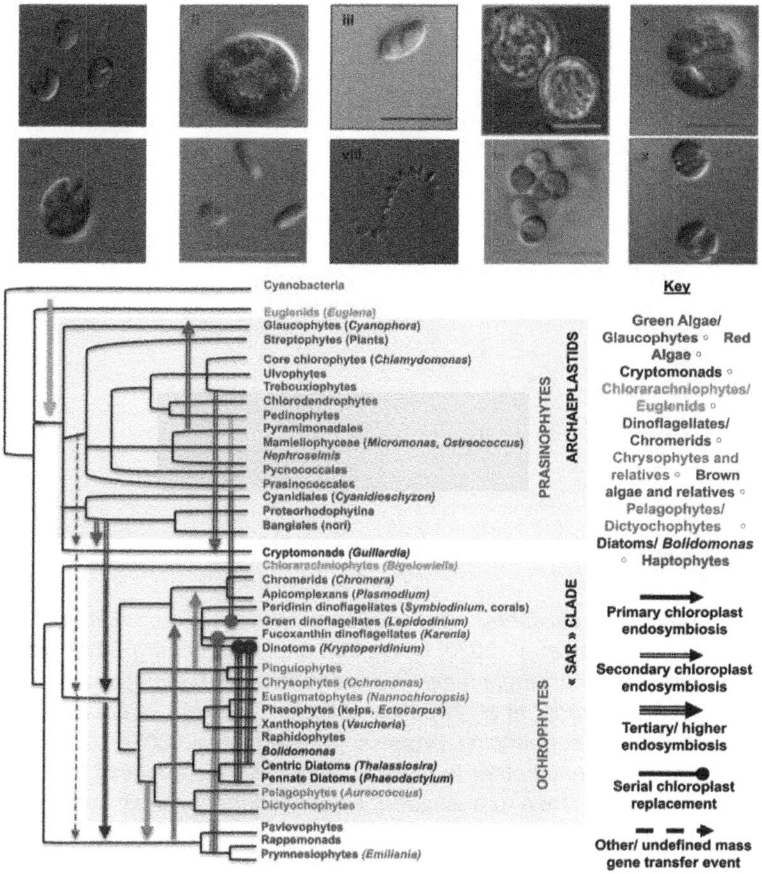

Figure 3.1. *Diversity of eukaryotic algae. Top: exemplar photos of eukaryotic algae. Images are (i)* Nephroselmis *sp. RCC1805 (prasinophyte); (ii)* Rhodella maculata *RCC655 (proteorhodophytinan red alga); (iii)* Hemiselmis *sp. RCC4215 (crypto-monad); (iv)* Lotharella reticulosa *RCC376 (chlorarachniophyte); (v)* Karlodinium veneficum *RCC2539 (fucoxanthin-containing dinoflagellate); (vi)* Heterosigma akashiwo *RCC1502 (raphidophyte); (vii)* Ochromonas *sp. RCC2350 (chrysophyte); (viii)* Asterionellopsis glacialis *RCC2636 (diatom); (ix)* Pelagococcus subviridis *RCC5644 (pelagophyte); and (x)* Isochrysis *sp. RCC1956 (haptophyte). Scale bars: 10 µm. All images kindly supplied by the Roscoff Culture Collection. Bottom: schematic tree of eukaryotic algae, summarizing key phylogenetic relationships and endosymbiotic gene transfers that underpin the evolution of algal genomes. Branch colors correspond to the box colors in the photo above. Data in this figure is summarized from Stiller* et al. *(2014), Dorrell* et al. *(2017) and Dorrell* et al. *(2021). For a color version of this figure, see www.iste.co.uk/maberly/photosynthesis.zip*

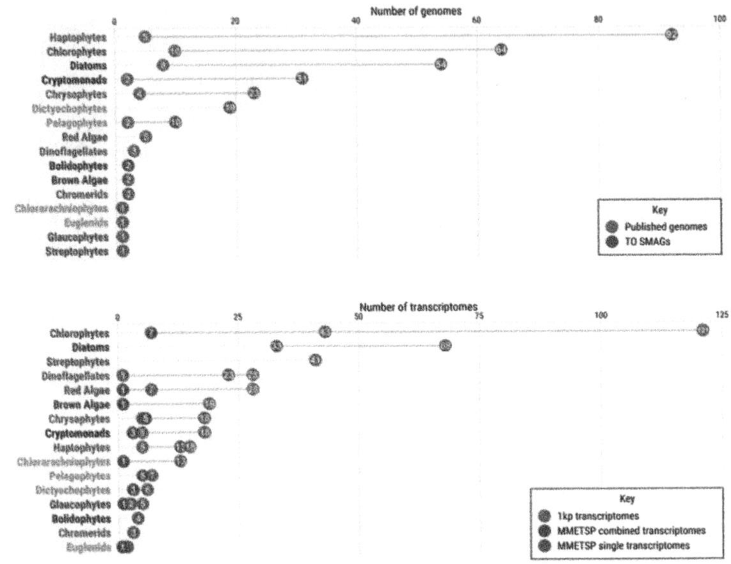

Figure 3.2. *Diversity of published algal sequence libraries. Top: number of published algal genomes (Grigoriev* et al. *2020; Hanschen and Starkenburg 2020; Nelson* et al. *2021) and* Tara *Oceans single amplified genomes and metagenome-assembled genomes (SMAGs, Delmont* et al. *2020); and bottom: number of equivalent from the MMETSP and OneKp sequencing projects (Keeling* et al. *2014; Carpenter* et al. *2019). Dots that shade each other (e.g. for chrysophyte and euglenid transcriptomes) denote a same value. Taxa are shaded by phylogenetic origin, using the same scheme as Figure 3.1. For a color version of this figure, see www.iste.co.uk/maberly/photosynthesis.zip*

3.2. Diversity of algae and their chloroplasts

The advent of genome sequencing has provided substantial insights into the taxonomy and diversity of eukaryotic algae (Figure 3.1). The earliest-evolving eukaryotic algal groups are found within the archaeplastids (glaucophytes and green and red algae), which are unified by the presence of chloroplasts surrounded by two membranes. Sequencing of archaeplastid chloroplast genomes confirmed the long-established hypothesis that they originated from the primary endosymbiosis of a free-living beta-cyanobacterium (Margulis 1975; Reith and Munholland 1995).

Archaeplastids notably include multicellular red algae that form important crop species (e.g. *Pyropia yezoensis*, or nori seaweed), the model single-celled freshwater alga *Chlamydomonas* within the chlorophyte lineage and land plants within the streptophyte lineage, both in the green algae (Merchant *et al.* 2006; Umen 2014; Skokan *et al.* 2019; Wang *et al.* 2020; Figure 3.1). The most abundant of the archaeplastid groups in the contemporary ocean are the prasinophytes, tiny single-celled chlorophytes considered particularly to dominate photosynthetic biomass and production in oligotrophic open ocean systems. Well-studied prasinophyte genera include *Ostreococcus*, considered to be the smallest (ca. 0.95 μm) free-living eukaryote (Derelle *et al.* 2006), and *Micromonas*, both of which are found worldwide.

The dominant eukaryotic algal groups in the oceans today, however, are distantly related to the archaeplastids, including groups such as cryptomonads, haptophytes, dinoflagellates and the various algal groups that comprise the ochrophytes (Figure 3.1). The majority of these algae contain chloroplasts that are red or brown in color (dependent on the accessory pigments they contain; typically phycobilisomes in red-colored cryptomonads, and fucoxanthin in brown-colored dinoflagellates, haptophytes and ochrophytes) and are surrounded by three or four membranes, as compared to the two found in archaeplastids (Dorrell and Smith 2011). Consideration of the membrane ultrastructure of these chloroplasts (Cavalier-Smith 2003), alongside phylogenetic analysis of their genomes (Kowallik *et al.* 1995), suggests that they have arisen through the secondary or higher (e.g. tertiary) endosymbioses of eukaryotic algae, as opposed to the primary endosymbiosis underpinning archaeplastid chloroplasts, and that the majority of algal groups in the ocean today possess chloroplasts of red algal origin. The secondary red chloroplasts are closely related to one another, although their underlying hosts are not (Dorrell and Smith 2011; Figure 3.1).

Specific groups of algae with secondary red chloroplasts have risen to positions of particular ecological importance. The diatoms, for example, which are members of the ochrophytes, are of unmatched dominance in marine ecosystems, probably performing well over one-fifth of total planetary photosynthesis (Bowler *et al.* 2010; Dorrell and Smith 2011). The characteristic highly elaborate siliceous cell walls of diatoms, known as frustules, are easily recognizable and are well preserved in the fossil record, and many species form chains of interlocking cells which are also easily recognizable (Figure 3.1). Diatoms tend to thrive in nutrient-rich turbulent systems such as those found in coastal upwelling regions and at high latitudes, and their life histories are often characterized by rapid proliferations, followed by die-offs, in spring and autumn following nutrient injection events (Malviya *et al.* 2016; Tréguer *et al.* 2018).

The haptophytes are mainly composed of small planktonic unicellular species, occurring either as single cells or colonies (Figure 3.1). This group includes the coccolithophores, which are covered by small regular calcareous plates (coccoliths) and are extremely important in biogeochemical cycles because they are responsible for about half of all modern precipitation of $CaCO_3$ in the ocean, forming the basis of calcium-containing rocks such as limestone (Milliman 1993). The most well-known representative is the coccolithophore *Emiliania huxleyi*, which can form giant blooms visible from space (Tyrrell and Merico 2004; Durak *et al.* 2017).

The dinoflagellates are a hyphenate species-rich group, with taxa inhabiting both planktonic and benthic environments, and with lifestyles ranging from photoautotrophy to heterotrophy, as well as symbionts and parasites (Steidinger and Jangen 1997; Simon *et al.* 2009). Some species can form highly resistant spores that are preserved in the geological record. Many others are endosymbionts of marine animals, the most well-known being *Symbiodinium*, the zooxanthellae of reef-building corals; while others can form toxic algal blooms, such as the red tide-causing *Alexandrium* (Cumbo *et al.* 2013; González-Pech *et al.* 2019). Dinoflagellates are known for their divergent and bizarre evolutionary trajectories, for example, not using histones for packaging their nuclear DNA, in contrast to all other known eukaryotic lineages, and possessing expanded nuclear genomes and shrunken chloroplast and mitochondrial genomes (Dorrell and Howe 2015; Gornik *et al.* 2019).

Alongside these three major groups are many others; for example, within the ochrophytes are the pelagophytes (typified by the harmful algal bloom-causing *Aureococcus anophagefferens*), chrysophytes (golden algae, e.g. *Ochromonas*) and phaeophytes (containing giant kelps) and the cryptomonads and chromerids, which also contain red-derived chloroplasts (Cumbo *et al.* 2013); euglenids and chlorarachniophytes (Figure 3.1). The euglenids and chlorarachniophytes differ from the other groups described, possessing chloroplasts derived from the secondary endosymbiosis of green algae, although analysis of euglenid and chlorarachniophyte chloroplast genomes indicate that they are derived from distantly related members of the green algae tree (Kamikawa *et al.* 2015; Jackson *et al.* 2018).

Finally, alongside other aspects of their unusual evolutionary history, dinoflagellates have acquired chloroplasts on multiple occasions, from multiple sources. These include the red-derived chloroplasts found in most dinoflagellate species, and also alternatives derived from green algae, and explicitly from the tertiary endosymbiosis of diatoms and haptophytes (themselves containing secondary, red-type chloroplasts) (Dorrell and Howe 2015; Sarai *et al.* 2020). This extreme promiscuity in chloroplast diversity is the result of rampant serial endosymbiosis, in which the original dinoflagellate chloroplast has been lost and replaced multiple times, and may be linked to the extreme propensity of dinoflagellates for phagotrophic and photo-mixotrophic life strategies (Dorrell and Howe 2015; Sarai *et al.* 2020; Cohen *et al.* 2021; Jeong *et al.* 2021) (Figure 3.1).

3.3. Genomic insights into algal evolution

The annotation of eukaryotic algal nuclear genome sequences, particularly within a phylogenetic context, has yielded an even more complex story. Large numbers of genes, which may be the footprints of endosymbiont-derived genes transferred to the nucleus (Bock 2017; Henze *et al.* 2002), phylogenetically link together different groups of algae with secondary chloroplasts, suggesting that several of them have in fact originated via sequential secondary, tertiary and even quaternary endosymbiosis (Stiller *et al.* 2014; Dorrell *et al.* 2017). For example, different phylogenetic investigations of algal genomes have suggested that the chloroplasts of ochrophytes originate from cryptomonad algae (Stiller *et al.* 2014); the chloroplasts of haptophytes may originate from pelagophytes within the ochrophytes (Dorrell *et al.* 2017, 2021); the chloroplasts of dinoflagellates may originate from haptophytes (Yoon *et al.* 2002; Dorrell *et al.* 2021) and the chloroplasts of chromerids may originate from chrysophytes within the ochrophytes (Ševčíková *et al.* 2015).

In addition, nuclear genome sequences have revealed cryptic genetic signals in many eukaryotic algae that are not related to their current chloroplasts. For instance, many of the algae with chloroplasts of purported red algal origin (e.g. cryptomonads, haptophytes and ochrophytes) contain large numbers of genes of green algal origin (Moustafa *et al.* 2009; Dorrell *et al.* 2017), whereas chlorarachniophytes and euglenids, which possess chloroplasts derived from the secondary endosymbioses of green algae, also contain genes derived from red algae, or their endosymbiotic progeny (Ponce-Toledo *et al.* 2018; Novák Vanclová *et al.* 2020). Many of these genes encode chloroplast-associated proteins, suggesting that their origin is in some way connected to chloroplast endosymbiosis (Dorrell *et al.* 2017); although it is not known whether these signals are the relics of ancient endosymbioses that have subsequently been replaced, similarly to the serial endosymbiotic replacements known in dinoflagellates (Dorrell and Howe 2015; Sarai *et al.* 2020), or may have other origins entirely, for example the endosymbiotic acquisition of algae that themselves contained mixtures of red and green genes (Dorrell *et al.* 2017; Ponce-Toledo *et al.* 2018).

Alongside endosymbiotic signals, algal nuclear genomes can contain large numbers of genes derived from non-endosymbiotic horizontal gene transfer events, which are known to occur between eukaryotic algae (Raymond and Kim 2012; Kazamia *et al.* 2018), and from bacterial and viral donors (Rossoni *et al.* 2019;

Vancaester *et al.* 2020; Dorrell *et al.* 2021; Nelson *et al.* 2021). These horizontally acquired genes may dramatically affect the functions of eukaryotic algae in the environment. For example, inspection of diatom genomes has identified unique combinations of genes of bacterial and eukaryotic origin that collectively encode non-canonical pathways of nutrient assimilation and metabolite management (Wilhelm *et al.* 2006), including a urea cycle that is integral to nitrogen metabolism (Allen *et al.* 2011; Smith *et al.* 2019; Dorrell *et al.* 2021), and high-affinity iron uptake systems not known in other eukaryotic groups (Kazamia *et al.* 2018; McQuaid *et al.* 2018).

Finally, the genomes of eukaryotic algae may also undergo complexification and simplification events pertaining to lifestyle transitions. Diatom genomes, for example, have been found to encode expanded sets of cyclins (Huysman *et al.* 2010), key regulators of cell division, as well as light-harvesting chlorophyll-binding proteins (Bailleul *et al.* 2010; Mock *et al.* 2017) that may be important in, respectively, driving proliferation and stress responses in the ocean environment. In contrast, many lineages of eukaryotic algae have secondarily reverted from photoautotrophic to exclusively heterotrophic lifestyles, accompanied by the loss of large numbers of genes associated with photosynthesis. This is particularly notable in the chrysophytes, within the ochrophytes, as well as the dinoflagellates, in which fewer than half of all documented species are photosynthetic (Sournia 1986; Gomez 2005; Simon *et al.* 2009; Dorrell *et al.* 2019).

3.4. Limitations of cultured algal sequencing projects

In Figures 3.2 and 3.3, we present the current (as of February 2021) phylogenetic and biogeographical diversity of available algal nuclear genomes and transcriptomes, based on the manually verified isolation sites of the source cultures sequenced (Vaulot *et al.* 2004; Gachon *et al.* 2013; Guiry *et al.* 2014; Boundy-Mills *et al.* 2015; Nelson *et al.* 2021). The greatest numbers of genomes have been sequenced from green algae and diatoms (Figure 3.2), reflecting their interest as models for plant evolution (in the case of green algae), aquaculture (both lineages) and ecological prominence in the contemporary ocean (diatoms in particular) (Bowler *et al.* 2010; Kumar *et al.* 2017; One Thousand Plant Transcriptomes Initiative 2019); however, multiple genome sequences have now been completed for many other algal groups, including haptophytes, pelagophytes and dinoflagellates (Grigoriev *et al.* 2020; Nelson *et al.* 2021).

Alongside this genomic diversity, hundreds of algal transcriptomes have been completed, either through individual projects or consortium efforts, as described above (Keeling *et al.* 2014; One Thousand Plant Transcriptomes Initiative 2019).

Transcriptome libraries carry some major advantages over genome assemblies, not least that coding sequences can be easily annotated without considering introns (which can be large and difficult to infer in some major algal groups, e.g. dinoflagellates (Gornik *et al.* 2019)), and that preliminary insights can be made into gene expression patterns. However, transcriptomes cannot provide insights into non-coding DNA, or genes typically removed by poly(A) selection (e.g. organelle transcripts, Smith and Sanitá Lima 2017), and may be more heavily impacted by biological contamination than genomes, in the absence of a possibility to assemble gene sequences into contigs (Carpenter *et al.* 2019). Nonetheless, transcriptomes used in partnership with genome sequences can radically expand our breadth of understanding of algal genome diversity, and allow exploration of algal lineages (e.g. dictyochophytes and bolidophytes; Figure 3.2) and biogeographical regions (the Arctic and Pacific Oceans; Figure 3.3) for which reference algal genomes remain relatively limited in number.

Despite this vast amount of resources, some biases and limitations are still visible. For example, the overwhelming majority of complete algal genomes and transcriptomes are from species isolated from North American and European oceanic regions (Figure 3.3), and many are from coastal and terrestrial environments, while open-ocean algae, which may be less easily accessible, are under-represented (Lommer *et al.* 2012) (Figure 3.3). Moreover, many of the algal species for which we have genome sequences, even though biologically interesting, may not be particularly representative of those that grow in the oceans. For example, the first sequenced red alga, *C. merolae*, was an extremophile isolated from acidic hot sulfur springs, with a growth temperature (up to 56°C) and pH (as low as 1.5) very different to most marine red algal relatives (Matsuzaki *et al.* 2004; Brawley *et al.* 2017), and may accordingly have very different core metabolic pathways to other red algal species (e.g. Helliwell *et al.* 2011; Rossoni *et al.* 2019).

Even within well-sampled oceanic regions, there may be limitations in what genomic information is available. Comparisons between observational (e.g. microscopy) and culturable data under a variety of environments have long pointed to the presence of an uncultured majority of marine micro-organisms (Razumov 1932; Jannasch and Jones 1959; Zimmermann *et al.* 1978), with only between 0.01 and 1% microbial taxa being cultivable (Amann *et al.* 1995; Rappé and Giovannoni 2003). Within the context of algae, several major groups of organisms have been identified from environmental sampling, such as MOCHs (Marine OCHrophytes), probably related to pelagophytes and dictyochophytes within the ochrophytes (Massana *et al.* 2014) and rappemonads, which form an early-diverging component of haptophytes (Kim *et al.* 2011; Kawachi *et al.* 2021); they have largely eluded or only recently been brought into successful culture in laboratory conditions (Absolon *et al.* 2019).

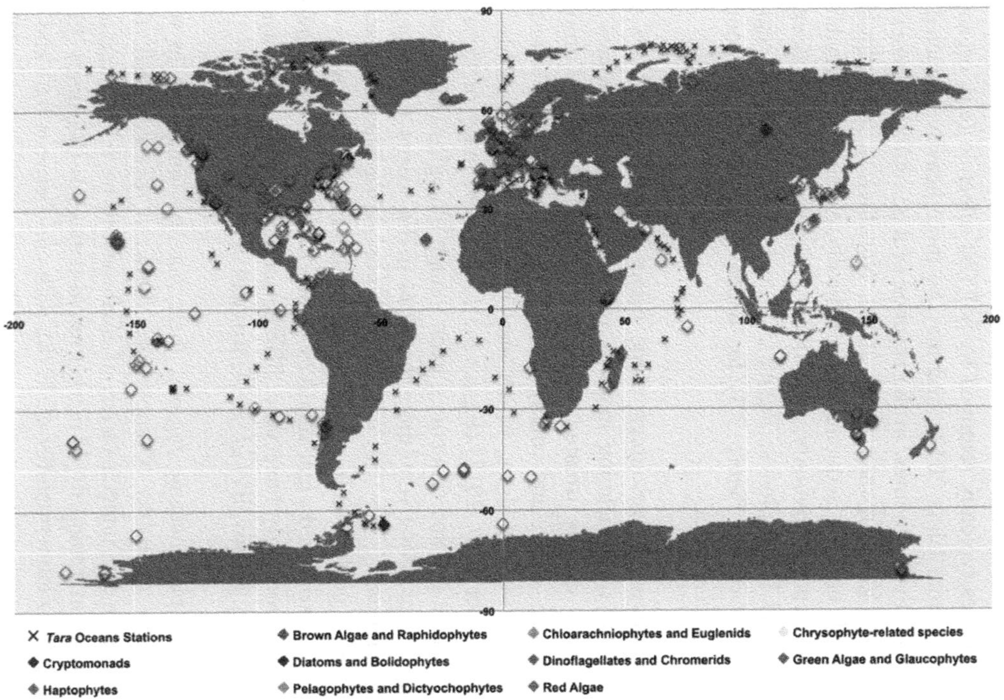

Figure 3.3. *Isolation sites of sequenced algal genomes (solid shapes) and MMETSP (Marine Microbial Eukaryotic Transcriptome Sequencing Project) transcriptomes (open shapes), compared to sampling sites for the* Tara Oceans *and* Tara Oceans Polar *Circle expeditions. Isolation sites were manually verified from culture collection accession for each strain (Vaulot et al. 2004; Gachon et al. 2013; Guiry et al. 2014; Boundy-Mills et al. 2015). Algal sequence libraries are colored by phylogenetic origin, using the same scheme as Figure 3.1. For a color version of this figure, see www.iste.co.uk/maberly/photosynthesis.zip*

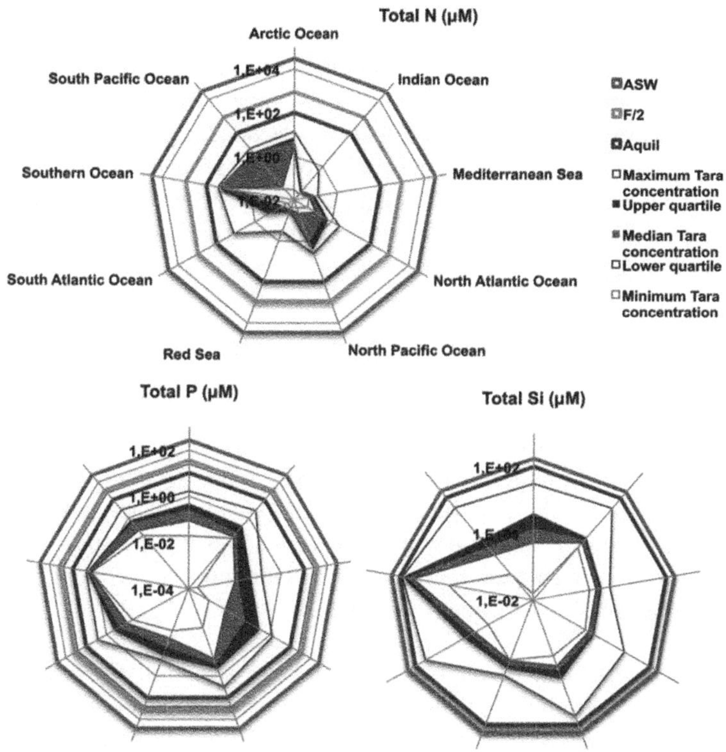

Figure 3.4. *Maximum and minimum values (blue lines) and interquartile ranges (blue sectors) of measured total concentrations at 5 m depth of nitrogen, phosphorus and silicon in different* Tara Oceans *regions; compared to supplemented concentrations of each nutrient in three growth media (ASW, f/2 and Aquil; red lines) commonly used for growth of laboratory algal cultures. The authors acknowledge Elena Kazamia (IBENS) for the calculation of growth media nutrient concentrations. For a color version of this figure, see www.iste.co.uk/maberly/photosynthesis.zip*

Finally, even the laboratory conditions in which algal genomes and transcriptomes are realized may be highly unnatural (Figure 3.4). Laboratory growth chambers may be unable to approach the extremely low environmental temperatures encountered in polar oceanic environments (Raymond and Kim 2012; Terrado *et al.* 2015). Furthermore, the growth media necessary to facilitate dense cultures for sequencing may contain nutrient concentrations many hundreds of times those experienced by algae in the open ocean (Figure 3.4), although admittedly the concentrations of these nutrients that are biologically available to algae in the lab and in the wild remain poorly understood (Lommer *et al.* 2012; Kazamia *et al.* 2018). Reference strains may have been cultured for many years or decades under nutrient replete conditions in the laboratory since isolation, and the somatic mutations that have occurred naturally to algae during their time in laboratory culture may impact our understanding of their physiology (Schaum and Collins 2014; Flowers *et al.* 2015; Rastogi *et al.* 2020). These biases, while necessary for the generation of algal sequence libraries in the first place, may limit our understanding of which genes truly underpin inducible responses to environmental stresses, or facilitate the acquisition of growth-limiting nutrients in the ocean (Browning *et al.* 2017).

3.5. History of omics-based approaches applied to environmental plankton samples

As discussed above, the application of next-generation sequencing to cultured algae has allowed us to understand algal diversity more accurately, but with less insight into functional biology in the environment. "-Omics" data generated directly from the natural environment may provide an interesting perspective, offering genomic contexts for taxa not represented in laboratory collections, and direct insights into organisms' experiences in the wild.

Community assessment of plankton samples has occurred for over 100 years (Pierella Karlusich *et al.* 2020b). These efforts started with data collection from the Western Channel Observatory, and the Continuous Plankton Recorder in the early 20th century (Reid *et al.* 2003; Southward *et al.* 2005), yielding large amounts of microscopic abundance and diversity data, that may even be transformable into genomic information via sequencing of preserved samples (Stern *et al.* 2018). Community assessment of plankton samples based on DNA sequencing itself started in the 1990s (Giovannoni *et al.* 1990; Fuhrman *et al.* 1995), but bloomed with the advent of high-throughput sequencing in the 2000s. A global-scale metagenomic sampling of the ocean began with J. Craig Venter's Global Ocean Sampling expeditions between 2003 and 2008, which included the collection of bacterial-enriched size fractions from surface waters from the North-Western Atlantic and Eastern Tropical Pacific Oceans from 2004 to 2006, generating a 6.1 million gene set using Sanger sequencing (Rusch *et al.* 2007). Later, in the context of the Census of Marine Life (http://www.coml.org/), a consortium of researchers deployed the International Census of Marine Microbes (Amaral-Zettler *et al.* 2010) to provide a detailed inventory of marine microbial diversity. Further projects of large spatial coverage came subsequently with the Malaspina expedition, led by Carlos Duarte, which targeted principally the deep ocean and also the epipelagic layers in a worldwide sampling campaign from 2010 to 2011 (Duarte 2015), and the Ocean

Sampling Day initiative, which began with a simultaneous global sampling campaign on June 21, 2014, at 191 different sites, mostly in coastal areas (Kopf *et al.* 2015).

From 2008, Eric Karsenti led a consortium of scientists that organized a circumglobal expedition on board the 36-m-long schooner SV *Tara* (Figure 3.3). The expedition was specifically designed for studying microscopic plankton ecosystems at global scale. Based on a holistic approach, the *Tara* Oceans pan-oceanic expedition sampled plankton ranging in size from viruses to small metazoans, coupled with comprehensive *in situ* biogeochemical measurements (Karsenti *et al.* 2011). A wide range of contrasting ecosystems were targeted, using sampling protocols that were highly standardized and consistent at each site. *Tara* Oceans is in fact derived from two research expeditions performed between 2009 and 2013. The first expedition (named *Tara* Oceans) lasted two years and eight months and sampled all of the principal ocean basins with the exception of the Arctic Ocean; the second (named *Tara* Oceans Polar Circle) lasted seven months and circumnavigated the Arctic Circle in 2013. An additional expedition denoted *Tara* Pacific, currently under analysis, retrieved sites across the Pacific Ocean, with a particular focus on the biology of coral ecosystems (Planes *et al.* 2019) (Figure 3.3). The *Tara* Oceans expeditions consisted of the collection of >35000 plankton samples from 210 sampling sites, which were used to generate >60 terabases of DNA and RNA sequences and ~7 million images (Sunagawa *et al.* 2020). Yet further expeditions are ongoing and planned, starting with the Mission Microbiomes expedition across both sides of the Atlantic Ocean, to understand how marine ecosystem processes are impacted by environmental perturbance and pollution (https://fondationtaraocean.org/expedition/mission-microbiomes/).

3.6. Biogeographical insights of algae from *Tara* Oceans metabarcoding

Environmental sequencing initiatives such as the *Tara* Oceans expedition have provided unprecedented insight into the biological diversity of ocean communities, placed within a geographical context. This diversity has been most clearly resolved by amplicon sequencing or barcoding, in which a fragment of the small subunit of the rRNA gene (16S for prokaryotes, 18S for eukaryotes) is universally amplified and massively sequenced from an environmental sample (Le Bescot *et al.* 2016). Most studies of eukaryotes have so far focused on hypervariable regions of the 18S rRNA gene, which are universal, taxonomically informative, flanked by conserved regions for primer binding, short enough for current sequencing technologies, and are by far the most represented in reference databases (Pawlowski *et al.* 2018), although alternatives exist (e.g. chloroplast *rbcL*, 16S and 23S rDNA; Djemiel *et al.* 2020). The V9 and V4 regions of the 18S rRNA gene are the most commonly used (Stoeck *et al.* 2010; Pernice *et al.*, 2016; Tragin *et al.*, 2018). This approach enabled a tremendous expansion of richness estimates in the microscopic realm, in spite of recurrent and unresolved problems (e.g. definition of species in the absence of data from sexual crosses, and artifacts introduced by the polymerase chain reaction and sequencing methods; Bhadury and Austen 2010; Berruti *et al.* 2017).

The first metabarcoding survey of eukaryotes from *Tara* Oceans consisted of the sequencing of >1.7 million reads belonging to the 18S rRNA gene (V9 region) from 334 samples from 47 different geographical sites and revealed the occurrence of approximately 110,000 distinct eukaryotic operational taxonomic units (OTUs, a proxy for species in molecular surveys), of which approximately 20% could be assigned to photosynthetic species (i.e. phytoplankton and hosts of photosynthetic symbionts (de Vargas *et al.* 2015; Le Bescot *et al.* 2016)). Among phytoplankton, the largest numbers of OTUs were assigned to dinoflagellates and diatoms, followed by prasinophytes within the green algae and haptophytes (Figure 3.5). Other photosynthetic groups, principally other members of the ochrophytes (dictyochophytes and chrysophytes) and cryptomonads were detected at >1% total abundance (Figure 3.5). In contrast, many other lineages of algae that have formed important references for genomic study, for example, red algae, glaucophytes and phaeophytes, represented relatively minor components of the marine microbial communities sampled (Figures 3.2 and 3.5).

In addition to biodiversity, molecular methods can give information into the habitat preferences and morphologies of different algal groups. For example, metabarcoding libraries generated from seawater samples filtered with different exclusion filters can be used to gain insight into the repartition of different algal groups into different size fractions (de Vargas *et al.* 2015), each of which may have very different functional physiologies and respond differently to environmental perturbations (Li *et al.* 2009). The largest size fractions considered by *Tara* Oceans (between 180 and 2,000 μm) are dominated by diatoms, due to the presence of chain-forming (e.g. *Hyalosira* and *Fragilaria*) and epizoic (e.g. *Pseudohimantidium*) species; while the phytoplankton in the 5–20 μm and 20–180 μm fractions principally consist of diatoms and dinoflagellates (Figure 3.5; Malviya *et al.* 2016). Abundance in the smaller 0.8–5-μm size fraction is much more heterogeneous, with haptophytes, dinoflagellates, diatoms (Leblanc *et al.* 2018), prasinophytes, pelagophytes, dictyochophytes and cryptomonads all making substantial contributions.

In relation to their biogeographical distribution, in general terms picophytoplankton (0.2–2 μm, mainly the cyanobacteria *Prochlorococcus* and *Synechococcus*, and also eukaryotes such as pelagophytes, some dinoflagellates, haptophytes and prasinophytes) are found in warm, nutrient-poor waters in the tropical and subtropical ocean, while nanophytoplankton (2–20 μm) can be detected potentially year-round in temperate regions (Figure 3.6). By contrast, microphytoplankton (20–200 μm) bloom mainly at higher latitudes at the beginning of spring and sometimes also in the late summer, as well as in upwelling regions (Figure 3.6) (Li *et al.* 2009). Diatoms are particularly prevalent at high latitudes and in upwelling environments (Figure 3.6), where they are able to bloom and outcompete other marine phytoplankton when nitrate and silicate are abundant (Malviya *et al.* 2016). The contrasting environmental distributions of diatoms and pelagophytes are particularly notable given their close relationship to one another within the ochrophytes (Figure 3.1).

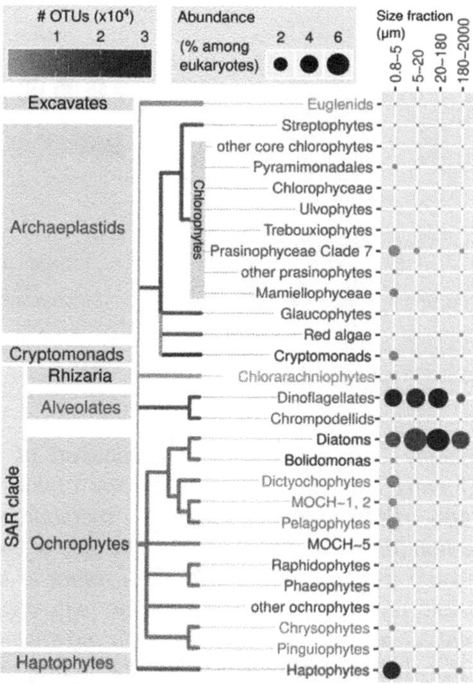

Figure 3.5. *Taxonomic distribution, abundance, and richness of marine eukaryotic phytoplankton based on 18S rRNA gene (V9) metabarcoding across different size fractionated* Tara *Oceans samples. The area of each circle corresponds to the average abundance relative to total eukaryotic read counts, while the color gradient varies according to the number of operational taxonomic units (OTUs). Data are derived from the complete* Tara *Oceans data set for 18S rRNA gene (V9 region) metabarcoding (de Vargas et al. 2015; Ibarbalz et al. 2019). Size-fractionated samples were generated via membrane filtration with different pore sizes. For a color version of this figure, see www.iste.co.uk/maberly/photosynthesis.zip*

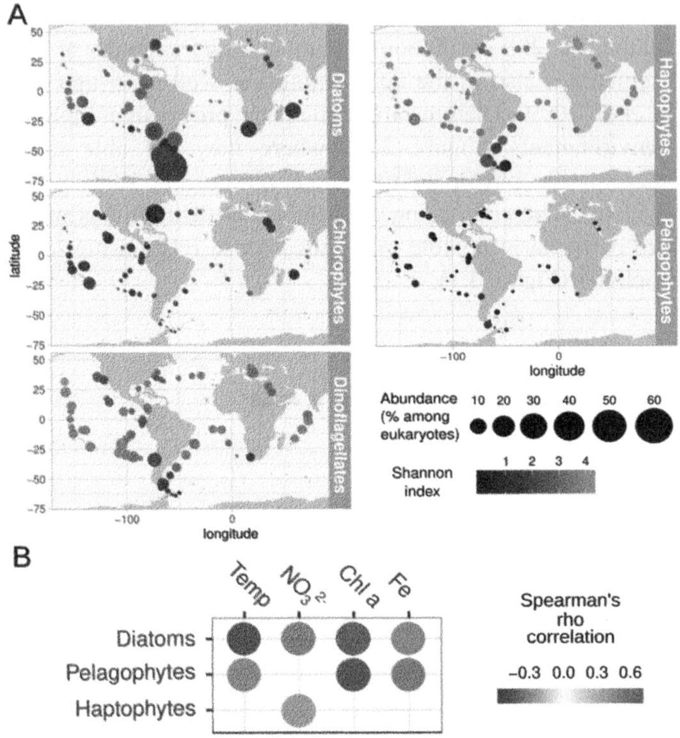

Figure 3.6. *Biogeography and environmental distribution of photosynthetic algae inferred from* Tara Oceans *samples. A: Biogeography of the major photosynthetic protists based on 18S rRNA gene (V9 region) metabarcoding data from the 0.8–2,000-µm size fraction (de Vargas* et al. *2015). The size of each circle corresponds to the abundance at each location, and the fill color refers to the Shannon diversity index (dark blue for low diversity and light blue for high diversity). B: Correlation analysis between relative abundances of the main phytoplankton groups and a selection of productivity-related parameters. Color represents Spearman's rho coefficients. Empty spaces refer to non-significant correlation values (p > 0.05). Chlorophytes and dinoflagellates are not shown due to the absence of significant correlations. For a color version of this figure, see www.iste.co.uk/maberly/ photosynthesis.zip*

In contrast to the clear spatial trends shown by picophytoplankton and diatoms, the abundance patterns of prasinophytes, dinoflagellates and haptophytes are less straightforward to contextualize, and may vary between different members of each group. For example, the haptophyte *Phaeocystis* is a dominant component of microalgal communities at high latitudes (shown in the corresponding map of

Figure 3.6a the larger circles with colors indicating lower diversity values), whereas other haptophytes (e.g. *Emiliania*) constituting diverse communities show predominantly temperate or even tropical distributions (Bendif *et al.* 2014; Pfaff *et al.* 2016). Within the prasinophytes, members of the Mamiellophyceae (e.g. *Ostreococcus*, *Micromonas*) are prominent in coastal waters (Monier *et al.* 2016; Vannier *et al.* 2016), while clade VII prasinophytes dominate oceanic waters (Lopes Dos Santos *et al.* 2017). The high diversity and broad biogeographical distribution of dinoflagellates, including coastal, oceanic, polar, temperate and tropical abundance (Le Bescot *et al.* 2016) (Figure 3.6) are indicative of the broad ecological strategies employed by different genera.

3.7. Functional studies of algae from *Tara* Oceans metagenomic and metatranscriptomic data

In addition to taxonomic analyses based on metabarcoding data, the functional analysis of natural phytoplankton communities can be carried out by metagenomics and metatranscriptomics. The current version of the *Tara* Oceans eukaryotic gene catalogue consists of 116 million non-redundant transcribed sequences mainly from eukaryotic plankton ranging from 0.8 to 2000 μm (Carradec *et al.* 2018). These sequences, termed meta-genes, were generated by assembling 441 poly-A+ metatranscriptomes from samples across the main ocean basins and has no apparent saturation (Carradec *et al.* 2018). Among the most abundant transcripts found in the catalogue encode chlorophyll *a*/*b*-binding proteins, involved in light harvesting and considered the most abundant membrane proteins on Earth (Dittami *et al.* 2010).

The variability of the gene and transcript levels of all these sequences across different environmental conditions can give clues about functional responses in phytoplankton communities. The environmental response of phytoplankton on short timescales (over the lifetime of the cell) is called acclimation and includes the modulation of gene expression. By contrast, adaptation to an environmental niche refers to genetic changes (e.g. in gene copy number) that accumulate over many generations. Responses to environmental cues also include shifts in the community structure, changes of broad taxonomic groups as well as changes of genotypes of the same species adapted to local conditions. These links were detected in diatoms, for example, with nitrogen (Busseni *et al.* 2019) and iron uptake systems (Caputi *et al.* 2019).

Figure 3.7 shows the correlations in *Tara* Oceans data between conditions of high productivity and the expression of gene families containing members involved in photosynthesis and carbon fixation. Among the displayed set of photosynthetic genes are those coding for ferredoxins, iron-containing proteins that can be replaced by flavodoxins, which are iron-free versions of less efficiency (Pierella Karlusich *et al.*,

2014). Although diatoms can concentrate up to 30–40% of their cellular iron into ferredoxin (Pierella Karlusich *et al.*, 2014), the expression of diatom ferredoxin genes shows an extreme positive correlation to environmental conditions indicative of primary production (Figure 3.7). While quantification of ferredoxin-encoding transcripts among poly-A+ metatranscriptomes can be affected by the fact that the gene is plastid encoded in many diatom species (Lommer *et al.* 2012; Dorrell *et al.* 2017), the strong positive correlation between ferredoxin and primary production might reflect diatom iron uptake and usage strategies (Carradec *et al.* 2018; Kazamia *et al.* 2018; Caputi *et al.* 2019).

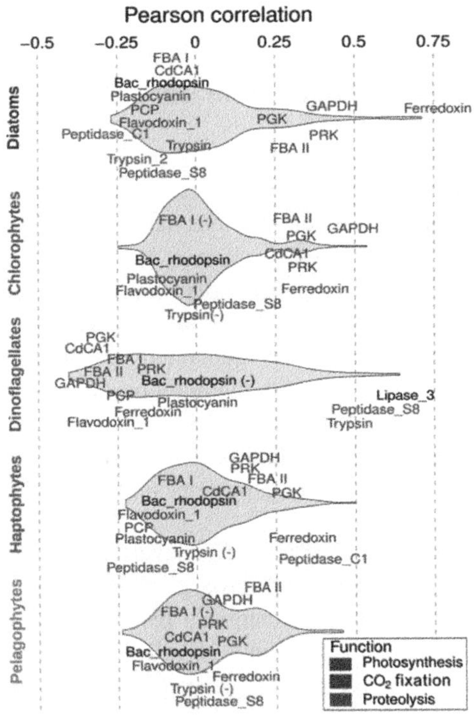

Figure 3.7. *Functional insights into global ocean communities of algae using* Tara *Oceans metatranscriptomes. This figure shows the distribution of Pearson's correlation values of Pfam domains with measurements of net primary productivity in the five major photosynthetic planktonic groups. The profiles illustrate the density distribution of all the correlation values of the Pfams having a relative expression value at least equal to 0.05% (relative expression value is indicated by the size of the grey dots). Pfam representatives of three main functional classes (CO_2 fixation through the Calvin–Benson–Bassham cycle, photosynthetic light reactions/light harvesting, and proteolysis) are colored. Adapted from Carradec* et al. *(2018). For a color version of this figure, see www.iste.co.uk/maberly/photosynthesis.zip*

Metatranscriptome data can even provide information regarding the broader life strategies used by different phytoplankton groups, for example, in the context of mixotrophy (Burkholder *et al.* 2008). Figure 3.7 shows that obligate photoautotrophs, such as diatoms and prasinophytes, exhibit high positive correlations between conditions of high productivity and the expression of gene families containing members involved in photosynthesis (i.e. light harvesting, photosynthetic electron transport and carbon dioxide assimilation through the Calvin–Benson–Bassham cycle); while haptophytes and dinoflagellates show weak or no correlations (Figure 3.7). Indeed, dinoflagellates show high positive correlations to primary production for the expression of gene families encoding cell lytic components, such as proteases and lipases, suggesting that dinoflagellates in the most productive marine communities function predominantly as heterotrophs and predators (Carradec *et al.* 2018). This might reflect switches in the trophic strategy of dinoflagellates under mixotrophic conditions, for example, the replacement of autonomous amino acid assimilation pathways with scavenging from prey or dissolved external organic material (Johnson 2015; Li *et al.* 2020; Cohen *et al.* 2021). Similar positive correlations to peptidases for haptophytes might also reflect previously reported facultative mixotrophy in this group (Unrein *et al.* 2014; Anderson *et al.* 2018).

3.8. Applying genome-resolved metagenomics to phototrophic eukaryotes

Initial studies of environmental datasets have focused either on the study of single marker genes (e.g. 18S rDNA) or functional genes, based on comparison to reference genomes and transcriptomes. While these data can provide deep insights into the diversity and functions of lineages with well annotated cultured representatives, they necessarily provide fewer insights into taxa for which reference libraries from cultured species do not exist, as discussed above (Absolon *et al.* 2019).

Recently, environmental sequence data has been harnessed to provide insights into the complete genome sequences of species and lineages that are not yet cultivable in the lab. Two major types of genomic data can be generated from natural samples: single amplified genomes (SAGs) and metagenome-assembled genomes (MAGs). The former are generated by isolating individual cells, amplifying their genomes using whole genome amplification (WGA) and sequencing their DNA, with the first environmental SAGs generated in 2007 for microbes of the human mouth (Marcy *et al.* 2007). Conversely, MAGs are obtained by assembling metagenomic reads into contigs (continuous DNA segment) with the help of specific algorithms, then grouped using nucleotide sequence signatures such as GC content, tetra-nucleotide frequencies, marker gene phylogenies, depth of DNA sequence coverage, and abundance across the samples (Iverson *et al.* 2012). The grouped samples are then assigned to individual populations using both their coverage and composition as proxies.

The first MAGs were reconstructed from prokaryotic communities in an acid mine drainage (Tyson *et al.* 2004). This technique has subsequently been widely applied to marine viral, archaeal and bacterial populations (Parks *et al.* 2017; Tully *et al.* 2017; Delmont *et al.* 2018; Tully *et al.* 2018; Gregory *et al.* 2019; Cao *et al.* 2020; Reji and Francis 2020), and for some heterotrophic eukaryotes (Yoon *et al.* 2011; Mangot *et al.* 2017; Seeleuthner *et al.* 2018); but very few examples have been available from eukaryotic algae. This bias reflects the size and complexity of eukaryotic genomes, which makes them difficult to assemble from limited environmental sequencing data (Tyson *et al.* 2004; Gregory 2005), particularly in the case of genomes without closely related reference genomes (Bowers *et al.* 2017; Nelson *et al.* 2020), and historical priority in funding biases for less easily cultivable heterotrophic protists (del Campo *et al.* 2014). Moreover, the absence of standardized meta-data for assembled metagenomes can limit their direct comparability to one another (Sibbald and Archibald 2017).

Nonetheless, new methods are emerging to improve the recovery of eukaryotic genomes from environmental data (West *et al.* 2018). For instance, among the MAGs generated for marine phototrophic eukaryotes are small species such as the prasinophytes *Micromonas, Bathycoccus* and *Ostreococcus* (Delmont *et al.* 2015, 2018; Duncan *et al.* 2020) and the haptophyte *Emiliania* (Delmont *et al.* 2018), alongside phylogenetically diverse MAG assemblies from specific marine ecosystems, for example, the Mediterranean Sea and the Arctic Ocean (Tully *et al.* 2018; Duncan *et al.* 2020). Such studies allow us to gain insight into the diversity and ecology of these groups based on their natural populations.

Very recently, an important database has been released that represents 713 marine eukaryotic SMAGs (both MAGs and a few SAGs) from a wide range of organisms, generated from the global diversity of *Tara* Oceans samples (Delmont *et al.* 2020). In this, more than 270 SMAGs (~30% of the database) have been assigned as marine eukaryotic algae, dramatically expanding the number of algal genomes available (Figure 3.2a). For instance, it increases by 18 times the number of available genomes for haptophytes, by 15 for cryptomonads, by 6 for prasinophytes, diatoms and chrysophytes, and enables access for the first time to genomic data for taxonomic groups (e.g. dictyochophytes and bolidophytes; Figure 3.2a) for which no genomes were previously available. Moreover, this *Tara* Oceans SMAGs database gathers highly contextualized data, is freely available, and the genomes have relatively elevated completeness (~40% on average and up to 94% BUSCO completion, a measure of genome completedness based on the detection of broadly conserved marker proteins; Simão *et al.* 2015, Hanschen and Starkenburg 2020). This new database will be a very important resource to gain insight into the genomic diversity of previously under-sampled branches of the eukaryotic tree of life.

3.9. Perspectives

In this chapter, we have explored the dramatic changes in eukaryotic algal genome resources that have become available within one generation of scientific research, starting with the completion of the first chloroplast genomes of cultured algae in the mid-1990s (Kowallik *et al.* 1995; Reith and Munholland 1995) and currently extending to ecologically contextualized genomes and transcriptomes of uncultured algae from environmental sequence metadata (Carradec *et al.* 2018; Delmont *et al.* 2020). This is revolutionizing our understanding of algal diversity, evolution and functional biology, but challenges nonetheless remain.

The quality of the assembly and annotation of cultured algal genomes and transcriptomes may vary substantially (Hanschen and Starkenburg 2020), and indeed assembly is only the start of the much more complex process of functional annotation of genome content, which may depend on the development of transformation strategies enabling reverse genetics (Faktorová *et al.* 2020). However, an increasing number of genomes from transformable algal species (e.g. the diatom *Phaeodactylum tricornutum*; Rastogi *et al.* 2018; Yang *et al.* 2018) have undergone multiple rounds of revision, integrating transcriptomic and even proteomic data to completely annotate the encoded complement of genes. An additional question to consider is the genomic representativeness of individual isolates for an entire species. Recent projects, focusing on within-species genomic diversity, have shown that genomic content and organization can vary dramatically even between strains of the same algal species, isolated from different environments (Read *et al.* 2013; Flowers *et al.* 2015; Rastogi *et al.* 2020). This micro-evolution may be as significant as the phylogenetic distances between major groups for underpinning genomic diversity in eukaryotic algae (Blanc-Mathieu *et al.* 2017).

The environmental sequence datasets available through projects such as *Tara* Oceans expand our insights into the biodiversity of previously unexplored oceanic regions, although with notable gaps (e.g. the Bay of Bengal; much of the southern

Indian Ocean; and the Atlantic African coast; Figure 3.3). We hope that future sequencing projects may be able to further redress some of the geographical discrepancies in what marine sequence data are available, and may involve more direct participation from researchers in transitional and developing economies, which may be highly phycologically diverse and disproportionately affected by anthropogenic pollution and climate change (Msuya and Hurtado 2017; Lee 2019). These projects may also form the starting point for groundbreaking explorations of algal diversity in freshwater and terrestrial environments, which have very different chemical properties (e.g. extreme nutrient limitations and enrichments, highly variable pH, lower salinity, and large diurnal and annual fluctuations in temperature), and in which very different patterns (e.g. a more dominant role for chrysophytes and cryptomonads, and of mixotrophic and obligately heterotrophic species) are observed (Dorrell *et al.* 2019; Djemiel *et al.* 2020; Singer *et al.* 2020; Nelson *et al.* 2021).

The identification and assembly of SMAGs raise new challenges concerning assembly and quality, for example, concerning the minimum information about single-amplified and metagenome-assembled genomes (MISAGs and MIMAGs) acceptable for deposition in public repositories (Bowers *et al.* 2017). Certain lineages, particularly dinoflagellates, are currently not represented at all in SMAGs (Figure 3.2), reflecting their large and highly complex nuclear genomes (Gornik *et al.* 2019). These may require deeper sequencing than currently feasible for genome-resolved metagenomics. Finally, the metagenomic identification of poorly cultivable organisms, some of which may correspond to novel branches of the algal tree of life (Delmont *et al.* 2020), may open up new chapters in laboratory research. SMAGs may be particularly important in this regard, revealing possible undocumented auxotrophies (e.g. for vitamins; Helliwell *et al.* 2011; Absolon *et al.* 2019) that may eventually allow us to bring some of the previously uncultured lineages, such as MOCHs and rappemonads (Kim *et al.* 2011; Massana *et al.* 2014), into the laboratory for more detailed observation.

The genome revolution has already transformed our understanding of eukaryotic algal biology, but this revolution is ongoing. Understanding the genomics and biodiversity of eukaryotic algae will be essential in meeting conservation and biofuels targets agreed within the COP21 climate agreement (Pires 2017), and is a central objective of the upcoming UN Decade for Ocean Science (Lee 2019; Nash *et al.* 2020). Further exploration, and new sequence assembly and analysis, may help expose new patterns of diversity and functions in marine algal communities.

Barcode: a short region of sequence (typically from the nuclear small subunit rDNA) that has been sequenced in many cultured organisms. Sequencing DNA barcodes amplified from environmental samples, and comparing to databanks of equivalent sequences from cultured species, allows us to profile the probable species diversity in the wild.

Diatoms: "the rainforests of the oceans", diatoms are typically rather large, sessile algae living inside silica shells, which contribute up to one-fifth of total planetary photosynthesis. Diatoms are particularly abundant at high latitudes (e.g. the Arctic and the Antarctic), where they can exploit abundant seawater nutrients to underpin the marine food chain.

Dinoflagellates: a highly heterogeneous and diverse group of algae, encompassing photosynthesizers, mixotrophs and heterotrophs. Dinoflagellates include the principal photosynthetic symbiont of coral (*Symbiodinium*) alongside harmful "red tide" species, which can produce toxic compounds that result in mass shellfish and marine vertebrate die-offs.

Endosymbiosis: the process by which mitochondria and chloroplasts originate, proposed in 1904 by Konstantin Mereschkowsky and revitalized in the 1960s by Lynn Margulis. Free-living organisms (photosynthetic in the case of chloroplasts; and either bacteria or single-celled eukaryotes) are swallowed up by a predatory eukaryote, and converted into permanent bioenergetic organelles. While only one mitochondrial endosymbiosis is known, at the origin of eukaryotes, chloroplasts have been acquired through endosymbiosis at least ten times across the eukaryotic tree.

Genome: an assembled sequence of all DNA found in a genome, including both coding and non-coding regions. After sequencing and assembly, genomes have to be annotated to define where genes are located, and predict probable coding content.

Haptophytes: typically small, eukaryotic algae that include coccolithophorids, a group of algae living inside calcium carbonate shells and that can form highly abundant seasonal "blooms" visible from space. The calcium shells, if deposited on the seafloor and compressed, can be converted over geological timescales into calcareous rocks, for example, limestone and chalk. In addition to coccolithophorids, non-calcifying haptophyte lineages such as the genus *Phaeocystis* are also abundant in the environment and some of them are associated with harmful algal bloom (HAB) events that negatively affect fish farming.

Horizontal Gene Transfer: also referred to as HGT or Lateral Gene Transfer; the non-sexual exchange of genetic information between cells, typically of different species. It may occur as a continuous process over evolution, and may involve both prokaryotic and eukaryotic partners, and may be a driving force in rapid environmental adaptations.

Meta-genome and **meta-transcriptome:** sequencing of all DNA or RNA in an environmental sample, for example, seawater. This will provide the sequences of not just one species, but a coexisting community of different organisms, which allows insights into ecological interactions at the expense of resolving sequence origin.

Photo-mixotrophy: a lifestyle in which an organism both derives energy through photosynthesis, and through the phagotrophic consumption of other organisms or the uptake of dissolved organic nutrients. This strategy, while rare in plants (see carnivorous plants) is widespread in algae, and is particularly important for the biology of dinoflagellates.

Phytoplankton and **Algae:** typically refer to single-celled or colonial (chain-forming) photosynthetic organisms, which can either include bacteria (i.e. cyanobacteria) or refer only to eukaryotes (e.g. diatoms, dinoflagellates and haptophytes). Algae are evolutionarily diverse, with many of the main eukaryotic groups distantly related to plants, to us, and even to one another, although all eukaryotic algae share chloroplasts, derived through endosymbiosis.

SMAGs: assembled genome sequences obtained from uncultured organisms, either by intensive sequencing of single isolated cells (SAGs) or the mathematical resolution of large amounts of sequence data from multiple metagenome datasets (MAGs).

Transcriptome: a survey of all expressed sequences in a cell, based on next-generation sequencing of cellular RNA (via cDNA). Transcriptomes can provide direct insight into the sequences of coding regions (i.e. genes) in eukaryotic genomes and (by allowing consideration of the levels of different gene expression) reveal physiological responses to environmental stimuli.

Box 3.1. *Glossary box*

3.10. Acknowledgments

RGD acknowledges a CNRS Momentum Fellowship (awarded 2019–2021). CB has received funding from the European Research Council (ERC) under the European Union's Horizon 2020 research and innovation program (Diatomic; grant agreement No. 835067). Additional funding is acknowledged from the FFEM – French Facility for Global Environment (Fonds Français pour l'Environnement Mondial), and the French Government "Investissements d'Avenir" Programmes MEMO LIFE (Grant ANR-10-LABX-54), Université de Recherche Paris Sciences et Lettres (PSL) (Grant ANR-1253 11-IDEX-0001-02), France Genomique (ANR-10-INBS-09), and OCEANOMICS (Grant ANR-11-BTBR-0008). JJPK acknowledges postdoctoral funding from the Fonds Français pour l'Environnement

Mondial. CN acknowledges postdoctoral funding from the European Research Council (ERC) under the European Union's Horizon 2020 research and innovation program through the project Diatomic (grant agreement No. 835067) to CB. The authors thank Daniel Vaulot and Ian Probert (Roscoff Culture Collection) for the reuse of images included in Figure 3.1; Elena Kazamia for assistance with the generation of Figure 3.4; and Quentin Carradec for assistance with the generation of Figure 3.7.

3.11. References

Absolon, D., Smith, A., Helliwell, K. (2019). Shining a light on microbial dark matter: A role for the forgotten B vitamins in marine algal communities? *Access Microbiol.*, 10, 1099.

Allen, A.E., Dupont, C.L., Oborník, M., Horák, A., Nunes-Nesi, A., McCrow, J.P., Zheng, H. *et al.* (2011). Evolution and metabolic significance of the urea cycle in photosynthetic diatoms. *Nature*, 473(7346), 203–207.

Amann, R.I., Ludwig, W., Schleifer, K.H. (1995). Phylogenetic identification and *in situ* detection of individual microbial cells without cultivation. *Microbiol. Rev.*, 59(1), 143–169.

Amaral-Zettler, L., Artigas, L.F., Baross, J., Loka Bharathi, P.A., Boetius, A., Chandramohan, D., Herndl, G. *et al.* (2010). Global census of marine microbes. In *Life in the World's Oceans: Diversity, Distribution and Abundance*, McIntyre, A.D. (ed.). Blackwell Publishing Ltd., Oxford.

Anderson, R., Charvet, S., Hansen, P.J. (2018). Mixotrophy in chlorophytes and haptophytes–effect of irradiance, macronutrient, micronutrient and vitamin limitation. *Front. Microbiol.*, 1704.

Armbrust, E.V., Berges, J.A., Bowler, C., Green, B.R., Martinez, D., Putnam, N.H., Zhou, S. *et al.* (2004). The genome of the diatom *Thalassiosira pseudonana*: Ecology, evolution, and metabolism. *Science*, 306(5693), 79–86.

Bailleul, B., Rogato, A., de Martino, A., Coesel, S., Cardol, P., Bowler, C., Falciatore, A., Finazzi, G. (2010). An atypical member of the light-harvesting complex stress-related protein family modulates diatom responses to light. *Proc. Natl. Acad. Sci. USA*, 107(42), 18214–18219.

Bendif, E.M., Probert, I., Carmichael, M., Romac, S., Hagino, K., de Vargas, C. (2014). Genetic delineation between and within the widespread coccolithophore morpho-species *Emiliania huxleyi* and *Gephyrocapsa oceanica* (Haptophyta). *J. Phycol.* 50(1), 140–148.

Berruti, A., Desirò, A., Visentin, S., Zecca, O., Bonfante, P. (2017). ITS fungal barcoding primers versus 18S AMF-specific primers reveal similar AMF-based diversity patterns in roots and soils of three mountain vineyards. *Envi. Microbiol. Rep.*, 9(5), 658–667.

Bhadury, P. and Austen, M.C. (2010). Barcoding marine nematodes: An improved set of nematode 18S rRNA primers to overcome eukaryotic co-interference. *Hydrobiologia*, 641(1), 245–251.

Blanc-Mathieu, R., Krasovec, M., Hebrard, M., Yau, S., Desgranges, E., Martin, J., Schackwitz, W. *et al.* (2017). Population genomics of picophytoplankton unveils novel chromosome hypervariability. *Sci. Adv.*, 3(7), e1700239.

Bock, R. (2017). Witnessing genome evolution: Experimental reconstruction of endosymbiotic and horizontal gene transfer. *Ann. Rev. Genet.*, 51(11), 1–22.

Bork, P., Bowler, C., de Vargas, C., Gorsky, G., Karsenti, E., Wincker, P. (2015). *Tara* Oceans studies plankton at a planetary scale. *Science*, 348(6237), 873.

Boundy-Mills, K., Hess, M., Bennett, A.R., Ryan, M., Kang, S., Nobles, D., Eisen, J.A. *et al.* (2015). The United States Culture Collection Network (USCCN), enhancing microbial genomics research through living microbe culture collections. *Appl. Env. Microbiol.*, 81 (17), 5671–5674.

Bowers, R.M., Kyrpides, N.C., Stepanauskas, R., Harmon-Smith, M., Doud, D., Reddy, T.B.K., Schulz, F. *et al.* (2017). Minimum information about a single amplified genome (MISAG) and a metagenome-assembled genome (MIMAG) of Bacteria and Archaea. *Nat. Biotechnol.*, 35(8), 725–731.

Bowler, C., Vardi, A., Allen., A.E. (2010). Oceanographic and biogeochemical insights from diatom genomes. *Ann. Rev. Mar. Sci.*, 2, 333–365.

Brawley, S.H., Blouin, N.A., Ficko-Blean, E., Wheeler, G.L., Lohr, M., Goodson, H.V., Jenkins, J.W. *et al.* (2017). Insights into the red algae and eukaryotic evolution from the genome of *Porphyra purpurea* (Bangiophyceae, Rhodophyta). *Proc. Natl. Acad. Sci. USA*, 114(31), 6361–6370.

Browning, T.J., Achterberg, E.P., Rapp, I., Engel, A., Bertrand, E.M., Tagliabue, A., Moore, C.M. (2017). Nutrient co-limitation at the boundary of an oceanic gyre. *Nature*, 551 (7679), 242–246.

Burkholder, J.M., Glibert, P.M., Skelton, H.M. (2008). Mixotrophy, a major mode of nutrition for harmful algal species in eutrophic waters. *Harmful Algae*, 8(1), 77–93.

Busseni, G., Vieira, F.R.J., Amato, A., Pelletier, E., Pierella Karlusich, J.J., Ferrante, M.I., Wincker, P. *et al.* (2019). Meta-omics reveals genetic flexibility of diatom nitrogen transporters in response to environmental changes. *Mol. Biol. Evol.*, 36,(11), 2522–2535.

del Campo, J., Sieracki, M.E., Molestina, R., Keeling, P., Massana, R., Ruiz-Trillo, I. (2014). The others: Our biased perspective of eukaryotic genomes. *Trends Ecol. Evol.*, 29(5), 252–259.

Cao, S., Zhang, W., Ding, W., Wang, M., Fan, S., Yang, B., McMinn, A. *et al.* (2020). Structure and function of the Arctic and Antarctic marine microbiota as revealed by metagenomics. *Microbiome*, 8(1), 47.

Caputi, L., Carradec, Q., Eveillard, D., Kirilovsky, A., Pelletier, E., Pierella Karlusich, J.J., Vieira, F.R.J. *et al.* (2019). Community-level responses to iron availability in open ocean plankton ecosystems. *Global Biogeochem. Cy.*, 33(3), 391–419.

Carpenter, E.J., Matasci, N., Ayyampalayam, S., Wu, S., Sun, J., Yu, J., Vieira, F.R.J. *et al.* (2019). Access to RNA-sequencing data from 1,173 plant species: The 1000 plant transcriptomes initiative (1KP). *GigaScience*, 8(10), 1093.

Carradec, Q., Pelletier, E., Da Silva, C., Alberti, A., Seeleuthner, Y., Blanc-Mathieu, R., Lima-Mendez, G. *et al.* (2018). A global ocean atlas of eukaryotic genes. *Nat. Commun.*, 9(1), 373.

Cavalier-Smith, T. (2003). Genomic reduction and evolution of novel genetic membranes and protein-targeting machinery in eukaryote-eukaryote chimaeras (meta-Algae). *Phil. Trans. R. Soc. Biol.*, 358 (1429), 109–134.

Cohen, N.R., McIlvin, M.R., Moran, D.M., Held, N.A., Saunders, J.K., Hawco, N.J., Brosnahan, M. *et al.* (2021). Dinoflagellates alter their carbon and nutrient metabolic strategies across environmental gradients in the central Pacific Ocean. *Nat. Microbiol.*, 6, 173–186.

Cumbo, V.R., Baird, A.H., Moore, R.B., Negri, A.P., Neilan, B.A., Salih, A., van Oppen, M.J.H. *et al.* (2013). *Chromera velia* is endosymbiotic in larvae of the reef corals *Acropora digitifera* and *A. tenuis*. *Protist*, 164(2), 237–244.

Delmont, T.O., Eren, A.M., Vineis, J.H., Post, A.F. (2015). Genome reconstructions indicate the partitioning of ecological functions inside a phytoplankton bloom in the Amundsen Sea, Antarctica. *Front. Microbiol.*, 6(10), 1090.

Delmont, T.O., Quince, C., Shaiber, A., Esen, Ö.C., Lee, S.T., Rappé, M.S., McLellan, S.L. *et al.* (2018). Nitrogen-fixing populations of Planctomycetes and Proteobacteria are abundant in surface ocean metagenomes. *Nat. Microbiol.*, 3(7), 804–813.

Delmont, T.O., Gaia, M., Hinsinger, D.D., Fremont, P., Fernandez Guerra, A., Eren, A.M. Vanni, C. *et al.* (2020). Functional repertoire convergence of distantly related eukaryotic plankton lineages revealed by genome-resolved metagenomics. *Cold Spring Harbor Laboratory* [Online]. Available at: https://doi.org/10.1101/2020.10.15.341214.

Derelle, E., Ferraz, C., Rombauts, S., Rouzé, P., Worden, A.Z., Robbens, S., Partensky, F. *et al.* (2006). Genome analysis of the smallest free-living eukaryote *Ostreococcus tauri* unveils many unique features. *Proc. Natl. Acad. Sci USA*, 103(31), 11647–11652.

Dittami, S.M., Michel, G., Collén, J., Boyen, C., Tonon, T. (2010). Chlorophyll-binding proteins revisited – A multigenic family of light-harvesting and stress proteins from a brown algal perspective. *BMC Evol. Biol.*, 10(11), 365.

Djemiel, C., Plassard, D., Terrat, D., Crouzet, O., Sauze, J., Mondy, S., Nowak, V. *et al.* (2020). µgreen-db: A reference database for the 23S rRNA gene of eukaryotic plastids and cyanobacteria. *Sci Rep.*, 10, 5915.

Dorrell, R.G., and Howe, C.J. (2015). Integration of plastids with their hosts: Lessons learned from dinoflagellates. *Proc. Natl. Acad. Sci USA*, 112(33), 10247–10254.

Dorrell, R.G., and Smith, A.G. (2011). Do red and green make brown? Perspectives on plastid acquisitions within chromalveolates. *Eukaryot. Cell*, 10(7), 856–868.

Dorrell, R.G., Gile, G., McCallum, G., Méheust, R., Bapteste, E.P., Klinger, C.M., Brillet-Guéguen, L. *et al.* (2017). Chimeric origins of ochrophytes and haptophytes revealed through an ancient plastid proteome. *eLife*, 6(5), 23717.

Dorrell, R.G., Azuma, T., Nomura, M., Audren de Kerdrel, G., Paoli, L., Yang, S., Bowler, C. *et al.* (2019). Principles of plastid reductive evolution illuminated by nonphotosynthetic chrysophytes. *Proc. Natl. Acad. Sci. USA*, 116(14), 6914–6923.

Dorrell, R.G., Villain, A., Perez-Lamarque, B., McCallum, G., Watson, A.K., Ait-Mohamed, O., Alberti, A. *et al.* (2021). Phylogenomic fingerprinting of tempo and functions of horizontal gene transfer within ochrophytes. *Proc. Natl. Acad. Sci. USA*, 118(4), e2009974118.

Duarte, C.M. (2015). Seafaring in the 21st century: The Malaspina 2010 circumnavigation expedition. *Limnol. Oceanogr. Bull.*, 24, 11–14.

Duncan, A., Barry, K., Daum, C., Eloe-Fadrosh, E., Roux, S., Tringe, S.G., Schmidt, K. *et al.* (2020). Metagenome-assembled genomes of phytoplankton communities across the Arctic Circle. *Cold Spring Harbor Laboratory* [Online]. Available at: https://doi.org/10.1101/2020.06.16.154583.

Dunthorn, M., Otto, J., Berger, S.A., Stamatakis, A., Mahé, F., Romac, S., de Vargas, C., *et al.* (2014). Placing environmental next-generation sequencing amplicons from microbial eukaryotes into a phylogenetic context. *Mol. Biol. Evol.*, 31(4), 993–1009.

Durak, G.M., Brownlee, C., Wheeler, G.L. (2017). The role of the cytoskeleton in biomineralisation in haptophyte algae. *Sci. Rep.*, 7(1), 15409.

Faktorová, D., Nisbet, R.E.R., Fernández Robledo, J.A., Casacuberta, E., Sudek, L., Allen, A.E., Ares Jr, M. *et al.* (2020). Genetic tool development in marine protists: Emerging model organisms for experimental cell biology. *Nat. Methods*, 17(5), 481–494.

Fernández-Castro, B., Mouriño-Carballido, B., Benítez-Barrios, V.M., Chouciño, P., Fraile-Nuez, E., Graña, R., Piedeleu, M. *et al.* (2014). Microstructure turbulence and diffusivity parameterization in the tropical and subtropical Atlantic, Pacific and Indian Oceans during the Malaspina 2010 Expedition. *Deep Sea Res. Part I Oceanogr. Res. Pap.*, 94, 15–30.

Flowers, J.M., Hazzouri, K.M., Pham, G.M., Rosas, U., Bahmani, T., Khraiwesh, B., Nelson, D.R. *et al.* (2015). Whole-genome resequencing reveals extensive natural variation in the model green alga *Chlamydomonas reinhardtii*. *Plant Cell*, 27(9), 2353–2369.

Fuhrman, J.A., McCallum, K., Davis, A.A. (1995). Phylogenetic diversity of subsurface marine microbial communities from the Atlantic and Pacific Oceans. *Appl. Env. Microbiol.*, 59(5), 1294–1302.

Gachon, C.M.M., Heesch, S., Küpper, F.C., Achilles-Day, U.E.M., Brennan, D., Campbell, C.N., Clarke, A. *et al.* (2013). The CCAP KnowledgeBase: Linking protistan and cyanobacterial biological resources with taxonomic and molecular data. *Systemat. Biodiv.*, 11(4), 407–413.

Giovannoni, S.J., Britschgi, T.B., Moyer, C.L., Field, K.G. (1990). Genetic diversity in Sargasso Sea bacterioplankton. *Nature*, 345(6270), 60–63.

Gomez, F. (2005). A list of free-living dinoflagellates species in the world's ocean. *Acta Botanica Croatica*, 64, 129–212.

González-Pech, R.A., Bhattacharya, D., Ragan, M.A., Chan, C.X. (2019). Genome evolution of coral reef symbionts as intracellular residents. *Trends Ecol. Evol.*, 34(9), 799–806.

Gornik, S.G., Hu, I., Lassadi, I., Waller, R.F. (2019). The biochemistry and evolution of the dinoflagellate nucleus. *Microorganisms*, 7(8), 245.

Gregory, T.R. (2005). Synergy between sequence and size in large-scale genomics. *Nat. Rev. Genet.*, 6(9), 699–708.

Gregory, A.C., Zayed, A.A., Conceição-Neto, N., Temperton, B., Bolduc, B., Alberti, A., Ardyna, M. *et al.* (2019). Marine DNA viral macro- and microdiversity from pole to pole. *Cell*, 177(5), 1109–1123.e14.

Grigoriev, I.V., Hayes, R.D., Calhoun, S., Kamel, B., Wang, A., Ahrendt, S., Dusheyko, S. *et al.* (2020). PhycoCosm, a comparative algal genomics resource. *Nucl. Acids Res.*, 49, D1004–D1011.

Guiry, M.D., Guiry, G.M., Morrison, L., Rindi, F., Valenzuela Miranda, S., Mathieson, A.C., Parker, B.C. *et al.* (2014). AlgaeBase: An on-line resource for algae. *Cryptogamie, Algologie.*, 35(2), 105–115.

Hanschen, E.R. and Starkenburg, S.R. (2020). The state of algal genome quality and diversity. *Algal Res.*, 50, 101968.

Helliwell, K.E., Wheeler, G.L., Leptos, K.C., Goldstein, R.E., Smith, A.G. (2011). Insights into the evolution of vitamin B12 auxotrophy from sequenced algal genomes. *Mol. Biol. Evol.*, 28(10), 2921–2933.

Henze, K., Martin, W., Schnarrenberger, C. (2002). Endosymbiotic gene transfer: A special case of horizontal gene transfer germane to endosymbiosis, the origins of organelles and the origins of eukaryotes. In *Horizontal Gene Transfer*, Syvanen, M. and Kado, C. (eds). Academic Press, London.

Hinder, S.L., Hays, G.C., Edwards, M., Roberts, E.C., Walne, A.W., Gravenor, M.B. (2012). Changes in marine dinoflagellate and diatom abundance under climate change. *Nat. Clim. Change*, 2, 271–275.

Huysman, M.J.J., Martens, C., Vandepoele, K., Gillard, J., Rayko, E., Heijde, M., Bowler, C. *et al.* (2010). Genome-wide analysis of the diatom cell cycle unveils a novel type of cyclins involved in environmental signaling. *Genom. Biol.*, 11(2), R17.

Ibarbalz, F.M., Henry, H., Brandão, M.C., Martini, S., Busseni G., Byrne, H., Coelho L.P. *et al.* (2019). Global trends in marine plankton diversity across kingdoms of life. *Cell*, 179(5), 1084–1097.e21

Iverson, V., Morris, R.M., Frazar, C.D., Berhiaume, C.T., Morales, R.L., Armbrust, E.V. (2012). Untangling genomes from metagenomes: Revealing an uncultured class of marine Euryarchaeota. *Science*, 335(6068), 587–590.

Jackson, C., Knoll, A.H., Chan, C.X., Verbruggen, H. (2018). Plastid phylogenomics with broad taxon sampling further elucidates the distinct evolutionary origins and timing of Secondary Green Plastids. *Sci. Rep.*, 8(1), 1523.

Jannasch, H.W. and Jones, G.E. (1959). Bacterial populations in sea water as determined by different methods of enumeration. *Limnol. Oceanogr.*, 4(2), 128–139.

Jeong, H.J., Kang, H.C., Lim, A.S., Jang, S.H., Lee, K., Lee, S.Y., Ok, J.H. *et al.* (n.d.). Feeding diverse prey as an excellent strategy of mixotrophic dinoflagellates for global dominance. *Sci Adv.*, 7(2), eabe4214.

Johnson, M.D. (2015). Inducible mixotrophy in the dinoflagellate *Prorocentrum minimum. J. Euk. Microbiol.*, 62(4), 431–443.

Kamikawa, R., Tanifuji, G., Kawachi, M., Miyashita, H., Hashimoto, T., Inagaki, Y. (2015). Plastid genome-based phylogeny pinpointed the origin of the green-colored plastid in the dinoflagellate *Lepidodinium chlorophorum. Genom. Biol. Evol.*, 7(4), 1133–1140.

Karsenti, E., Acinas, S.G., Bork, P., Bowler, C., de Vargas, C., Rae, J., Sullivan, M. *et al.* (2011). A holistic approach to marine eco-systems biology. *PLoS Biology*, 9(10), e1001177.

Kawachi, M., Nakayama, T., Kayama, M., Nomura, M., Miyashita, H., Bojo, O., Rhodes, L. *et al.* (2021). Rappemonads are haptophyte phytoplankton. *Curr Biol.*, 31(11), 2395–2403.

Kazamia, E., Sutak, R., Paz-Yepes, J., Dorrell, R.G., Vieira, F.R.J., Mach, J., Morrissey, J. *et al.* (2018). Endocytosis-mediated siderophore uptake as a strategy for Fe acquisition in diatoms. *Sci. Adv.*, 4(5), 4536.

Keeling, P.J., Burki, F., Wilcox, H.M., Allam, B., Allen, E.E. Amaral-Zettler, L.A., Armbrust, E.V. *et al.* (2014). The marine microbial eukaryote transcriptome sequencing project (MMETSP), illuminating the functional diversity of eukaryotic life in the oceans through transcriptome sequencing. *PLoS Biol.*, 12(6), e1001889.

Kim, E., Harrison, J.W., Sudek, S., Jones, M.D.M., Wilcox, H.M., Richards, T.A., Worden, A.Z. *et al.* (2011). Newly identified and diverse plastid-bearing branch on the eukaryotic tree of life. *Proc. Natl. Acad. Sci. USA*, 108(4), 1496–1500.

Kopf, A., Bicak, M., Kottmann, R., Schnetzer, J., Kostadinov, I., Lehmann, K., Fernandez-Guerra, A. *et al.* (2015). The ocean sampling day consortium. *GigaScience*, 4(6), 27.

Kowallik, K.V., Stoebe, B., Schaffran, I., Kroth-Pancic, P., Freier., U. (1995). The chloroplast genome of a chlorophylla *c*-containing alga, *Odontella sinensis. Plant Mol. Biol. Rep.*, 13, 336–342.

Kumar, D., Singh, B., Sharma, Y.C. (2017). Challenges and opportunities in commercialization of algal biofuels. In *Algal Biofuels*, Gupta, S., Malik, A., Bux, F. (eds). Springer Switzerland, Cham.

Le Bescot, N., Mahé, F., Audic, S., Dimier, C., Garet, M.J., Poulain, J., Wincker, P. *et al.* (2016). Global patterns of pelagic dinoflagellate diversity across protist size classes unveiled by metabarcoding. *Env. Microbiol.*, 18(2), 609–626.

Leblanc, K., Quéguiner, B., Diaz, F., Cornet, V., Michel-Rodriguez, M., Durrieu de Madron, X., Bowler, C. *et al.* (2018). Nanoplanktonic diatoms are globally overlooked but play a role in spring blooms and carbon export. *Nat. Commun.*, 9(1), 953.

Lee, Y.H. (2019). Ocean cultural heritage and ocean literacy programs in the UN Decade of Ocean Science for Sustainable Development (2021–2030). *J. Ocean Cult.*, 2, 136–146.

Li, W.K.W., McLaughlin, F.A., Lovejoy, C., Carmack, E.C. (2009). Smallest algae thrive as the arctic ocean freshens. *Science*, 326(5952), 539.

Li, H., Li, L., Yu, L., Yang, X., Shi, X., Wang, J., Li, J. *et al.* (2020). Transcriptome profiling reveals versatile dissolved organic nitrogen utilization, mixotrophy, and N conservation in the dinoflagellate, *Prorocentrum shikokuense* under N deficiency. *Sci. Tot. Env.*, 143103.

Lommer, M., Specht, M., Roy, A.S., Kraemer, L., Andreson, R., Gutowska, M.A., Wolf, J. *et al.* (2012). Genome and low-iron response of an oceanic diatom adapted to chronic iron limitation. *Genom. Biol.*, 13(7), R66.

Lopes Dos Santos, A., Gourvil, P., Tragin, M., Noël, MH., Decelle, J., Romac, S., Vaulot, D. (2017). Diversity and oceanic distribution of prasinophytes Clade VII, the dominant group of green algae in oceanic waters. *ISME J.*, 11(2), 512–528.

Malviya, S., Scalco, E., Audic, S., Vincent, F., Veluchamy, A., Poulain, J., Wincker, P. *et al.* (2016). Insights into global diatom distribution and diversity in the world's ocean. *Proc. Natl. Acad. Sci. USA*, 113(11), 1516–1525.

Mangot, J.F., Logares, R., Sánchez, P., Latorre, F., Seeleuthner, Y., Mondy, S., Sieracki, M.E. *et al.* (2017). Accessing the genomic information of unculturable oceanic picoeukaryotes by combining multiple single cells. *Sci. Rep.*, 7(1), 41498.

Marcy, Y., Ouverney, C., Bik, E.M., Lösekann, T., Ivanova, N., Martin, H.S., Szeto, E. *et al.* (2007). Dissecting biological "dark matter" with single-cell genetic analysis of rare and uncultivated TM7 microbes from the human mouth. *Proc. Natl. Acad. Sci. USA*, 104(29), 11889–11894.

Margulis, L. (1975). Symbiotic theory of the origin of eukaryotic organelles; criteria for proof. *Symposia Soc. Exp. Biol.*, 29, 21–38.

Massana, R., del Campo, J., Sieracki, M.E., Audic, S., Logares, R. (2014). Exploring the uncultured microeukaryote majority in the oceans: Reevaluation of ribogroups within stramenopiles. *ISME J.*, 8(4), 854–866.

Matsuzaki, M., Misumi, O., Shin-I, T., Maruyama, S., Takahara, M., Miyagishima, S.Y., Mori, T. *et al.* (2004). Genome sequence of the ultrasmall unicellular red alga *Cyanidioschyzon merolae* 10D. *Nature*, 428(6983), 653–657.

McQuaid, J.B., Kustka, A.B., Oborník, M., Horák, A., McCrow, J.P., Karas, B.J., Zheng, H. *et al.* (2018). Carbonate-sensitive phytotransferrin controls high-affinity iron uptake in diatoms. *Nature*, 555(7697), 534–537.

Merchant, S.S., Prochnik, S.E., Vallon, O., Harris, E.H., Karpowicz, S.J., Witman, G.B., Terry, A., *et al.* (2007). The *Chlamydomonas* genome reveals the evolution of key animal and plant functions. *Science*, 318(5848), 245–250.

Milliman, J.D. (1993). Production and accumulation of calcium carbonate in the ocean: Budget of a nonsteady state. *Global Biogeochem. Cycles*, 7(4), 927–957.

Mock, T., Otillar, R.P., Strauss, J., McMullan, M., Paajanen, P., Schmutz, J., Salamov, A. *et al.* (2017). Evolutionary genomics of the cold-adapted diatom *Fragilariopsis cylindrus*. *Nature*, 541(7638), 536–540.

Monier, A., Worden, A.Z., Richards, T.A. (2016). Phylogenetic diversity and biogeography of the Mamiellophyceae lineage of eukaryotic phytoplankton across the oceans. *Env. Microbiol. Rep.*, 8(4), 461–469.

Moustafa, A., Beszteri, B., Maier, U.G., Bowler, C., Valentin, K., Bhattacharya, D. (2009). Genomic footprints of a cryptic plastid endosymbiosis in diatoms. *Science*, 324(5935), 1724–1726.

Msuya, F.E. and Hurtado, A.Q. (2017). The role of women in seaweed aquaculture in the western Indian Ocean and south-east Asia. *Eur. J. Phycol.*, 52(4), 482–494.

Nash, K.L., Alexander, K., Melbourne-Thomas, J., Novioaglio, C., Sbrocchi, C., Villanueva, C., Pecl, G.T. (2020). Developing achievable alternate futures for key challenges during the UN decade of ocean science for sustainable development. *Rev. Fish Biol. Fish.* [Online]. Available at: https://doi.org/10.1007/s11160-020-09629-5.

Nelson, W.C., Tully, B.J., Mobberley, J.M. (2020). Biases in genome reconstruction from metagenomic data. *PeerJ*, 8(10), e10119.

Nelson, D.R., Hazzouri, K.M., Lauersen, K.J., Jaiswal, A., Chaiboonchoe, A., Mystikou, A., Fu, W. *et al.* (2021). Large-scale genome sequencing reveals the driving forces of viruses in microalgal evolution. *Cell Host Microbe*, 29(2), 250–266.

Novák Vanclová, A.M.G., Zoltner, M., Kelly, S., Soukal, P., Záhonová, K., Füssy, Z., Ebenezer, T.E. *et al.* (2020). Metabolic quirks and the colourful history of the *Euglena gracilis* secondary plastid. *New Phytol.*, 225(4), 1578–1592.

Ohta, N., Sato, N., Kuroiwa, T. (1998). Structure and organization of the mitochondrial genome of the unicellular red alga *Cyanidioschyzon merolae* deduced from the complete nucleotide sequence. *Nucl. Acids Res.*, 26(22), 5190–5198.

One Thousand Plant Transcriptomes Initiative (2019). One thousand plant transcriptomes and the phylogenomics of green plants. *Nature*, 574(7780), 679–685.

Parks, D.H., Rinke, C., Chuvochina, M., Chaumeil, P.A., Woodcroft, B.J., Evans, P.N., Hugenholtz, P. *et al.* (2017). Recovery of nearly 8,000 metagenome-assembled genomes substantially expands the tree of life. *Nat. Microbiol.*, 2(11), 1533–1542.

Pawlowski, J.W., Kelly-Quinn, M., Altermatt, F., Apothéloz-Perret-Gentil, L., Beja, P., Boggero, A., Borja, A. *et al.* (2018). The future of biotic indices in the ecogenomic era: Integrating (e)DNA metabarcoding in biological assessment of aquatic ecosystems. *Sci Total Env.*, 637–638, 1295–1310.

Pernice, M.C., Giner, C.R., Logares, R., Perera-Bel, J., Acinas, S.G., Duarte, C.M., Gasol, J.M., *et al.* (2016). Large variability of bathypelagic microbial eukaryotic communities across the world's oceans. *ISME J.*, 10, 945–958.

Pfaff, S., Gustavs, L., Reichardt, A., Jaskulke, R., Gäbler-Schwarz H.E., Karsten, E. (2016). Ecophysiological plasticity in the Arctic phytoplankton species *Phaeocystis pouchetii* (Prymnesiophyceae, Haptophyta). *Algol. Stud.*, 151–152, 87–102.

Pierella Karlusich, J.J., Lodeyro, A.F., Carrillo, N. (2014). The long goodbye: The rise and fall of flavodoxin during plant evolution. *J. Exp. Bot.*, 65(18), 5161–5178.

Pierella Karlusich, J.J., Ibarbalz, F.M., Bowler, C. (2020a). Phytoplankton in the *Tara* Ocean. *Ann. Rev. Mar. Sci.*, 12(1), 233–265.

Pierella Karlusich, J.J., Ibarbalz, F.M., Bowler, C. (2020b). Exploration of marine phytoplankton: From their historical appreciation to the omics era. *J. Plankton Res.*, 52 (10), 595–612.

Pires, J.C.M. (2017). COP21: The algae opportunity? *Renew. Sust. Energy Rev.*, 79, 867–877.

Planes, S., Allemand, D., Agostini, S., Banaigs, B., Boissin, E., Boss, E., Bourdin, G. *et al.* (2019). The *Tara* Pacific expedition – A pan-ecosystemic approach of the "-omics" complexity of coral reef holobionts across the Pacific Ocean. *PLoS Biol.*, 17(9), e3000483.

Ponce-Toledo, R.I., Moreira, D., López-García, P., Deschamps, P. (2018). Secondary plastids of euglenids and chlorarachniophytes function with a mix of genes of red and green algal ancestry. *Mol. Biol. Evol.*, 35(9), 2198–2204.

Rappé, M.S. and Giovannoni, S.J. (2003). The uncultured microbial majority. *Ann. Rev. Microbiol.*, 57, 369–394.

Rastogi, A., Maheswari, U., Dorrell, R.G., Vieira, F.R.J., Maumus, F., Kustka, A., McCarthy, J. *et al.* (2018). Integrative analysis of large scale transcriptome data draws a comprehensive landscape of *Phaeodactylum tricornutum* genome and evolutionary origin of diatoms. *Sci. Rep.*, 8(1), 4834.

Rastogi, A., Vieira, F.R.J., Deton-Cabanillas, A.F., Veluchamy, A., Cantrel, C., Wang, G., Vanormelingen, P. *et al.* (2020). A genomics approach reveals the global genetic polymorphism, structure, and functional diversity of ten accessions of the marine model diatom *Phaeodactylum tricornutum*. *ISME J.*, 14, 347–363.

Raymond, J.A., and Kim H.J. (2012). Possible role of horizontal gene transfer in the colonization of sea ice by algae. *PLOS ONE*, 7(5), 35968.

Read, B.A., Kegel, J., Klute, M.J., Kuo, A., Lefebvre, S.C., Maumus, F., Mayer, C. *et al.* (2013). Pan genome of the phytoplankton *Emiliania* underpins its global distribution. *Nature*, 499(7457), 209–213.

Reid, P.C., Colebrook, J.M., Matthews, J.B.L., Aiken, J. (2003). The continuous plankton recorder: Concepts and history, from plankton indicator to undulating recorders. *Prog. Oceanogr.*, 58 (2–4), 117–173.

Reith, M. and Munholland, J. (1995). Complete nucleotide sequence of the *Porphyra purpurea* chloroplast genome. *Plant Mol. Biol. Report.*, 13, 333–335.

Reji, L. and Francis. C.A. (2020). Metagenome-assembled genomes reveal unique metabolic adaptations of a basal marine *Thaumarchaeota* lineage. *ISME J.*, 14(8), 2105–2015.

Rossoni, A.W., Price, D.C., Seger, M., Lyska, D., Lammers, P., Bhattacharya, D., Weber, A.P. (2019). The genomes of polyextremophilic cyanidiales contain 1% horizontally transferred genes with diverse adaptive functions. *eLife*, 10, 45017.

Rusch, D.B., Halpern, A.L., Sutton, G., Heidelberg, K.B., Williamson, S., Yooseph, S., Wu, D. *et al.* (2007). The Sorcerer II global ocean sampling expedition: Northwest Atlantic through Eastern Tropical Pacific. *PLoS Biol.*, 5(3), e77.

Sarai, C., Tanifuji, G., Nakayama, T., Kamikawa, R., Takahashi, K., Yazaki, E., Matsuo, E. *et al.* (2020). Dinoflagellates with relic endosymbiont nuclei as models for elucidating organellogenesis. *Proc. Natl. Acad. Sci. USA*, 117(10), 5364–5375.

Schaum, C.E. and Collins, S. (2014). Plasticity predicts evolution in a marine alga. *Phil. Trans. R. Soc. Biol.*, 281, 20141486.

Seeleuthner, Y., Mondy, S., Lombard, V., Carradec, Q., Pelletier, E., Wessner, M., Leconte, L. *et al.* (2018). Single-cell genomics of multiple uncultured stramenopiles reveals underestimated functional diversity across oceans. *Nat. Commun.*, 9(1), 310.

Ševčíková, T., Horák, A., Klimeš, V., Zbránková, V., Demir-Hilton, E., Sudek, S., Jenkins, J. *et al.* (2015). Updating algal evolutionary relationships through plastid genome sequencing: Did alveolate plastids emerge through endosymbiosis of an ochrophyte? *Sci. Rep.*, 5(5), 10134.

Sibbald, S.J. and Archibald, J.M. (2017). More protist genomes needed. *Nat. Ecol. Evol.*, 1(5), 145.

Simão, F.A., Waterhouse, R.M., Ioannidis, P., Kriventseva, E.V., Zdobnov. E.M. (2015). BUSCO: Assessing genome assembly and annotation completeness with single-copy orthologs. *Bioinformatics*, 31(19), 3210–3212.

Simon, N., Cras, AL., Foulon, E., Lemée, R. (2009). Diversity and evolution of marine phytoplankton. *C. R. Biol.*, 332(2–3), 159–170.

Singer, D., Seppey, C.V.W., Lentendu, G., Dunthorn, M., Bass, D., Belbahri, L., Blandenier, Q. *et al.* (2020). Protist taxonomic and functional diversity in soil, freshwater and marine ecosystems. *Env. Int.*, 146(11), 106262.

Skokan, R., Medvecká, E., Viaene, T., Vosolsobě, S., Zwiewka, M., Müller, K., Skůpa, P. *et al.* (2019). PIN-driven auxin transport emerged early in streptophyte evolution. *Nature Plants*, 5(11), 1114–1119.

Smith, D.R. and Sanitá Lima, M. (2017). Unraveling chloroplast transcriptomes with ChloroSeq, an organelle RNA-Seq bioinformatics pipeline. *Brief. Bioinformat.*, 18(6), 1012–1016.

Smith, S.R., Dupont, C.L., McCarthy, J.K., Broddrick, J.T., Oborník, M., Horák, A., Füssy, Z. *et al.* (2019). Evolution and regulation of nitrogen flux through compartmentalized metabolic networks in a marine diatom. *Nat. Commun.*, 10(1), 4552.

Sournia, A. (1986). *Atlas du phytoplancton marin : introduction, cyanophycées, dictyochophycées, dinophycées et raphidophycées*. Centre National de la Recherche Scientifique, Paris.

Southward, A.J., Langmead, O., Hardman-Mountford, N.J., Aiken, J., Boalch, G.T., Dando, P.R., Genner, M.J. *et al.* (2005). Long-term oceanographic and ecological research in the western English Channel. *Adv. Mar. Biol.*, 47, 1–105.

Steidinger, K.A. and Jangen, K. (1997). Dinoflagellates. In *Identifying Marine Phytoplankton*, Tomas, C.R. (ed.). Elsevier, New York.

Stern, R., Kraberg, A., Bresnan, E., Kooistra, W.H.C., Lovejoy, C., Montresor, M., Morán, X. A.G. *et al.* (2018). Molecular analyses of protists in long-term observation programmes – Current status and future perspectives. *J. Plankton Res.*, 40(5), 519–536.

Stiller, J.W., Schreiber, J., Yue, J., Guo, H., Ding, Q., Huang, J. (2014). The evolution of photosynthesis in chromist algae through serial endosymbioses. *Nat. Commun.*, 5(12), 5764.

Sunagawa, S., Tara Oceans Coordinators, Acinas, S.G., Bork, P., Bowler, C., Eveillard, D., Gorsky, G. *et al.* (2020). *Tara* Oceans: Towards global ocean ecosystems biology. *Nat. Rev. Microbiol.*, 18, 424–445.

Terrado, R., Monier, A., Edgar, R., Lovejoy, C. (2015). Diversity of nitrogen assimilation pathways among microbial photosynthetic eukaryotes. *J. Phycol.*, 51(3), 490–506.

Tréguer, P., Bowler, C., Moriceau, B., Dutkiewicz, S., Gehlen, M., Aumont, O., Bittner, L. *et al.* (2018). Influence of diatom diversity on the ocean biological carbon pump. *Nat. Geosci.*, 11, 27–38.

Tully, B.J., Sachdeva, R., Graham, E.D., Heidelberg, J.F. (2017). 290 Metagenome-assembled genomes from the Mediterranean Sea: A resource for marine microbiology. *PeerJ*, 5(7), e3558.

Tully, B.J., Graham, E.D., Heidelberg, J.F. (2018). The reconstruction of 2,631 draft metagenome-assembled genomes from the global oceans. *Sci. Data*, 5(1), 170203.

Tyrrell, T. and Merico, A. (2004). *Emiliania huxleyi*: Bloom observations and the conditions that induce them. In *Coccolithophores*, Thierstein, H.R. and Young, J.R. (eds). Springer, Berlin.

Tyson, G.W., Chapman, J., Hugenholtz, P., Allen, E.E., Ram, R.J., Richardson, P.M., Solovyev, V.V. *et al.* (2004). Community structure and metabolism through reconstruction of microbial genomes from the environment. *Nature*, 428(6978), 37–43.

Umen, J.G. (2014). Green algae and the origins of multicellularity in the plant kingdom. *Cold Spring Harb. Perspect. Biol.*, 6(11), 016170.

Unrein, F., Gasol, J.M., Not, F., Forn, I., Massana, R. (2014). Mixotrophic haptophytes are key bacterial grazes in oligotrophic coastal waters. *ISME J.*, 8, 164–176.

Vancaester, E., Depuydt, T., Osuna-Cruz, C.M., Vandepoele, K. (2020). Comprehensive and functional analysis of horizontal gene transfer events in diatoms. *Mol. Biol. Evol.*, 37(11), 3243–3257.

Vannier, T., Leconte, J., Seeleuthner, Y., Mondy, S., Pelletier, E., Aury, J.M., de Vargas, C. *et al.* (2016). Survey of the green picoalga *Bathycoccus* genomes in the global ocean. *Sci. Rep.*, 6(11), 37900.

de Vargas, C., Audic, S., Henry, N., Decelle, H.J., Mahé, F., Logares, R., Lara, E. *et al.* (2015). Eukaryotic plankton diversity in the sunlit ocean. *Science*, 348(6237), 1261605.

Vaulot, D., Le Gall, F., Marie, D., Guillou, L., Partensky, F. (2004). The Roscoff Culture Collection (RCC): A collection dedicated to marine picoplankton. *Nova Hedwigia*, 79(1–2), 49–70.

Wang, D., Yu, X., Xu, K., Bi, G., Cao, M., Zelzion, E., Fu, C. *et al.* (2020). *Pyropia yezoensis* genome reveals diverse mechanisms of carbon acquisition in the intertidal environment. *Nat. Commun.*, 11(1), 4028.

West, P.T., Probst, A.J., Grigoriev, I.V., Thomas, B.C., Banfield, J.F. (2018). Genome reconstruction for eukaryotes from complex natural microbial communities. *Genom. Res.*, 28(4), 569–580.

Willhelm, C., Büchel, C., Fisahn, J., Goss, R., Jakob, T., Laroche, J., Lavaud, J. *et al.* (2006). The regulation of carbon and nutrient assimilation in diatoms is significantly different from green algae. *Protist*, 157(2), 91–124.

Wolff, G., Plante, I., Lang, B.F., Kück, U., Burger, G. (1994). Complete sequence of the mitochondrial DNA of the chlorophyte alga *Prototheca wickerhamii*. Gene content and genome organization. *J. Mol. Biol.*, 237(1), 75–86.

Yang, M., Lin, X., Liu, X., Zhang, J., Ge, F. (2018). Genome annotation of a model diatom *Phaeodactylum tricornutum* using an integrated proteogenomic pipeline. *Mol. Plant*, 11(10), 1292–1307.

Yoon, H.S., Hackett, J.D., Bhattacharya, D. (2002). A single origin of the peridinin- and fucoxanthin-containing plastids in dinoflagellates through tertiary endosymbiosis. *Proc. Natl. Acad. Sci. USA*, 99(18), 11724–11729.

Yoon, H.S., Price, D.C., Stepanauskas, R., Rajah, V.D., Sieracki, M.E., Wilson, W.H., Yang, E.C. *et al.* (2011). Single-cell genomics reveals organismal interactions in uncultivated marine protists. *Science*, 332(6030), 714–717.

Zimmermann, R., Iturriaga, R., Becker-Birck, J. (1978). Simultaneous determination of the total number of aquatic bacteria and the number thereof involved in respiration. *Appl. Env. Microbiol.*, 36, 926–935.

4

Inorganic Carbon Acquisition by Aquatic Primary Producers

Sebastian D. ROKITTA[1], Sven A. KRANZ[2] and Björn ROST[1,3]

[1] *Alfred-Wegener-Institute – Helmholtz Center for Polar and Marine Research, Bremerhaven, Germany*

[2] *Florida State University, Department of Earth, Ocean and Atmospheric Science, Tallahassee, USA*

[3] *University of Bremen, FB Biology/Chemistry, Germany*

4.1. Overview

The acquisition of dissolved inorganic carbon by aquatic primary producers became increasingly challenging with higher structural complexity of algae, and with simultaneously declining atmospheric CO_2 partial pressure. The seemingly easy diffusive supply of CO_2 to Rubisco turned into a bottleneck for photosynthesis, which consequently required alternative inorganic carbon acquisition processes and pathways to evolve. In order to ensure sufficient CO_2 supply to Rubisco, aquatic photosynthesizing organisms started to employ facilitated CO_2 uptake, active HCO_3^- trafficking across multiple membranes as well as carbonic anhydrases, located at the outer cell membrane and in several cellular compartments. The modes of these so-called CO_2-concentrating mechanisms (CCMs) are very diverse, non-canonical even within phylogenetic groups, and possess differently efficient CO_2 accumulation capacities, depending on the requirements of Rubisco, the physico-chemical conditions in the boundary layer, membrane properties and cellular architecture. However, different independently evolved CCMs also exhibit a high degree of functional similarity,

Blue Planet, Red and Green Photosynthesis,
coordinated by Stephen Christopher MABERLY and Brigitte GONTERO.

owing to the functional similarity of the photosynthetic process. To introduce the topic to the reader, this chapter starts with a brief outline of Rubisco's properties and the reasons why CCMs are required (section 4.2). Then, the principle chemical nature of dissolved inorganic carbon in water is described (section 4.3): its speciation and kinetic behavior, as well as relevant co-determinants of carbonate chemistry. We furthermore touch upon the physico-chemical basis of carbon availability in aquatic environments (section 4.4), and subsequently elaborate on the known transport modes of different inorganic carbon species. Subsequently, the current state of knowledge on existing strategies in main algal groups is presented (sections 4.5–4.9). Finally, we consider the operation of CCMs in the context of co-occurring cellular processes (section 4.10), such as calcification and N_2 fixation, which rely on the provision of ample inorganic carbon and/or energy and, in the case of calcification, can have important consequences for compartmental pH homeostasis.

4.2. Rubisco and the problem of its own success

With only a few exceptions, microbial primary producers rely on the key enzyme Ribulose-1,5-bisphosphate carboxylase/oxygenase (Rubisco) to catalyze the fixation of CO_2 into biomass (see Chapters 1 and 5). This enzyme evolved in cyanobacteria about three billion years ago at times when the atmosphere contained large amounts of CO_2 and literally no O_2 (Berner 2006). As a result, the enzyme exhibits a generally low affinity, that is, a high K_m value for its prime substrate CO_2. In addition, the enzyme was (and still is) prone to a competing, futile reaction with O_2 that initiates the process of photorespiration and lowers the overall carbon fixation efficiency. Rubisco's preference for carboxylation over oxygenation is described by its so-called "specificity" ($S_{(C/O)}$), which is defined as $V_{CO2}*K_{m(O2)}/V_{O2}*K_{m(CO2)}$, where V and K are the maximum reaction rates and the half-saturation constants, respectively (Watson and Tabita 1997; Badger *et al.* 1998; Tortell 2000). Under ancient, O_2-free conditions, low specificities had no relevance, but over the following billions of years, primary producers began to draw down atmospheric CO_2 and to simultaneously enrich the oceans and atmosphere with O_2, the side-product of photosynthetic light reactions. As a result of their own evolutionary success, primary producers changed the composition and chemistry of the atmosphere and the oceans, and the shifted abundances of CO_2 and O_2 turned into bottlenecks for photosynthetic CO_2 fixation.

Research consensually indicates that a positive selection for higher carboxylation rates and lower half-saturating concentrations occurred, which improved the kinetic properties of Rubisco over the temporal evolution of aquatic microorganisms. Consequently, the "evolutionary youngest" microalgae from the "red lineage" (Falkowski *et al.* 2004), including red algae, haptophytes and diatoms, nowadays possess the most efficient Rubisco forms with lowest half-saturating concentrations and highest specificities. Nevertheless, data from multiple studies (Read and Tabita 1994; Whitney and Andrews 2001; Haslam *et al.* 2005; Whitney *et al.* 2011a, 2011b; Young *et al.* 2016) show that K_m values of Rubisco for CO_2 are typically above surface ocean concentrations. A comprehensive across-the-spectrum analysis of the carboxylation rates, half-saturation concentrations and specificities of different Rubisco forms was recently published (Iñiguez *et al.* 2020). Figure 4.1 depicts known ranges of K_m, V_{CO2} (carbon fixation rates per active Rubisco site) and specificities. Given the properties of their Rubisco, aquatic primary producers had to develop strategies to accumulate CO_2 at the active site of Rubisco to maximize carboxylation over oxygenation. To this end, aquatic primary producers developed the so-called CO_2-concentrating mechanisms (CCMs). These include transport of CO_2 and/or HCO_3^- into the cell, biochemical pathways to concentrate CO_2 around Rubisco (see Chapter 5), measures to prevent CO_2 leakage from the cell, and often the expression of carbonic anhydrase (CA), an enzyme that accelerates the interconversion between CO_2 and HCO_3^- (see Chapter 6). Even though all aquatic photoautotrophs are principally facing the same need for a CCM imposed by Rubisco's properties, different algal groups have specific, often cell-architectural challenges, and thus developed different solutions to the problem, that is, different modes of carbon acquisition. This evolution was a gradual process, evidenced by the fact that those microalgal groups having evolved early during periods with high CO_2 and low O2 levels not only have a lower Rubisco specificities, but also exhibit higher CCM activities than more recently evolved groups (Tortell 2000; Whitney and Andrews 2001; Van de Waal *et al.* 2019).

Before addressing further details on CCMs, however, we need to introduce some background on aquatic carbonate chemistry to illustrate the challenges of carbon uptake for aquatic primary producers. The reader who wishes to delve more into Rubisco's properties and evolution is advised to explore further the according primary literature, such as Badger *et al.* (1998), Whitney and Andrews (2001), Haslam *et al.* (2005), Young *et al.* (2012, 2016), Iñiguez *et al.* (2020) as well as Read and Tabita (1994) and Chapters 1 and 5.

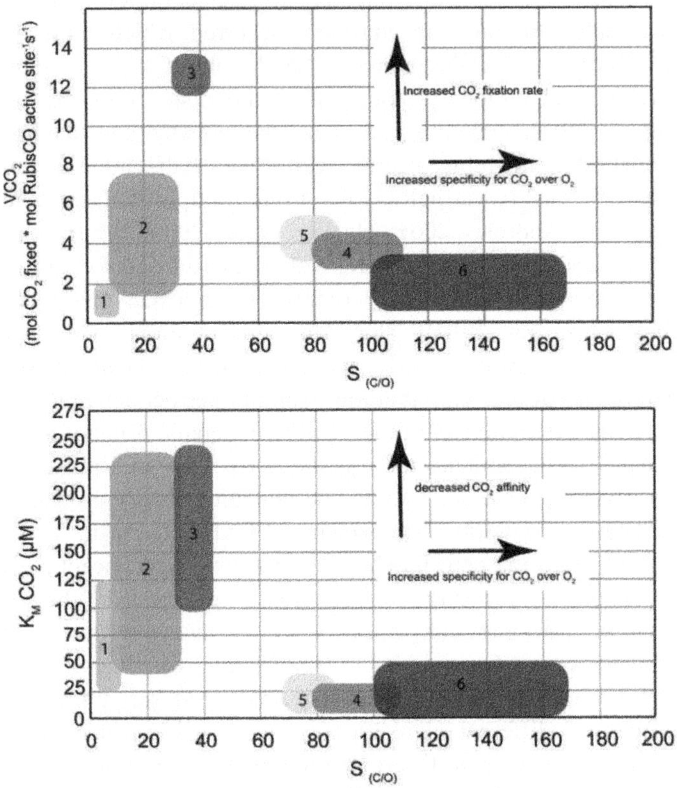

Figure 4.1. *Ranges of carboxylation rates (upper panel) and CO_2 affinities (lower panel) of different Rubisco types correlated with the according specificities; 1: Form III (archaea); 2: Form II (proteobacteria); 3: Form I (green; cyanobacteria); 4: Form I (C_3-plants); 5: Form I (C_4-plants); 6: Form I (red lineage-"on green algae"); redrawn and simplified from Whitney* et al. *(2011a). For a color version of this figure, see www.iste.co.uk/maberly/photosynthesis.zip*

4.3. Dissolved inorganic carbon and its behavior in water

Like all other gases, atmospheric CO_2 dissolves in water, following Henry's law (Henry 1803). Accordingly, the concentration of CO_2 in the water depends on its partial pressure in the overlying atmosphere and its solubility, which in turn depends on salinity, pressure and temperature (equation [4.1]):

$$[CO_2]_{aq} = K * pCO_{2;atm} \qquad [4.1]$$

where $[CO_2]_{aq}$ is the concentration of CO_2 in water (in mol kg^{-1}), K is the Henry constant (in mol kg^{-1} Pa^{-1}), and $pCO_{2;atm}$ is the partial pressure of CO_2 in the atmosphere (in Pa). Being a linear function, a doubling of atmospheric pCO_2 will cause a doubling of $[CO_2]_{aq}$, provided that salinity, pressure and temperature remain constant. Unlike most other gases, however, CO_2 can react with water (H_2O) to form "true carbonic acid" (H_2CO_3) that rapidly dissociates into bicarbonate (HCO_3^-), carbonate (CO_3^{2-}) and hydrogen ions (H^+). CO_2 can also react with OH^- to form HCO_3^-. Irrespective of the specific reaction pathway, all of the named species are connected by a chemical reaction equilibrium (equation [4.2]), so that both reactions of CO_2 increase the concentration of H^+ and thus lower the pH.

$$CO_{2(aq)} + H_2O \leftrightarrow H_2CO_3 \leftrightarrow H^+ + HCO_3^- \leftrightarrow 2\,H^+ + CO_3^{2-} \qquad [4.2]$$

Consequently, in- and out-gassing of CO_2 alters the pH. *Vice versa*, artificially or biologically driven pH changes will alter the speciation of the carbonate system. For example, an acidification can increase $[CO_2]$ to conditions above the "Henry-equilibrium", causing an outgassing to the overlying atmosphere. CO_2, H_2CO_3, HCO_3^- and CO_3^{2-} are collectively referred to as "dissolved inorganic carbon" (DIC). Due to its rapid dissociation, the proportion of H_2CO_3 is negligibly small, so that in chemical notations it is typically subsumed in the CO_2 term. While CO_2 is the dominant species at pH values *below* the first dissociation point (pK_{a1} ~6.1 in seawater), CO_3^{2-} dominates *above* the second dissociation point (pK_{a2} ~9.1 in seawater) and HCO_3^- dominates in between (Figure 4.2; see Chapter 1). This speciation into bicarbonate and carbonate ions allows water to contain a higher concentration of dissolved inorganic carbon than would otherwise be possible due to the solubility of CO_2 alone. This capacity for high "CO_2 uptake" in the form of DIC is further enhanced by "alkalinity", a distinct chemical property of water. Total alkalinity (TA) may be described operationally, as an amount of negative charges present in the water that is able to accept the H^+ released in the dissociation reactions of carbonic acid.

This removal of "reaction products" of equation [4.2] facilitates further uptake of CO_2 into the water: If idealized (TA = 0), saline water is left to equilibrate with the ambient atmosphere (~400 µatm CO_2; temperature = 20°C, salinity = 32), it exhibits DIC concentrations of ~0.016 mmol kg^{-1}. If seawater with a typical TA background of 2.35 mmol kg^{-1} is used instead, however, the water will contain ~2.1 mmol kg^{-1} DIC, that is, more than 130-fold the amount that non-alkaline water can take up.

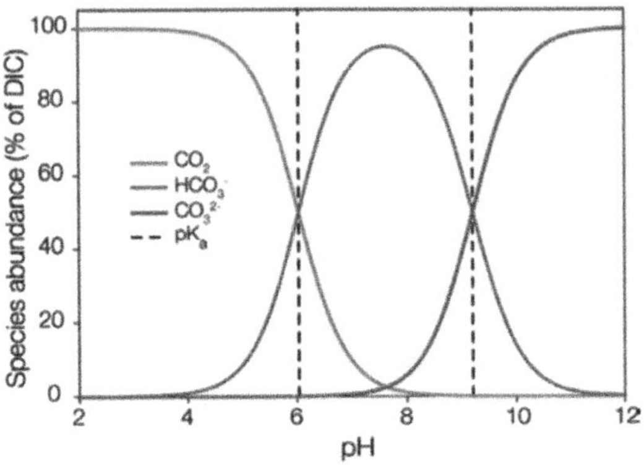

Figure 4.2. *Bjerrum Plot; pKa values taken from Mehrbach* et al. *(1973) refitted by Dickson and Millero (1987); temperature = 20°C, pressure = 1dbar, salinity = 32). For a color version of this figure, see www.iste.co.uk/maberly/photosynthesis.zip*

The origins of alkalinity in natural waters are weathering processes, caused by the slightly acidic rain that forms in an atmosphere containing CO_2 (Middelburg *et al.* 2020; Chapter 1). The subsequent "carbonic acid-supported" dissolution of minerals such as olivine (Mg_2SiO_4; equation [4.3]) or limestone ($CaCO_3$; equation [4.4]) will set free the constituent ions:

$$Mg_2SiO_4 + 4H^+ + 4\,HCO_3^- \rightarrow 2\,Mg^{2+} + 4\,HCO_3^- + Si(OH)_4 \qquad [4.3]$$

$$CaCO_3 + H^+ + HCO_3^- \rightarrow Ca^{2+} + 2\,HCO_3^- \qquad [4.4]$$

Having donated their original protons, the resulting bicarbonate ions are now present with the cations released from the mineral, that is, overall electroneutral. In both cases, due to their lower or equal pK_a value, the weak bases that are released from the mineral can accept H^+ that originate from the dissolution of atmospheric carbonic acid. Consequently, the oceans contain an excess of proton acceptors (i.e. bases derived from weak acids with $pK_a \geq 4.5$) over proton donors. This is in fact the traditional definition of TA as given by Dickson (1981; equation [4.5])

$$TA = [HCO_3^-] + 2\,[CO_3^{2-}] + [OH^-] + [B(OH)_4^-] + [HPO_4^{2-}] + 2[PO_4^{3-}]$$
$$+[H_3SiO_4^-] + [NH_3] + [HS^-] - [H^+] - [HSO_4^-] - [HF] - H_3PO_4 \qquad [4.5]$$

TA (in mol kg^{-1}) is a conservative parameter, because the number of relevant negative charges per 1 kg of seawater is independent of volume changes caused by varying temperature or pressure. However, single constituents *within* the TA expression (e.g. HCO_3^-, CO_3^{2-} or $B(OH)_4^-$; equation [4.5]) are themselves *not* conservative, because pK_a values of their acid-base systems are influenced by temperature or pressure. To overcome this, Dickson's definition was re-expressed in an *explicit conservative* way by Wolf-Gladrow *et al.* (2007; equation [6.6]):

$$TA_{ec} = [Na^+] + 2[Mg^{2+}] + 2[Ca^{2+}] + [K^+] + 2[Sr^{2+}] + \cdots$$
$$- [Cl^-] - [Br^-] - [NO_3^-]$$
$$- \cdots TPO_4 + TNH_3 - 2\,TSO_4 - THF - THNO_2 \qquad [4.6]$$

where TPO_4 represents the sum of the concentrations of all species of phosphoric acid, TNH_3 represents the sum of $[NH_3]$ and $[NH_4^+]$, TSO_4 represents the sum of the concentrations of $[SO_4^{2-}]+[HSO_4^-]$, THF represents the sum of $[F^-]+[HF]$, and $THNO_2$ represents the sum of $[NO_2^-]+[HNO_2]$. TA is thereby expressed based on an overall charge balance. DIC species do not appear at all, because CO_2 invasion or release does not affect the concentrations of conservative ions or uncharged species (such as total phosphoric acid, TPO_4). This expression also makes it possible to better grasp effects of biological processes in terms of their effect on alkalinity. It can be seen, for example, that the production of 1 unit $CaCO_3$ reduces TA by 2 units because two charges are removed from solution (see equation [4.6]), whereas only 1 unit of C is required. This and other biological effects on carbonate chemistry are illustrated in Figure 4.3 (based on Zeebe and Wolf-Gladrow 2001; Rost *et al.* 2008).

In summary, at equilibrium with the atmosphere the concentration of CO_2 in water is exclusively governed by Henry's law. Although the basic principles of carbonate chemistry apply equally to freshwater and seawater, oceanic environments contain significantly different alkalinity backgrounds. Therefore, these environments strongly differ in their overall inorganic carbon content, with fresh water containing more spatially and temporally variable DIC concentrations than seawater (see Chapter 1). The invasion of CO_2 into water changes its pH and *vice versa*, pH changes shift the speciation of DIC, leading to either a further ingassing or outgassing of CO_2. Processes such as calcification and $CaCO_3$ dissolution, as well as biomass production and remineralization, likewise affect the carbonate system. Despite this rather short summary, carbonate chemistry holds many more fascinating aspects and relates to a surprisingly large number of scientific disciplines. For further reading, we refer the interested reader to the book *CO_2 in Seawater: Equilibrium, Kinetics, Isotopes* by Zeebe and Wolf-Gladrow (2001), in which the authors give a comprehensive and practical introduction to marine carbonate chemistry, related topics and scientific applications.

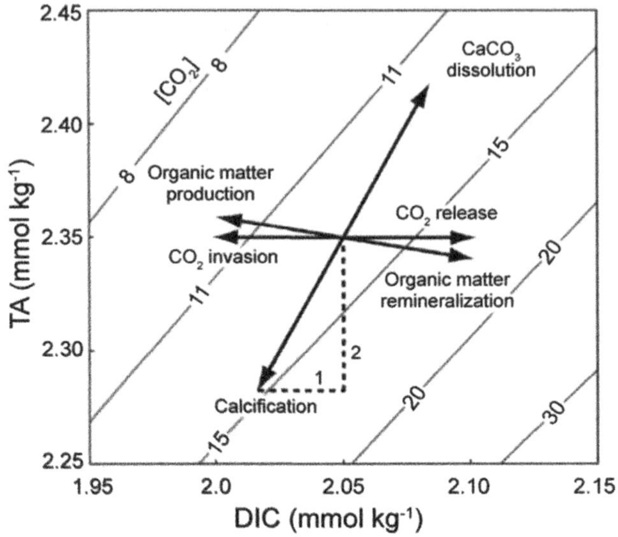

Figure 4.3. *Processes affecting DIC and TA (background isoclines indicate [CO_2] in μmol kg^{-1}): Whereas CO_2 invasion and release increase or decrease [DIC], the production and dissolution of $CaCO_3$ decreases or increases both, [DIC] and TA, in a 1:2 stoichiometry. Biomass production and remineralization decrease or increase [DIC], respectively, and also slightly increase or decrease TA due to the uptake or release of nitrate and phosphate (see equation [4.6]). Redrawn from Zeebe and Wolf-Gladrow (2001)*

4.4. Disequilibrium situations and implications of transport processes

Generally, the utilization of one or more DIC species, for example, for photosynthetic biomass buildup, results in the drawdown of the respective inorganic carbon species in the aqueous environment. On a closer look, the cellular uptake of DIC species from the vicinity of the cell is often faster than can be replenished by diffusion from the bulk seawater or by re-equilibration between carbon species (see, for example, Wolf-Gladrow *et al.* 1999; Kranz *et al.* 2010; Hopkinson *et al.* 2013). As a result, the concentration of inorganic carbon in the vicinity of the cell can be significantly decreased compared to the bulk-seawater, forming a so-called diffusive boundary layer around the cell. Molecular diffusion of the consumed species towards the cell is one way to replenish the inorganic carbon species in this boundary layer. However, diffusion rates in water are about 10.000 times slower than in air, and the diffusion coefficients of DIC species in water are strongly temperature-dependent. At a temperature of 10°C, for instance, the diffusion rate of CO_2 is only ~25% compared to the rate at 20°C (Boudreau 1997; Zeebe and

Wolf-Gladrow 2001). Re-equilibration from the chemically "connected" pools of DIC species is the second way to replenish inorganic carbon species in the boundary layer. Here, it must be noted that while the equilibration between the HCO_3^- and CO_3^{2-} pools is quasi-instantaneous (Zeebe and Wolf-Gladrow 2001), the equilibration between the CO_2 and HCO_3^- pools are comparably slow and temperature- as well as pH-dependent (Figure 4.4), so that the issue of slow replenishment primarily affects the CO_2 re-equilibration in the cellular boundary layer.

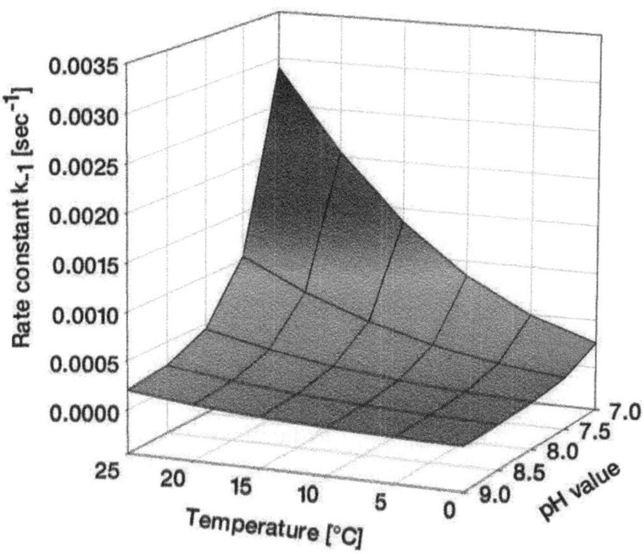

Figure 4.4. *Temperature- and pH-dependence of the rate constant k_{-1} for the reaction $HCO_3^- + H^+ \rightarrow CO_2$ in seawater (salinity 32). It can be seen that with increasing pH and decreasing temperature, the conversion rate is drastically decreased. Rate constant was calculated based on Zeebe and Wolf-Gladrow (2001), using pK_a-values of carbonic acid derived by Roy* et al. *(1993). For a color version of this figure, see www.iste.co.uk/maberly/photosynthesis.zip*

For example, if CO_2 is taken up by a cell, the depletion of CO_2 in the boundary layer will be a function of many factors: First, the cell's CO_2 uptake rate, which naturally scales with the photosynthetic performance and thus, also with factors such as light. Second, the cell size, and thereby the surface:volume ratio, which determines the water volume from which CO_2 is removed. Third, kinetic aspects, that is, CO_2 diffusion rates in the water and the re-equilibration rates from the "connected" HCO_3^- and CO_3^{2-} pools, which determine the CO_2 re-supply. And finally, TA and pH, which determine the water's equilibrium CO_2 and HCO_3^- concentration (Egleston *et al.* 2010). From these few points, it becomes clear that many interacting factors determine whether cells fall short in carbon supply with regard to their carbon demand, both factors determining the need for CCM activity. Based on these aspects of environmental and ecophysiological properties, it is clear that large cells with high uptake rates establish thicker diffusive boundary layers, in which the utilized carbon species is depleted. This is additionally amplified under low temperatures due to slow diffusion and re-equilibration. On the other side of the spectrum, smaller cells in a warmer environment with lower uptake rates are likely to experience less depletion of the utilized carbon species in their thinner boundary layer. To learn more about such "disequilibrium" situations in the cells' boundary layers, we refer the reader to the works of Johnson (1982), Schulz *et al.* (2006), Flynn *et al.* (2012) and Eigen (1964).

Overall, because of the above-mentioned characteristics of the carbonate system, inorganic carbon in general, and CO_2 in particular, can become significantly depleted around an actively photosynthesizing cell (Sweeney 2003). To deal with this shortage in CO_2 supply and the already mentioned high CO_2 demand imposed by Rubisco, aquatic primary producers evolved diverse CCMs. While the so-called "biophysical CCMs" enrich DIC in the cell by trafficking inorganic carbon species across membranes, the "biochemical CCMs" employ specialized metabolic pathways (C4 or CAM; see Chapter 5) to facilitate a directed flux of DIC species to the site of carbon fixation, that is, where Rubisco is located (see Chapter 7). In any case, DIC species must cross membranes, and due to their very different chemical properties, also their transport modalities differ considerably.

CO_2 is a small uncharged molecule and, due to its symmetric architecture with two opposing dipoles, it can be well hydrated. However, it has no overall dipole moment that would facilitate any polar interactions with larger molecules. Consequently, it can readily diffuse through biological membranes, following its concentration gradient, that is, obeying Fick's law of diffusion (Fick 1855). At a pH of 7.5, diffusion of CO_2 across artificial, two-dimensional, protein-free membranes is primarily rate-limited by the diffusive boundary layers that automatically establish when CO_2 permeates a membrane and its concentration decreases on the source-side (Missner *et al.* 2008). This exponential decrease of CO_2 concentration towards the membrane follows kinetic theory, but the thickness of the boundary layer and the degree of CO_2 depletion can be significantly affected by the presence or absence of carbonic anhydrases (CA, see Chapter 6). CAs increase the rate of re-equilibration with HCO_3^- and thus replenish the CO_2 pool much quicker than would be possible without them.

The influence of membrane composition on CO_2 diffusion has been investigated with a multitude of techniques using artificial, protein-free membranes, membranes containing a diversity of proteins as well as whole cells (see Figure 4.5 for references). Analytical models have also been used to determine the diffusion limitations originating from the diffusive boundary layer. These simulations revealed that membrane bilayers are responsible for about 85%–95% of the resistance against CO_2 transition, which is in line with previous findings (Missner *et al.* 2008). *In vivo* measurements combined with numerical modeling showed that diffusion rates over the cytoplasmic membrane can vary widely between species (ranging from 0.001 to 3.2 cm s^{-1}; Figure 4.5). Based on these observations, the cytoplasmic membrane is *per se* not a significant barrier to CO_2. Consequently, a "CCM" that simply allows transition of CO_2 molecules across the plasma membrane would be highly inefficient, because any CO_2 accumulated to above-ambient concentrations would leak back to the medium. Still, there is ongoing debate about physiologically relevant CO_2 uptake mediated by protein channels like human or plant aquaporins (Uehlein *et al.* 2003; Endeward *et al.* 2006) or the RhaG protein, a gas channel that allows passage of CO_2 in addition to NH_3 (Endeward *et al.* 2008). In most cases, the data indeed suggest an eased diffusion of CO_2 through the transporter pores that ceases when the pores are blocked by an irreversibly binding inhibitor (Prasad *et al.* 1998).

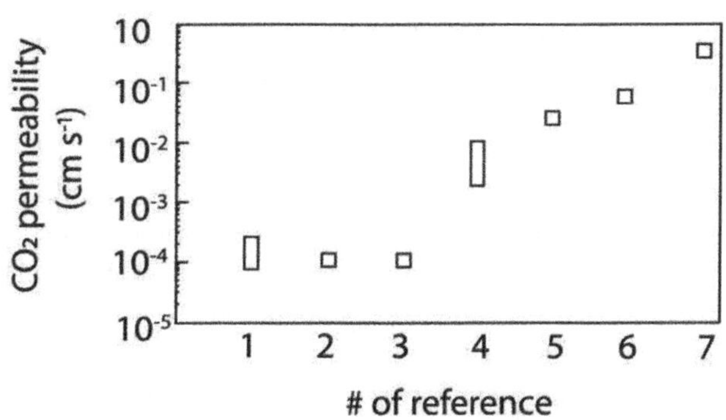

Figure 4.5. *CO_2 membrane permeabilities (in cm s-1); 1: Sültemeyer and Rinast (1996), Blank and Ehmke (2003) 2: Prasad et al. (1998) 3: Yang* et al. *(2000) 4: Hopkinson* et al. *(2011) 5: Itel* et al. *(2012) 6: Gutknecht* et al. *(1977) 7: Missner* et al. *(2008). Please note that most measurements have been conducted with artificial membranes*

Cases of enzyme-facilitated or "semi-active" transport of CO_2 across membranes have been shown, for example, in cyanobacteria. Here, the so-called NDH1 proteins at the thylakoid and plasma membranes (see section 4.4) convert cytosolic CO_2 into

HCO_3^-. The resulting lower cytoplasmic CO_2 concentration consequently facilitates diffusive uptake. In the eukaryotic green alga *Chlamydomonas* (see section 4.5), a transmembrane HCO_3^- transporter closely cooperates with a CA, which apparently functions in unidirectional transport so that CO_2 in one compartment can be hydrated faster to HCO_3^- shortly before passing into another compartment. Such cooperativity mechanisms of enzymes indeed facilitate an apparent "active CO_2 transport", while the existence of physiologically relevant, true active CO_2 pumps remains to be proven.

In contrast to the neutral CO_2 molecule, the charged and more "bulky" HCO_3^- ion cannot freely diffuse over membranes and hence principally requires transport mechanisms to fulfill the cells' demand for inorganic carbon. For this reason, it also constitutes the ideal "accumulation product" inside cells. The inward transport of HCO_3^- necessitates the transport of further ions: First, positive or negative ions (e.g. Na^+ or Cl^-) need to be sym- or antiported to ensure overall electroneutrality of the cell. Second, facing the massive fluxes of carbon, these co-transported ions would either accumulate *in*, or be massively depleted *from* the cell, so that further transport processes, the so-called rectifying currents, are required to maintain ion composition of the cytoplasm within certain boundaries. Some HCO_3^- transporters have been described as uniporters that facilitate the transport of only HCO_3^- into the cell; however, antiporters (that transport HCO_3^- inwards and a negative ion outwards) as well as symporters (that transport both HCO_3^- and a positive ion inwards) have been described as well. Typical cations and anions involved in anti- and symport are Cl^- and Na^+, respectively (see later sections for details).

Understanding the structure and function of transmembrane HCO_3^- transport has recently gained new traction as researchers aim to bioengineer components of microalgal CCMs into higher plants to increase biomass production (e.g. Rae *et al.* 2017). While in cyanobacteria, three independent HCO_3^- transporter systems at the plasma membrane have been identified so far, eukaryotic microalgae, like the green alga *Chlamydomonas*, or the diatom *Phaeodactylum*, employ HCO_3^- transporters at diverse membranes, which often are complemented by additional enzymes like CAs, highlighting the diversity of CCMs among aquatic photoautotrophic organisms (Colman *et al.* 2002).

The CO_3^{2-} ion, at the current state of knowledge, is unlikely to be transported across membranes. CO_3^{2-} carries two negative charges, so that electroneutral transport would require a sym- or antiport of two positive or negative ions, respectively. According to kinetic theory, the likelihood of three substrates binding the transport enzyme in a sufficiently narrow time window is low, and thus, such a transporter would be inefficient, especially when compared to HCO_3^- or CO_2 transport. Moreover, if electroneutrality was assured by cotransport of double-negative or double-positive ions (e.g. Mg^{2+}, Ca^{2+} and SO_4^{2-}), all of these large ions would, like the CO_3^{2-} itself, have to be stripped of their hydrate hull prior to transport, which massively increases the energy barrier for the transport. Finally, in today's ocean with pH values of ~8.1, the concentration of HCO_3^- is about an order of magnitude higher than that of CO_3^{2-}, making it much more efficient to transport the former species of DIC. Results from green macroalgae also show that even at high pH (>10), HCO_3^- and not CO_3^{2-} is the major source of DIC (Maberly 1992). The aspects of unfavorable kinetics of substrate binding, costs imposed by maintaining electroneutrality, and the comparably low substrate concentrations are supposedly the reasons why no CO_3^{2-} transporter has been characterized in aquatic primary producers to date.

We have now presented properties of DIC species and their disequilibria, as well as implications of their transport, but to understand the operation of CCMs mechanistically, a more holistic picture has to be drawn. In the following sections, we will briefly describe different carbon acquisition strategies and summarize proposed mechanisms for prokaryotic and eukaryotic photoautotrophs, often focusing on key model organisms.

4.5. CCM operation in cyanobacteria

Cyanobacteria, prokaryotic photoautotrophs, inhabit diverse and often extreme habitats that have evolutionarily shaped their CCMs. Genomic analyses have shown that distinct cyanobacterial CCM strategies exist in connection with different lifestyles and habitats (Badger *et al.* 2002). A separation of cyanobacteria into α- and β-cyanobacteria, based on their carboxysome structure (see Chapter 7), shows that most of the α-cyanobacteria (e.g. *Prochlorococcus* and *Synechococcus*) inhabit the marine environment while most of the β-cyanobacteria (e.g. *Synechocystis* and *Anabaena*) mainly inhabit freshwater environments, with exceptions like *Trichodesmium* as a prominent marine β-cyanobacterium. More complex CCMs, involving multiple components (e.g. high-affinity HCO_3^- transporters), have been mainly found in β-cyanobacteria likely because freshwater environments exhibit larger variability in pH and DIC. Marine species, in contrast, appear to lack some of these DIC uptake systems (Badger *et al.* 2006).

Regarding carbon acquisition, an exclusive diffusive supply of CO_2 would be insufficient to saturate the cyanobacterial Rubisco with its very high half-saturation constant for the carboxylation reaction (Price and Badger 1989a, 1989b; Figure 1). Hence, cyanobacteria utilize a number of active transport systems that enrich DIC in the cytosol (up to 1.000-fold enrichment compared to ambient conditions; Coleman 1991). Five distinct, and seemingly independent, transport systems have so far been identified in cyanobacteria, three of which are the plasma membrane-spanning HCO_3^- transporters BCT1, BicA and SbtA (Price *et al.* 2004, 2008; Price and Howitt 2011), while the other two involve special, reductant-driven "carbonic anhydrase tunnels", NDH-1_4 and NDH-1_3, that facilitate unidirectional conversion of CO_2 to HCO_3^- at the thylakoid lumen and/or the plasma membrane. In the following section, we briefly describe these systems (Figure 4.6).

BCT1 was the first unambiguously identified HCO_3^- transporter (Omata *et al.* 1999), discovered in the globally abundant marine cyanobacterium *Synechococcus* sp. (strain PCC7942). This transporter is characterized as a "low-DIC inducible", ATP-binding cassette (ABC) uniporter (Omata *et al.* 1999) and is considered to be a high-affinity transport system with half-saturation concentrations for HCO_3^- below 15 μmol L^{-1} (Omata *et al.* 1999). Despite its first discovery in the marine α-cyanobacterium *Synechococcus*, BCT1 seems to be mostly found in freshwater β-cyanobacteria as well as some extremophiles, such as endolithic and thermophilic cyanobacteria (Omata *et al.* 1999; Badger *et al.* 2006). As a high-affinity transporter, BCT1 apparently facilitates sufficient carbon uptake under adverse environmental conditions, like the low DIC concentrations in freshwater, the low atmospheric CO_2 exchange inside rocks that is typically associated with high pH, and the generally low CO_2 solubility in hot-springs.

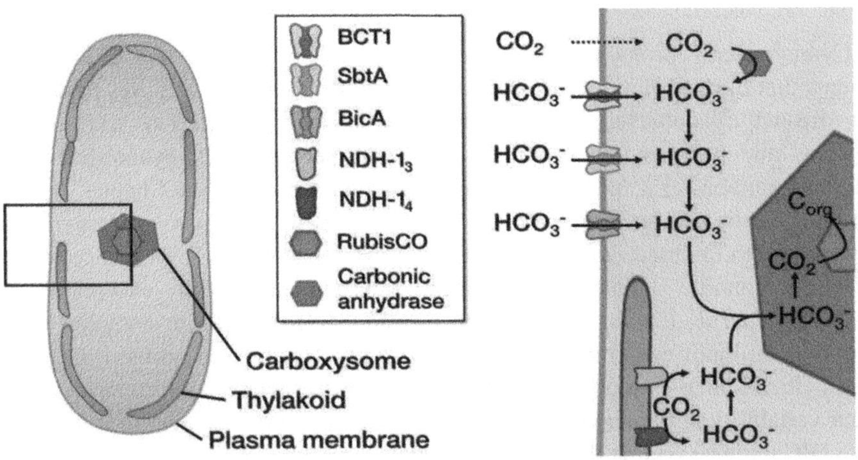

Figure 4.6. *Schematic diagram of cyanobacterial CCM components, redrawn based on Price et al. (2008); the embedding of Rubisco into a CA-containing carboxysome causes a flux of HCO_3^- to the carboxysome. Cytosolic HCO_3^- can be replenished by active uptake, CA-assisted re-equilibration from diffusively imported CO_2, or from NDH1-assisted HCO_3^- generation. For a color version of this figure, see www.iste.co.uk/maberly/photosynthesis.zip*

The HCO_3^- transporter BicA was also first identified in *Synechococcus*, and is characterized as a Na^+-dependent, low-affinity/high-flux transporter with an ATP-binding cassette (Omata *et al.* 1999; Price *et al.* 2004). This HCO_3^- transporter is a member of the sulfate permease/solute carrier (SulP/SLC26) family of anion transporters found in many prokaryotes and eukaryotes. The kinetic properties of BicA transporters vary considerably, with apparent half-saturation concentrations ranging from ~75-350 μmol L^{-1}, supporting maximum photosynthesis rates ranging from ~450 to ~1500 μmol O_2 (mg chlorophyll)$^{-1}$ h^{-1} (Price and Howitt 2011). BicA has been described to possess a catalytic trade-off where lower affinity coincides with higher V_{max} and *vice versa* (Price and Howitt 2011). Stoichiometrically, it was estimated to require the energy equivalent of ~1 ATP for the transport of 2 HCO_3^- ions across the plasma membrane (Raven *et al.* 2014).

SbtA, first discovered in *Synechococcus* as well, is characterized as an Na^+-dependent ATP-binding cassette (ABC) transporter. SbtA is considered to be a high-affinity uptake system with half-saturation concentrations below 15 μmol L^{-1} (Price and Howitt 2011; Price *et al.* 2011; Du *et al.* 2014). Its gene expression responds to DIC supply (Schwarz *et al.* 2011), with higher expression levels under low DIC supply or high carbon demand due to high light intensities, respectively. It is not clear whether the transporter activity is exclusively driven by the inward-directed Na^+-gradient, thus functioning as HCO_3^-/Na^+ symporter. Like for the aforementioned HCO_3^- pumps, BCT1 and BicA, the exact role of the ATP-binding cassette in the transport processes is unclear. The cassette might directly provide energy (by ATP consumption) for transport, or might fulfill a sensing function, so that the transport is controlled by cellular ATP availability. Interestingly, the SbtA transporter is co-transcribed with SbtB, a sensor protein that binds cellular messenger molecules (cAMP and AMP) and also seems to fulfill redox-sensory function (Selim *et al.* 2018). AMP in particular is an indicator of ceasing glycolysis and overall carbon shortage, and is apparently perceived by SbtB as "low DIC concentrations".

An apparent "active CO_2" uptake (i.e. utilization of CO_2 following Michaelis–Menten kinetics) has been measured (Kranz *et al.* 2009). Such uptake requires an actively maintained gradient across the plasma membrane with a significantly lower CO_2 concentration in the cytosol compared to the cellular diffusive boundary layer (Hopkinson *et al.* 2011), which can only be achieved by an instantaneous and efficient removal of cytosolic CO_2. It was hypothesized that this can be facilitated by the NDH-1 complexes that are located at the cytosolic side of the thylakoidal plasma membranes. Two different NDH-1 complexes have been described (Ohkawa *et al.* 2001; Price *et al.* 2002). NDH-1_3 is inducible under DIC limitation, and its expression correlates with a high apparent CO_2 uptake affinity. This system seems to be driven by photosynthetic electron transport (Ohkawa *et al.* 2000a; 2000b). NDH-1_4 appears to be constitutively expressed and was described to be associated with the cyanobacterial respiratory electron transport at the plasma membrane, although it is suspected to be driven by both photosynthetic and respiratory electron transport (please note that cyanobacteria share aspects of their photosynthetic and respiratory electron transport chain). Multiple studies proposed a diverse distribution of these two CO_2 uptake systems in α- and β-cyanobacteria (Badger *et al.* 2002, 2006; Badger and Price 2003). A detailed model of NDH-1 facilitated CO_2 uptake has been proposed (Price *et al.* 2002; Badger and Price 2003), by which NDH-1 catalyzes a combined hydration/reduction reaction of CO_2 to create HCO_3^- via oxidation of NAD(P)H and ferredoxin that originates from photosynthesis or respiration (Ohkawa *et al.* 2000a; Shibata *et al.* 2001; Maeda *et al.* 2002). Mechanistically, the NDH-1 complexes are "CA-like", but function unidirectionally, using reductant pressure to create HCO_3^- from cytosolic CO_2, and thereby establish the strong transmembrane CO_2 gradient required for the diffusive uptake. As a "side" reaction, one additional H^+ gets channeled through the thylakoid membrane, contributing to the transmembrane H^+ gradient, which can drive ATP production (Price *et al.* 2002). Hence, the conversion of CO_2 to HCO_3^- can simultaneously supply inorganic carbon and enhance the cellular ATP generation.

The reader who wishes to know more about the diversity and functionality as well as ecology and genomics of cyanobacterial CCMs is referred to the studies and reviews of Badger *et al.* (2006), Price *et al.* (2008), and Raven and Beardall (2016).

4.6. CCM operation in green algae

In eukaryotic microalgae, CCMs differ from those of cyanobacteria because eukaryotic cells have organelles, some of which (e.g. chloroplasts) contain multiple membranes. This striking evolutionary feature of compartmentalization improves control over metabolic processes but also adds challenges to supply inorganic carbon to Rubisco efficiently. To date, the structure and function of inorganic carbon transporters have been covered most intensively in the biotechnologically relevant model green alga *Chlamydomonas reinhardtii.*

C. reinhardtii lives in diverse ecosystems that exhibit wide ranges of CO_2 concentration (Spalding 2008; Sasso *et al.* 2018). To thrive in these habitats, it has evolved different CO_2 acclimation phenotypes that are under transcriptional and ultimately environmental control. Mass spectrometry studies demonstrated early on that both HCO_3^- and CO_2 can be taken up by *Chlamydomonas* during photosynthesis, and that these uptake systems are upregulated under limiting DIC conditions (Sültemeyer *et al.* 1989, 1991). More recent research, including gene expression studies, has shown that *Chlamydomonas* possesses an inducible CCM that has at least two prominent acclimation states that are widely recognized: a "low CO_2 state" (0.03–0.5% CO_2, including air-equilibrated pCO_2) and a "very low CO_2 state" (<0.03% CO_2; Mackinder 2018). Intermediate CO_2/DIC-availability is likely to produce intermediate states. Although the gene expression profiles under the respective acclimation states involve almost 25% of *Chlamydomonas*' genome, the acclimation states are commonly defined by different DIC affinities and growth rates. In addition, different localizations of particular CCM components are involved. Hence, the different DIC acclimation states account for more than the expression of specific genes involved in carbon acquisition (Fang *et al.* 2012).

During photosynthesis in the "low CO_2 state", a thylakoid-membrane located anion transporter complex, BST1-3 (Mukherjee *et al.* 2019), conveys HCO_3^- along its concentration gradient from the alkalized chloroplast stroma to the acidic thylakoid lumen (Figure 4.7). In the lumen, the carbonic anhydrase CAH3 accelerates the dehydration of the HCO_3^- to CO_2, a reaction that is "pulled" by the low luminal pH during photosynthesis (Sinetova *et al.* 2012). The gradient driven flux of CO_2 out of the thylakoid lumen into the pyrenoid area subsequently supplies Rubisco with the required CO_2 and efficiently fuels the carboxylation reaction (see Chapter 7). The pyrenoid structure, however, cannot prevent leakage of unfixed CO_2, since it does not constitute a physical barrier to the stroma. To minimize the resulting loss, the stromal carbonic anhydrase LCIB scavenges leaking CO_2 and, exploiting the high stromal pH, produces HCO_3^-, which can be redirected back to the lumen by BST1-3. Additional CCM components closely associated with the gene-product LCI1 are localized at the plasma membrane (Burow *et al.* 1996; Ohnishi *et al.* 2010; Kono *et al.* 2020). These components, however, are not well characterized to date and their preferably transported DIC species are still unknown. While LCI1 is expressed at air-levels of CO_2, its contribution to DIC uptake in the "very low CO_2" state is minimal (Kono and Spalding 2020), making it seemingly the "standard" plasma membrane uptake system for DIC at air-equilibrated CO_2 levels with an apparent half-saturation concentration of 40–100 μmol L^{-1} DIC over a pH range of 7.0–7.8 (Ohnishi *et al.* 2010).

Under the "very-low CO_2 state", when DIC becomes limiting, cells use a retrograde Ca^{2+} signaling (Wang *et al.* 2016) to induce expression of the two HCO_3^- transporters, HLA3 and LCIA (Figure 4.7; Wang and Spalding 2014a), and other, yet uncharacterized components (Fang *et al.* 2012). HLA3 is a plasma membrane protein classified as a putative ATP-binding cassette (ABC) type transporter in the multidrug resistance-related protein family. LCIA, a member of the formate-nitrite transporter family, is localized in the chloroplast envelope (Duanmu *et al.* 2009b; Wang and Spalding 2014b; Gao *et al.* 2015; Yamano *et al.* 2015). With this constellation, cells can efficiently take up HCO_3^- from the environment into the cytoplasm and enhance the uptake of HCO_3^- into the stroma. A dual role is taken by the carbonic anhydrase LCIB in the two acclimation states: whereas under "low CO_2", it localizes in the bulk stroma, in the "very-low CO_2" state, it associates with LCIA around the pyrenoid. In this state, it hydrates the leaked CO_2 to HCO_3^-, which is then directly pumped back into the bulk DIC pool of the plastid stroma (Duanmu *et al.* 2009a; Yamano *et al.* 2010; Wang and Spalding 2014b). The inability of LCIB-defective mutants to accumulate DIC also in "low CO_2", the epistatic relation between the carbonic anhydrases LCIB and CAH3 (Duanmu *et al.* 2009a), and recent protein structure data (Jin *et al.* 2016) suggest that LCIB serves as a primary active CO_2 uptake system in both "very low" and "low" CO_2 states, as well as an efficient CO_2 recapture system in "very low CO_2" states.

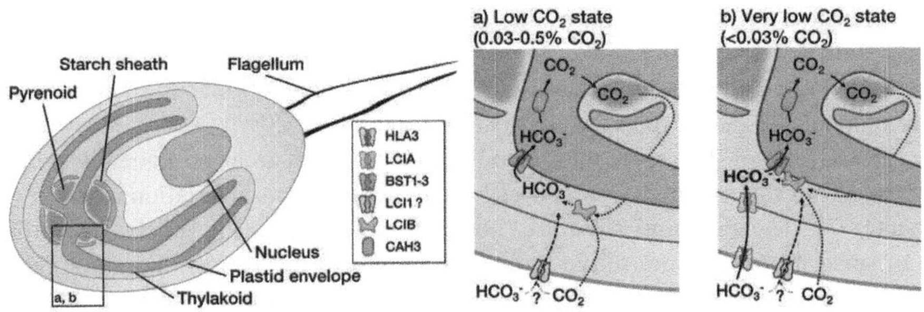

Figure 4.7. *Schematic diagram of a* Chlamydomonas reinhardtii *cell including the two acclimation phenotypes of the "low CO2 state" and the "very low CO2 state" and the tentative expression states of CCM components. Figure redrawn and modified based on Mackinder (2018). For a color version of this figure, see www.iste.co.uk/maberly/photosynthesis.zip*

Readers interested in structure and function of the CCM of *Chlamydomonas* are referred to the recent literature, especially the reviews by Mackinder (2018) and Hennacy and Jonikas (2020).

4.7. CCM operation in diatoms

Diatoms are a very successful group of microalgae contributing about 20% of annual global primary production (Field *et al.* 1998). They originated from secondary endosymbiosis and harbor a chloroplast surrounded by four membranes (Keeling 2013; Chapters 2 and 3), unlike green algae, which are primary-endosymbiotic photoautotrophs with only two membranes around the chloroplast.

This morphological feature complicates the facilitation of an efficient flux of DIC towards Rubisco. Numerous experiments have shown that diatoms utilize both CO_2 and HCO_3^- to cover their carbon demand (Rotatore *et al.* 1995; Burkhardt *et al.* 2001; Rost *et al.* 2003; Tortell *et al.* 2008a; Kranz *et al.* 2015), and that species exhibit certain preferences for particular proportions of DIC forms (Clement *et al.* 2016). This diversity enables diatoms to inhabit many niches in the aquatic and also the terrestrial environment. Experiments with natural communities exposed to environmentally relevant ranges of CO_2 concentrations clearly showed that changes in pH, and thus, CO_2/HCO_3^- availability causes shifts in species composition (Tortell *et al.* 2008b; Trimborn *et al.* 2013; Petrou *et al.* 2019), highlighting the importance of DIC supply and uptake capacity for the functioning of ecosystems. Despite the distant relationship and different cellular architecture, diatom and green algal CCMs share important principles and exhibit considerable similarity. For instance, both groups possess species-specific, constitutively expressed CCM modules alongside inducible, responsive CCM components that incrementally strengthen the CCM under CO_2 or DIC deprivation (Clement *et al.* 2016; Kikutani *et al.* 2016; Tsuji *et al.* 2017b). Also, both diatom and green algal CCMs feature an apparent membrane uptake of CO_2 and HCO_3^-, a transport across the chloroplast membranes, an efficient CA-assisted CO_2 conversion within the thylakoids and a CA-based "recycling system" to avoid CO_2 leakage (Tortell *et al.* 2008a; Moroney *et al.* 2011; Nakajima *et al.* 2013; Wang and Spalding 2014b; Yamano *et al.* 2015; Hopkinson *et al.* 2016; Kikutani *et al.* 2016; Mackinder *et al.* 2016).

A major breakthrough in characterizing diatom CCM components on a genetic basis was the report of a plasma membrane located, low-CO_2-inducible mammalian-type SLC4 transporter in *Phaeodactylum tricornutum*. This transporter was shown to pump HCO_3^- and to require high concentrations of Na^+ ions with a saturation level of about 100 mmol L^{-1} Na^+ (Nakajima *et al.* 2013). DIC uptake in *P. tricornutum* was maximal at a typical seawater pH of 8.2, where HCO_3^- is the dominant form of DIC, while it was suppressed to 40% at pH 9.5, where CO_3^{2-} is the dominant DIC form (Nakajima *et al.* 2013). Inhibition of the SLC4 transporter greatly decreased overall DIC affinity, underlining its function in maintaining cellular HCO_3^- influx under limiting CO_2 conditions. To date, six more SLC4-type and three more SLC26-type genes were found in *P. tricornutum* (Nakajima *et al.* 2013), which are known to be homologues of HCO_3^- transporters found in mammalian cells (e.g. Cordat and Casey 2009; Nakajima *et al.* 2013). Orthologous SLC4 and SLC26 genes are also present in the centric diatom *Thalassiosira pseudonana* (Nakajima *et al.* 2013) and the pennate freshwater diatom *Asterionella formosa* (Maberly *et al.* 2021), suggesting the widespread use of transmembrane HCO_3^- uptake in phylogenetically distant centric and pennate diatom taxa.

Based on observations in *P. tricornutum* and in *T. pseudonana* (Harada and Matsuda 2005' Harada *et al.* 2006a, 2006b' Tanaka *et al.* 2016), it is believed that the second messenger cAMP mediates CO_2 responses in many or most marine diatoms (Tsuji *et al.* 2017b), which might be the purpose of the ATP-binding cassette motif carried by many CCM components. Interestingly, despite facilitating similar processes in similar ways, the HCO_3^- transporters identified in cyanobacteria and the green alga *C. reinhardtii* do not share homology with the SLC4 transporters of diatoms, indicating multiple origins of algal HCO_3^- transporters and strong pressure for evolutionary convergence (Tsuji *et al.* 2017b).

Diatoms also exhibit an apparent "active" CO_2 uptake, which is likely a facilitated diffusion through CO_2 permeable aquaporins (Tu *et al.* 1978; Johnston and Raven 1996; Burkhardt *et al.* 2001; Rost *et al.* 2003; Tortell *et al.* 2008a; Hopkinson *et al.* 2011) that results from low cytoplasmic CO_2 concentrations. The cytoplasmic "CO_2 deficit" required for such uptake is, like in green algae, created through the active transport of HCO_3^- into the chloroplast that causes a low cytoplasmic HCO_3^- concentration and subsequent CA-mediated CO_2 conversion in the cytoplasm (Hopkinson *et al.* 2011; Hopkinson 2014; Sethi *et al.* 2020).

The four-layered chloroplast membrane constitutes a challenge for DIC trafficking and it has been suggested that an efficient flux requires HCO_3^- transporters in conjunction with CAs located between membranes (Figure 4.8; Tachibana *et al.* 2011; Hopkinson 2014; Samukawa *et al.* 2014). However, chloroplast membrane HCO_3^- transporters have yet to be confirmed and characterized in diatoms: Promising candidates are some SLC26 genes found in the genome of *P. tricornutum* (Nakajima *et al.* 2013). SLC26 transcription was not CO_2 responsive (Nakajima *et al.* 2013), suggesting that it is a constitutively expressed CCM component that maintains DIC flux from the cytosol into the plastid. In *P. tricornutum*, chloroplastic SLC4 transporters presumably function cooperatively with SLC4 transporters in the plasma membrane to deliver DIC from seawater to the stroma, although this awaits experimental verification (Tsuji *et al.* 2017b). In accordance with this idea, numerous CAs have been identified in diatom genomes (see Chapter 8; Roberts *et al.* 1997; Lane and Morel 2000; Lane *et al.* 2005; Xu *et al.* 2008; Kikutani *et al.* 2016) and experiments in which gene products have been tagged and localized in the diatoms *P. tricornutum* and *T. pseudonana* showed a high diversity of types and subcellular localizations of these CAs (Tachibana *et al.* 2011; Samukawa *et al.* 2014).

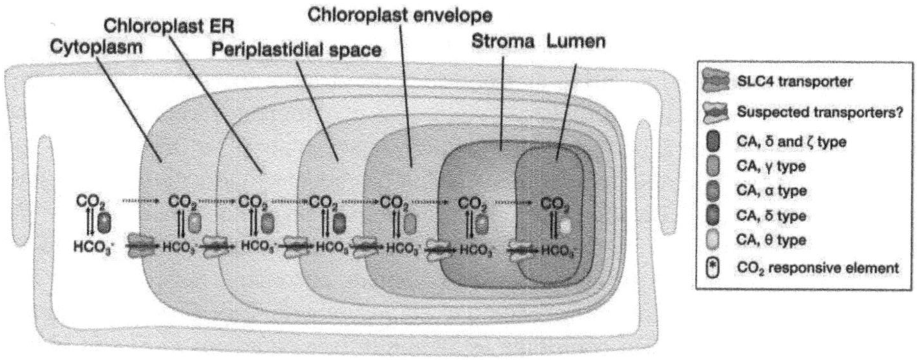

Figure 4.8. *Schematic of confirmed and suspected HCO_3^- transporters and CAs that together facilitate the transport of DIC into the thylakoid lumen in diatoms. Please note that this scheme includes elements found independently in* Phaeodactylum tricornutum *and* Thalassiosira pseudonana, *which exhibit different CO_2 acquisition strategies, as well as suspected elements that yet need to be confirmed. Figure redrawn and modified based on Tsuji et al. (2017b). For a color version of this figure, see www.iste.co.uk/maberly/photosynthesis.zip*

Owing to the structural complexity and phylogenetic diversity of diatoms, the field of diatom CCM research is rapidly expanding, and our state of knowledge is continuously changing. The interested reader is referred to the recent reviews by Tsuji *et al.* (2017b) and Sethi *et al.* (2020), who sum up the recent progress made regarding the localization and function of CCM elements as well as their applicability and potential for genetic engineering.

4.8. CCM operation in the coccolithophore *Emiliania huxleyi*

Coccolithophores are marine microalgae that surround themselves with elaborate calcium carbonate shells formed in an internal calcification compartment, the coccolith vesicle. It is estimated that their photosynthetic activity contributes up to 10% of the overall oceanic primary production of about 55 Pg yr^{-1} (Field *et al.* 1998), while their calcification in the pelagic is estimated to be less than ~3 Pg yr^{-1} (see Hopkins and Balch 2018). The dense calcite shells cause ballasting in marine aggregates and thereby enhance the efficiency of the biological carbon pump (Klaas and Archer 2002). Coccolithophores originated from secondary endosymbionts and harbor a chloroplast with four membranes (Keeling 2013), so that the maintenance of an efficient DIC flux towards Rubisco likely requires a multi-component, as well as multi-compartment CCM, similar to those of diatoms (Figure 4.9).

The globally most abundant coccolithophore species is *Emiliania huxleyi*. Unlike most coccolithophores, it has the ability to form large blooms, extending up to eight million km^2 (Moore *et al.* 2012) and pervades the oceans up to the northern and southern polar fronts. It is also very "young" in evolutionary terms and may for those reasons not be a perfect cell-biological representative for the majority of coccolithophores. Still, due to its prominence, it has become a model organism, and its photosynthesis, calcification and responses to environmental drivers are by far the best investigated among the coccolithophores. Even though *E. huxleyi* and its CCM have been studied in the context of carbonate chemistry, most attention has been given to calcification and its relation to photosynthesis. Calcification consumes CO_3^{2-} and Ca^{2+} ions, which contribute to alkalinity (see equations [4.5] and [4.6]). As a result, this process lowers the pH or respectively increases [CO_2] at the site of calcification, leading to speculations about calcification functioning as a CCM that supports photosynthesis (e.g. Sikes *et al.* 1980; Nimer and Merrett 1993; Anning *et al.* 1996; Buitenhuis *et al.* 1999). More recent studies, however, have shown that calcification does not enhance photosynthesis analogous to a CCM, because photosynthesis is unaffected even when calcification ceases (Herfort *et al.* 2002; Trimborn *et al.* 2007; Leonardos *et al.* 2009). In fact, both processes compete for inorganic carbon rather than facilitating one another (Rokitta and Rost 2012).

A number of studies focusing on CCM aspects found that *E. huxleyi* cells are principally able to take up both CO_2 and HCO_3^- (Nimer and Merrett 1996; Rost *et al.* 2003; Rokitta and Rost 2012). Lacking evidence for active CO_2 pumps suggests that observed CO_2 uptake functions similarly to that of diatoms, chlorophytes and cyanobacteria, that is, cytoplasmic HCO_3^- is pumped into the inner compartments, and the CA-mediated CO_2 deficit in the cytosol in turn results in a diffusive uptake. With a K_m of 10–40 µmol L^{-1}, the apparent affinity of this CO_2 uptake is quite low, for example, compared to diatoms (Rost *et al.* 2003, 2006b). In addition to CO_2 uptake, *E. huxleyi* operates Cl^-/HCO_3^- transporters of the SLC4 family (Figure 4.9; Herfort *et al.* 2002; von Dassow *et al.* 2009; Mackinder *et al.* 2011; Rokitta *et al.* 2011). For a long time, the results were inconclusive about which DIC species is predominantly taken up and how the preference and uptake affinity for CO_2 and HCO_3^- are related to instantaneous and acclimation conditions (e.g. Nielsen 1995; Herfort *et al.* 2002; Rokitta and Rost 2012; Stojkovic *et al.* 2013). Using ^{14}C tracers, it was shown that the relative CO_2 vs. HCO_3^- uptake of *E. huxleyi* is primarily determined by the ambient pH during the measurement, irrespective of acclimation to pH/CO_2 regimes or incident light intensity. Under low pH of about 7.9, cells predominantly take up CO_2, while the contribution of HCO_3^- uptake gradually increases with pH, reaching a CO_2:HCO_3^- uptake ratio of 1:1 at about pH 8.7 (Kottmeier *et al.* 2014). This immediate pH effect explains many of the diverse findings from assays that applied different pH settings. A pH-dependent regulation of the carbon fluxes in *E. huxleyi* has also been confirmed using mass spectrometry (Kottmeier *et al.* 2016a, 2016b). Testing the effect of [CO_2] vs. [H^+] in "uncoupled carbonate chemistry", the observed regulation patterns could be specifically attributed to the inhibiting effect of [H^+] on the HCO_3^- uptake rather than the effect of high [CO_2] (Kottmeier *et al.* 2016a, 2016b).

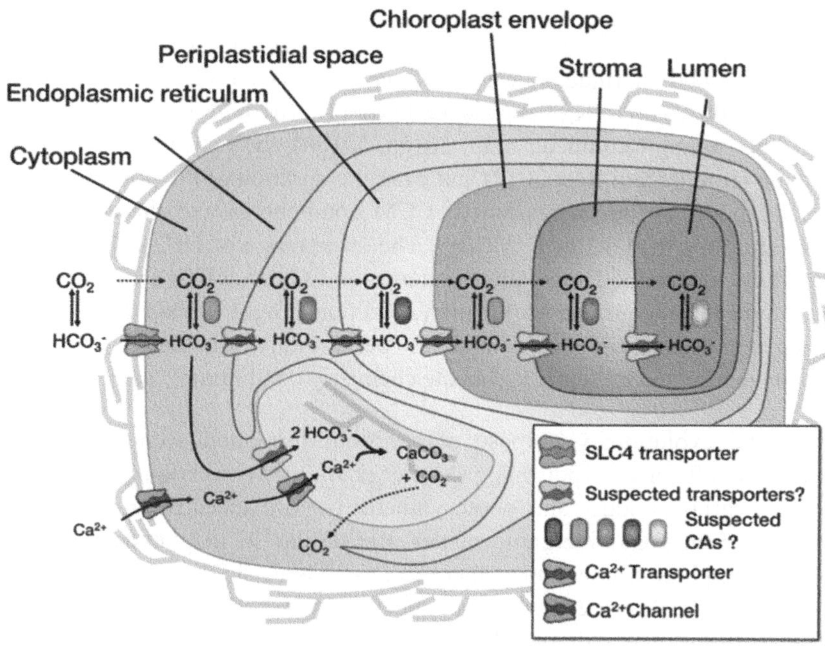

Figure 4.9. *Schematic diagram of the coccolithophore* Emiliania huxleyi. *The proposed scheme of carbon acquisition principally follows that of diatoms, which are also secondary endosymbiotic microalgae. The flux of DIC through the multiple membranes is putatively facilitated by a series of HCO_3^- transporters and CAs are likely present in the various compartments to facilitate re-equilibration. The coccolith vesicle imports HCO_3^- and Ca^{2+} ions and the following calcification reaction causes an increase in [CO_2]. The subsequent diffusion of CO_2 out of the coccolith vesicle is likely, but does not act as a CCM. For a color version of this figure, see www.iste.co.uk/maberly/photosynthesis.zip*

While pH determines the proportions of CO_2 vs. HCO_3^- uptake in *E. huxleyi*, the rate of DIC transport is apparently coupled to the photosynthesis rate, that is, cellular energization: when light intensity is kept static in assays, HCO_3^- uptake appears as a steady background flux, irrespective of external substrate concentration

(Rost *et al.* 2003). When light intensity is, however, altered during assays, the cellular HCO_3^- uptake responds instantly and scales with photosynthesis (Kottmeier *et al.* 2016a), confirming that instantaneous energization determines the CCM activity of *E. huxleyi*. In contrast to DIC uptake rates, DIC affinities during steady-state photosynthesis are not responsive to energization (Schulz *et al.* 2007b). However, they fall into a very wide range, making it unlikely that this phenomenon can be explained with "strain specific differences". Some studies observed comparably high DIC affinities with K_m values of ~100 μmol L^{-1} (Rokitta and Rost 2012); others found intermediate values (Nimer and Merrett 1996; Rost *et al.* 2006b; Schulz *et al.* 2007a) up to K_m values >1000 μmol L^{-1} (Rost *et al.* 2003). Apparently, the $[CO_2]$/pH setting (Kottmeier *et al.* 2016a, 2016b), the light intensity during the acclimation (Rokitta and Rost 2012) and even the duration of the photoperiod (Rost *et al.* 2006b) affect the expression of CCM components and thereby affect the apparent affinities for DIC in *E. huxleyi*. The large range of DIC affinities and the still not fully understood regulation patterns of CO_2 and HCO_3^- uptake rates in response to instantaneous and acclimation conditions might explain the observations that biomass production can sometimes be strongly stimulated or otherwise completely unaffected by future carbonate chemistry conditions.

While the evolutionary purpose of coccolith formation seems to lie in the alleviation of cellular Ca^{2+} poisoning (Müller *et al.* 2015), it certainly interacts with the overall carbon budget and photosynthetic performance. Because the carbon fluxes into organic and inorganic matter are similar in magnitude, the ratio of calcification to photosynthesis may also serve as an intrinsic short-term pH stabilizer (Sikes *et al.* 1980; Raven 2011a). More specifically, when imported HCO_3^- is used for photosynthesis, one additional H^+ is required per fixed CO_2, and when imported HCO_3^- is used for calcification, one H^+ is released per produced unit of $CaCO_3$ (Holtz *et al.* 2015; Kottmeier *et al.* 2016a, 2016b; Gafar *et al.* 2019). Thus, operating both photosynthesis and calcification in parallel might dampen negative effects that arise when only one of these processes makes use of HCO_3^- alone. Independent of the external carbonate chemistry and light conditions, the need to exchange H^+ with the environment seems to be generally lower in calcifying primary producers, and might give them an advantage over non-calcifying phytoplankton, which have to assure constant H^+ uptake to compensate for alkalization during HCO_3^--fueled photosynthesis (Raven 1986, 2011b). This advantage may add to the success of calcifying *E. huxleyi*, especially under bloom conditions, when seawater H^+ concentrations can become low.

4.9. CCM operation in macroalgae, seagrasses and freshwater plants

Macroalgae, seagrasses and freshwater plants face similar challenges with respect to supply and acquisition of inorganic carbon as phytoplankton do. Next to factors like light and nutrient availability, which primarily determine the carbon demand of a phototroph, macroalgae are further affected by secondary effects, for example, exposure to air due to tidal fluctuations, or strong changes in temperature and concentrations of O_2 or DIC, especially in freshwater. For example, in shallow seabeds and rock pools, dense macrophyte populations can reduce [CO_2] close to zero for extended periods of time (Middelboe and Hansen 2007). In addition, being much larger than phytoplankton, the diffusion limitation in macroalgae and seagrass is much more pronounced, leading to a substantially larger diffusive boundary layer (20 µm up to several millimeters; Johnston *et al.* 1992; James and Larkum 1996; Invers *et al.* 2001), which restricts the supply of DIC to the surface of the leaf. The transport of inorganic carbon through often multiple layers of cells or even tissues furthermore restricts the overall supply of CO_2 to Rubisco. The "red" chloroplast lineage of macroalgae that encompasses brown and red algae possesses form I Rubisco, which has a significantly higher CO_2 affinity than that of the "green" chloroplast lineage found in seagrasses (see Figure 4.1 as well as Chapters 2 and 5). Within this group, the K_m (CO_2) of brown algae tends to be lower than the K_m (CO_2) of red algae (Iñiguez *et al.* 2018, 2020). Because of the above-mentioned characteristics of their habitat as well as Rubisco kinetics, macroalgae and seagrasses have likely different demands regarding their CCMs.

Very few studies attempted to measure directly CO_2 and HCO_3^- uptake as well as CA activities in macroalgae and seagrasses, and most evidence stems from indirect approaches. For instance, the natural $^{13}C/^{12}C$ ratio of macroalgal and seagrass biomass or their behavior in pH drift experiments has often been used to describe the presence and regulation of CCMs. The ^{13}C isotope discrimination during biomass production, although driven by the intrinsic fractionation of Rubisco, is strongly modulated by the relative CO_2 vs. HCO_3^- uptake as well as CO_2 leakage in relation to total carbon uptake (Sharkey and Berry 1985; Raven *et al.* 2002). Observed changes in isotopic composition can be used as a proxy for the preferred carbon source, since a preference of isotopically heavier HCO_3^- over the isotopically lighter CO_2 should cause a heavier isotope signal in the algal biomass. However, uncertainties persist in this method, because the assumption of unaltered CO_2 leakage is not always true, the intrinsic fractionation factors of different Rubisco forms are mostly unknown, and disequilibrium situations during transport processes can blur the isotopic patterns substantially and cause strong deviations from the standard model (Sharkey and Berry 1985). In pH drift experiments, another common experimental approach, the measured compensation point indicates the pH at which the DIC uptake rate of the algae equals the rate of CO_2 release into the medium by its respiration and/or photorespiration. Low compensation pH values therefore indicate a higher dependence on CO_2 as a carbon source, while higher pH values can only be explained by significant direct HCO_3^- uptake (see pH-dependent changes in the relative proportion of CO_2 and HCO_3^-; Figure 4.2). Also this technique is to some extent error-afflicted, since it does not distinguish between effects of [CO_2] and direct effects of H^+ on the DIC uptake behavior (Middelboe and Hansen 2007).

Most macroalgae are able to use both CO_2 and HCO_3^- (Maberly 1990; Raven and Hurd 2012) even though the preference differs between species and the type of habitat (Maberly *et al.* 1992; Raven *et al.* 2002). HCO_3^- can either be used by CA-mediated conversion of HCO_3^- to CO_2 at the cell membrane and subsequent diffusive uptake, or by direct HCO_3^- uptake through the plasma membrane via an anion exchange protein (Figure 4.10; Drechsler *et al.* 1993; Axelsson *et al.* 1999, 2000; Fernández *et al.* 2014). While the first mechanism seems to be the most prevalent for HCO_3^- utilization (Israel and Hophy 2002), at high pH (>8.4), direct HCO_3^- uptake plays an increasingly dominant role (Axelsson *et al.* 1995; Fernández *et al.* 2014). The transported HCO_3^- can be accumulated within the cell, but eventually needs to be converted to CO_2 via an intracellular CA to fuel carbon fixation by Rubisco. Apparent affinities of photosynthesis for DIC ranged between ~200 and 1850 μmol L^{-1} among four red algae species, and photosynthesis rates saturated earlier under low irradiance, suggesting that energization plays a key role also in macroalgal DIC uptake (Johnston *et al.* 1992).

While multiple lines of evidence confirmed CCM activity in macroalgae, comparative studies highlighted habitat- and group-specific characteristics in carbon acquisition. For mostly red algae, which inhabit subtidal habitats where light is presumably limiting, Maberly *et al.* (1992) reported low isotopic values ($\leq$ -30‰) as well as low pH compensation points ($\leq$ pH 9.0), both indicating that these species rely primarily on diffusive CO_2 supply. Further up in the tidal zone, with exposure to higher irradiance, or even in the intertidal zone, where the time for photosynthesis is restricted during low tide, brown and red algal species apparently developed a more efficient form of photosynthesis based on HCO_3^- acquisition with higher affinities for CO_2 and HCO_3^- and lower fractionation values (Johnston 1991; Johnston *et al.* 1992). This illustrates that the carbon demand is an important driver behind the evolution and regulation of CCMs.

Some macrophytes, such as coralline red and green algae, and also some brown algae, precipitate $CaCO_3$ inside or outside their cell walls. While the physiological and ecological function of calcification is not fully understood, a few hypotheses have been put forward, such as protection against high light or defense against herbivory. An additional hypothesis, more relevant to this chapter, is the idea that calcification changes the carbonate chemistry in the close proximity of the cell wall. The precipitation of $CaCO_3$ is thought to facilitate uptake of HCO_3^- and other ions, especially nutrients (McConnaughey and Whelan 1997). Mechanistically, the active uptake of inorganic carbon can raise the pH in the close proximity of the cell wall, thereby favoring $CaCO_3$ precipitation (Borowitzka and Larkum 1976a, 1976b, Borowitzka 1981, 1984). Calcification might also benefit photosynthesis by releasing CO_2 directly (McConnaughey 1991; McConnaughey and Falk 1991) or by lowering total alkalinity in the ambient water, thus shifting the equilibrium towards higher proportions of CO_2 and HCO_3^- (Stepien *et al.* 2016). In summary, the reasons for and effects of calcification in macrophyte ecophysiology are still a topic left to be fully understood (Cornwall *et al.* 2013, 2017).

Seagrasses have evolved from terrestrial "green" lineage angiosperms that have re-colonized saltwater habitats (Larkum *et al.* 2017). All seagrasses have developed similar adaptations for living submerged in aqueous environments. Importantly, they possess photosynthetic cells mostly in the epidermis rather than in the mesophyll tissue and have strongly reduced the cutin wax-layer that is otherwise typical for related terrestrial plants. All tissues have a well-structured gas space system (aerenchyma/lacunae; Raven *et al.* 2002), which facilitates gas exchange between the various parts of the plant. During light exposure, photosynthetic O_2 evolution, the lower solubility of O_2 than of CO_2, and thus, the outward resistance of the diffusive boundary layer (Larkum *et al.* 1989; Koch *et al.* 2013) and the low permeability for O_2, cause the aerenchyma of leaves to increase in pressure. This causes leaves to become more buoyant and more upright, and to leak gas through small adventitious cracks in the leaf tissue. The increased pressure and concentration gradients within the lacunae can also cause gaseous movement between the leaves and roots. These adaptations partly alleviate the problem of slow diffusion of DIC across the large diffusive boundary layers in an aquatic medium (Maberly and Madsen 2002).

The structure and function of the CCM in seagrasses is still a matter of debate (Larkum *et al.* 2017). It is assumed that the enzyme kinetics of seagrass Rubisco are similar to those of the form IB Rubisco found in related flowering plants (Tcherkez *et al.* 2006; Galmés *et al.* 2016; Orr *et al.* 2016), and that most species operate a C_3 photosynthesis (Koch *et al.* 2013). Thus, a concentration of CO_2 around Rubisco seems to be required in some form, but no published data has so far clearly resolved details on the mode of action of seagrass CCMs (Larkum *et al.* 2017). Presumably, this is a result of the complexity of seagrass leaf structure with several photosynthetic and non-photosynthetic cell types as well as intercellular gas spaces, or originates from truly distinct modes of carbon acquisition, which the three families of seagrasses might have obtained by taking different evolutionary pathways.

Similar to macroalgae, pH drift experiments showed that many examined seagrass species exhibited pH compensation values that were higher than expected from diffusive CO_2 entry alone (Borum *et al.* 2016), and also the comparison of the affinity of photosynthesis for CO_2 with the affinity of terrestrial Rubisco for CO_2 suggests that CCMs are operative in seagrasses (Koch *et al.* 2013; Borum *et al.* 2016; Larkum *et al.* 2017). The analyses of carbon isotopic composition of seagrass biomass delivered diverse results. Assuming a constant isotopic DIC ratio in seawater, organic matter isotopic values more negative than –30‰ should indicate exclusive diffusive CO_2 entry (Benedict *et al.* 1980; Maberly *et al.* 1992; Raven *et al.* 2002,; Stepien 2015). No seagrass species unequivocally exhibits such low isotopic values, so that purely diffusive CO_2 uptake in conjunction with strict C_3 photosynthesis can be ruled out (Stepien 2015). In contrast, isotopic values more positive than –10‰ can only be explained by the exclusive use of HCO_3^- with or without C_4 photosynthesis (Benedict *et al.* 1980; Maberly *et al.* 1992; Raven *et al.* 2002; Stepien 2015), which is the case in roughly half of the data sets acquired from seagrasses (Larkum *et al.* 2017). The remaining data exhibit values between –30‰ and –10‰, which is indicative of mixed CO_2 and HCO_3^- uptake. Most of the available data on seagrass isotopic fractionation values tend towards the more positive end of the -30‰ to -10‰ range, suggesting high contribution of HCO_3^-, little leakage and respective internal accumulation of ^{13}C, and/or C_4 biochemistry. Nevertheless, the difficulties of accurately measuring isotopic values of DIC in the diffusive boundary layer as well as the uncertainties of Rubisco fractionation makes a firm judgment difficult (Larkum *et al.* 2017). For further details, the interested reader is referred to recent works on CCMs in macroalgae and seagrasses (Stepien *et al.* 2016; Larkum *et al.* 2017).

Like the case for seagrasses, information on HCO_3^- utilization mechanisms in freshwater plants, that also evolved from land plants, has yet to receive the attention given to microalgae. About 43% of the 131 species tested have an ability to use HCO_3^- (Iversen *et al.* 2019). The prevalence is lower in inland waters where CO_2 concentrations are high, such as in many streams and rivers, or in sites where HCO_3^- concentrations are low. This corresponds to the energy cost involved in active HCO_3^- use as well as an apparent reduced permeability for CO_2 in species able to use HCO_3^- compared to species restricted to CO_2 (Maberly and Madsen 1998).

In some species at least, the ability to use HCO_3^- is associated with spatial patterns of low and high pH at the leaf surface. These "polar leaves" export protons at the lower surface of their leaf, reducing pH and converting HCO_3^- to CO_2 that can then diffuse into the leaf. At the upper surface, there is a net extrusion of hydroxyl ions to maintain ionic balance, producing a high pH and often leading to the precipitation of calcium carbonate (Prins *et al.* 1980), which due to the reduction in alkalinity, sets free CO_2 (McConnaughey 1991). The green freshwater macroalgae, the charophytes, that use HCO_3^- have a similar system with bands of low pH and high pH around their cylindrical cells (Lucas and Smith 1973). The freshwater monocot *Ottelia alismoides* has the facultative ability to perform C_4 photosynthesis and crassulacean acid metabolism (see Chapter 5) and also has the constitutive ability to use HCO_3^- (Zhang *et al.* 2014; Shao *et al.* 2017). In this species, HCO_3^- utilization involves a periplasmic alpha carbonic anhydrase and an anion exchange protein of the SLC4 family (Huang *et al.* 2020).

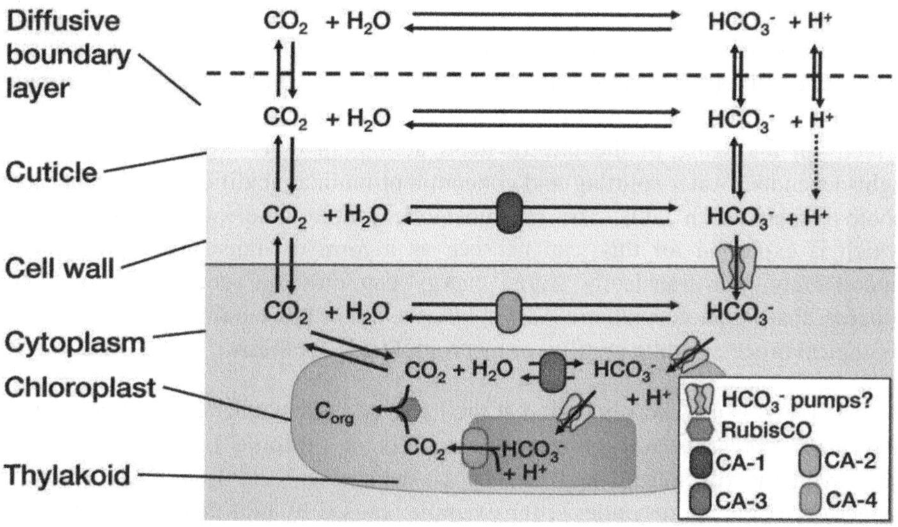

Figure 4.10. *Simplified scheme of assumed possible transport pathways for DIC at the epidermis of marine and freshwater macrophytes (redrawn and modified based on Larkum et al. 2017); CA-1, carbonic anhydrase excreted from epidermal cells; CA-2, cytoplasmic carbonic anhydrase; CA-3, carbonic anhydrase of the chloroplast stroma; CA-4 carbonic anhydrase of the thylakoid lumen. Note that while this diagram shows HCO_3^- uniports for simplicity, active movement of HCO_3^- will be connected to the exchange of Cl^-, Na^+ and so on, for which coupled systems are known. For a color version of this figure, see www.iste.co.uk/maberly/photosynthesis.zip*

4.10. CCM operation and its coupling with co-occurring processes

The functionalities of CCMs outlined above emphasize that the acquisition of inorganic carbon is metabolically highly integrated, with energetic and chemical feedbacks affecting many physiological processes. Consequently, the responses of CCM activity to environmental changes (like other physiological responses) require cellular information exchange and balancing of processes to achieve optimal growth (see Chapters 8 and 9). A striking overarching feature in this context that is conserved across distant algal taxa is that the pyrenoid (see Chapter 7) uses the luminal acidity to create the required massive flux of CO_2 towards the stromal Rubisco (Figures 4.7–4.10; Kikutani *et al.* 2016): during photosynthesis, the high pH in the chloroplast stroma establishes high concentrations of HCO_3^- while the low pH in the lumen permits only low concentrations of HCO_3^-, creating a steep gradient directed into the thylakoid lumen. Upon transfer of HCO_3^- into the thylakoid lumen, the acidic milieu favors its conversion into CO_2, which is not only strongly accelerated by luminal CAs (Tsuji *et al.* 2017a), but also continuously "reactant-driven" in the sense of the law of mass action. In other words, the continuous light-dependent water splitting and concomitant luminal acidification "turbocharge" photosynthesis with CO_2. The chemiosmotic proton gradient of photosynthesis, which is exploited for this, can be seen as a form of transiently stored energy. Interestingly, this transiently stored energy can only be accessed by one other process, that is, photosynthetic ATP synthesis, but is unavailable to most other cell biological processes, like motility or macromolecule synthesis.

The CCM is thus deeply rooted at the heart of photosynthetic energy generation, and also represents a regulatory circuit (or a "throttle") that can dissipate photosynthetic (over-)load and thereby co-determines ATP and NADPH output. Under excessive photosynthesis, for example, caused by high light intensities, cells can allow higher HCO_3^- uptake into the lumen, and thereby partially "buffer away" the chemiosmotic gradient, resulting (at worst) in an increased CO_2 leakage from the cell. This supposedly futile carbon pumping therefore does not only serve as a quenching process to dissipate excess light energy, but also the CO_2 burst may enhance activity of Rubisco and the Calvin–Benson–Bassham cycle, thereby allowing the short-term replenishment of ADP, which is required to keep the H^+ gradient in check and to avoid photodamage (Huertas *et al.* 2000; Tchernov *et al.* 2003). The high degree of integration, and *energization* as a common denominator are the reasons why environmental changes often cause responses of the CCM, or respectively, why changes in CCM-related environmental factors (e.g. CO_2/DIC concentration) can feed forward on cellular processes that, *a priori*, have little or nothing to do with inorganic carbon acquisition. In other words, environmental changes that affect the cellular energy budget cause a so-called re-allocation of cellular energy among the remaining energy demanding processes (Kranz *et al.* 2011; Rokitta and Rost 2012; Bercel and Kranz 2019).

A commonly observed phenomenon among many different microalgae is that an increase in [CO_2] and associated drop in pH (so-called "Ocean Acidification") enhance biomass production (e.g. Kroeker *et al.* 2013; Doney *et al.* 2020; Seifert *et al.* 2020). This effect has often been attributed to lowered costs for CCM operation and the fact that the liberated energy can be spent for other important processes like resource acquisition and cell division in general. In coccolithophores, this response could even be mechanistically explained by the inhibition of HCO_3^- pumps under high [H^+] that apparently enables cells to invest more effort into biomass production and growth rates (Kottmeier *et al.* 2016a, 2016b). In this specific example, energy was re-allocated from the CCM towards other processes.

A contrasting example can be seen in some diazotrophic cyanobacteria, which under N-scarcity operate the energy-intensive N_2 fixation via the enzyme nitrogenase. If these organisms are supplied with readily available inorganic N sources like NO_3^-, N_2 fixation becomes obsolete and is downregulated. Consequently, more energy is available for other processes, including the CCM (Eichner *et al.* 2014). Such an effect of energy re-allocation among physiological processes was also observed in the chlorophyte microalga *Dunaliella salina*, where even different inorganic nitrogen sources caused differential responses. While NH_4^+ can readily be assimilated into amino acids, 8 electrons per NO_3^- are required to reduce it to NH_4^+ before assimilation. If *Dunaliella* utilizes NH_4^+ as an N source, they have significantly more energy available that can be spent for CCM activity and biomass buildup (Giordano and Bowes 1997). In these specific examples, energy that was liberated by eased N acquisition was re-allocated towards the CCM.

Some studies also showed that elevated [CO_2] did not result in enhanced growth or productivity despite the CCM being downregulated (Kranz *et al.* 2009; Hoppe *et al.* 2018; Bercel and Kranz 2019; Luo *et al.* 2019). This apparently counterintuitive finding suggests that the response of cells to the possibility to downregulate the CCM (e.g. at high CO_2) depends on the instantaneous energy provision (i.e. light intensity), availability of resources (including all nutrients) and their respective allocation to cellular processes in order to balance cellular demands. In other words, energy- and resource re-allocation will only be observed if a re-adjusted state has significant benefits over the original situation. For example, the allocation of cellular energy among key processes in the cyanobacterium *Trichodesmium* in response to elevated [CO_2] was quantitatively analyzed in an optimality-based cellular model (Luo *et al.* 2019). The results indicated that under certain environmental conditions, CCM downregulation due to increased [CO_2] yields only a very limited energetic benefit that could not support significantly enhanced growth. Cells would have to spend the liberated energy to keep internal pH constant and to express more nitrogenase protein, which would at the same time increase their energy requirements.

It is generally assumed that the performance advantages conferred by a CCM outweigh the metabolic costs associated with facilitated DIC acquisition (Raven and Lucas 1985). While this is certainly true in general, the specific benefit of a CCM depends strongly on life styles and the biotic and abiotic environmental conditions, which strongly modulate cellular carbon demand. It should thus be kept in mind that efficient energy and resource (re-)allocation determine the overall efficiency of an organism's growth in any environment. Consequently, in some situations, facilitated DIC uptake will yield an ecological advantage while in others, the energy demanding process might lead to ecological disadvantages.

4.11. Future research foci

Scientists continue to work on CCM topics, combining new and traditional technologies to understand better and even improve the productivity of higher plants and microorganisms under diverse environmental conditions. Here, we briefly pinpoint some of the current and future foci for CCM research.

One field, as mentioned earlier, is the investigation of diffusive and facilitated CO_2 transport across membranes and through pore-like channels. We suggest that more work must be done to understand the relation between the nature of biological membranes and the wide range of resulting permeability coefficients. Reliable measurements of such coefficients are required to model CO_2 fluxes in cells and tissues, not only in plants and algae, but also in animal and human cells, in order to understand and cure diseases related to gas exchange.

A currently growing research direction is the biotechnological application of CCMs in which researchers aim to transplant CCM-related genes into higher plants and algae (Rae *et al.* 2017). This is relevant to improve crop yields and enhance global food security, and also to produce sustainable biofuels and other products of high commercial value like pigments and nutraceuticals (Sethi *et al.* 2020). The modular nature of many CCMs promises that particular enzymes or transporters could be transplanted in a straightforward way, but more research is necessary to understand their functioning in the larger scope of cellular energization, gene regulation and especially protein–protein interaction that seems to be critical for many CCMs to function properly.

The cell-biological aspect of energy distribution and provision and its correlation with CCM activity is an important field of interest. It has often been stated that photosynthetic electron transport ultimately provides the energy, in the form of ATP, to fuel all transport processes (Kaplan and Reinhold 1999). However, today we know that the DIC fluxes within the plastid are largely driven by the chemiosmotic gradient, and that ATP is not a primary export product of plastids. Thus, the utilization of energy generated by the mitochondria for potentially energy-demanding, non-plastidial parts of the CCM seems very plausible (Goyal and Tolbert 1990). In addition, the involvement of mitochondria in supplying energy to the CCM seems widespread as it has been observed in *Nannochloropsis* spp., *Vischeria stellata* and *Monodus subterraneus*, which are marine, freshwater and soil species, respectively (Huertas *et al.* 2000; Colman *et al.* 2002). How such a cooperative energy provision by chloroplasts and mitochondria, or rather how a crosstalk that tunes mitochondrial activity to sugar output from the chloroplasts is facilitated, remains to be investigated in future research.

Regarding our understanding of the modular structure and function of CCMs, puzzling aspects remain. For example, why do some microalgae typically not express certain CCM modules (specifically external CA) under "optimal" culture conditions (Sikes and Wheeler 1982; Sekino and Shiraiwa 1994; Rost *et al.* 2003), but actually do express them when cultures are aged (Nimer *et al.* 1994) and experience longer-term DIC depletion (Herfort *et al.* 2002; Martin and Tortell. 2006)? Such an interplay of external CA activity with direct HCO_3^- uptake mechanisms at high pH would hint towards an inducible "DIC recycling mechanism" (Trimborn *et al.* 2008). According to this mechanism, the active carbon uptake would increase $[CO_2]$ within the cytoplasm and result in a CO_2 efflux. The general high pH in the boundary layer would then favor an external CA-assisted conversion of leaked CO_2 into HCO_3^-, which could be readily taken up. This speculation still awaits physiological verification and especially ecological explanation, because under high pH, HCO_3^- would be the dominant DIC species anyway and the expression of additional enzymes seems costly and practically useless facing the comparably small advantage of "DIC recycling mechanism" under these circumstances.

It is important to note also that modes of inorganic carbon acquisition in the vast phylogenetic diversity of photosynthetic organisms (see Chapter 3) have not yet been extensively investigated. This might have to do with their irrelevance regarding biotechnical exploitation, small contribution to global biogeochemical fluxes, or their existence in specific and/or under-investigated ecosystems. A few studies have been conducted in eustigmatophytes (Huertas *et al.* 2000; 2002), picoeukaryotes (Iglesias-Rodriguez *et al.* 1998) or species that exhibit no obvious CCM activity at all (Raven *et al.* 2005), but that research is still far away from the well-founded conceptual frameworks as is the case, for example, in *Chlamydomonas*. The same is true for dinoflagellates, a diverse and highly relevant phytoplankton group that can develop toxic blooms ("red-tides") and affect ecosystem health and coastal economy (Brandenburg *et al.* 2019). Dinoflagellates are especially interesting as some species contain the inefficient type II Rubisco (Morse *et al.* 1995) while others, which underwent tertiary endosymbiosis, contain type I Rubisco (Yoon *et al.* 2002, 2005). Relatively few studies focused systematically on the CCM of dinoflagellates (Leggat *et al.* 1999; Dason and Colman 2004; Dason *et al.* 2004; Rost *et al.* 2006a; Bercel and Kranz 2019). However, more work on their CCMs seems warranted, because sedimented cysts and their isotopic signatures might serve in future paleoceanographic applications (Hoins *et al.* 2016a, 2016b) and because the understanding and projection of harmful algal blooms become more crucial with ongoing global change (Ji *et al.* 2020).

Generally, global change is affecting the physico-chemical environments of the marine and freshwater habitats (see Chapter 9). To project the responses of the planet's ecosystems by means of mathematical models, the particular responses of phytoplankton to changing environmental drivers need to be included, since they constitute the base of food webs, play important roles in the biogeochemical cycling of nutrients and (co-)determine the efficiency of the biological carbon pump that regulates Earth's climate. Ideally, future research on CCMs should involve more ecologically relevant species and apply environmentally realistic experimental conditions, so that gained data can be used to constrain the behavior of microalgae in a future, changing environment and increase the confidence in model projections.

4.12. Acknowledgments

The authors acknowledge the help of their colleagues Luke Mackinder and Kai Schulz, who critically read this chapter, gave important comments, and enabled them to shape the focus of this contribution. The authors also thank Stephen Christopher Maberly and Brigitte Gontero for helpful comments, text contributions and coordination during the writing process.

4.13. References

Anning, T., Nimer, N., Merrett, M.J., Brownlee, C. (1996). Costs and benefits of calcification in coccolithophorids. *Journal of Marine Systems*, 9, 45–56.

Axelsson, L., Ryberg, H., Beer, S. (1995). Two modes of bicarbonate utilization in the marine green macroalga *Ulva lactuca*. *Plant, Cell & Environment*, 18(4), 439–445.

Axelsson, L., Larsson, C., Ryberg, H. (1999). Affinity, capacity and oxygen sensitivity of two different mechanisms for bicarbonate utilization in *Ulva lactuca* L. (Chlorophyta). *Plant, Cell & Environment*, 22(8), 969–978.

Axelsson, L., Mercado, J., Figueroa, F. (2000). Utilization of HCO_3^- at high pH by the brown macroalga *Laminaria saccharina*. *European Journal of Phycology*, 35(1), 53–59.

Badger, M.R. and Price, G.D. (2003). CO_2 concentrating mechanisms in cyanobacteria: Molecular components, their diversity and evolution. *Journal of Experimental Botany*, 54, 609.

Badger, M.R., Andrews, T.J., Whitney, S.M., Ludwig, M., Yellowlees, D.C., Leggat, W., Price, G.D. (1998). The diversity and coevolution of Rubisco, plastids, pyrenoids, and chloroplast-based CO_2-concentrating mechanisms in algae. *Canadian Journal of Botany*, 76(6), 1052–1071.

Badger, M.R., Hanson, D., Price, G.D. (2002). Evolution and diversity of CO_2 concentrating mechanisms in cyanobacteria. *Functionmal Plant Biology*, 29, 161.

Badger, M.R., Price, G.D., Long, B.M., Woodger, F.J. (2006). The environmental plasticity and ecological genomics of the cyanobacterial CO_2 concentrating mechanism. *Journal of Experimental Botany*, 57(2), 249–265.

Benedict, C.R., Wong, W.W.L., Wong, J.H.H. (1980). Fractionation of the stable isotopes of inorganic carbon by seagrasses. *Plant Physiology*, 65(3), 512–517.

Bercel, T.L. and Kranz, S.A. (2019). Insights into carbon acquisition and photosynthesis in *Karenia brevis* under a range of CO_2 concentrations. *Progress in Oceanography*, 172, 65–76.

Berner, R.A. (2006). GEOCARBSULF: A combined model for Phanerozoic atmospheric O_2 and CO_2. *Geochimica et Cosmochimica Acta*, 70(23), 5653–5664.

Blank, M.E. and Ehmke, H. (2003). Aquaporin-1 and HCO_3^- Cl^- transporter-mediated transport of CO_2 across the human erythrocyte membrane. *Journal of Physiology*, 550, 419.

Borowitzka, M.A. (1981). Photosynthesis and calcification in the articulated coralline red algae *Amphiroa anceps* and *A. foliacea*. *Marine Biology*, 62(1), 17–23.

Borowitzka, M.A. (1984). Calcification in aquatic plants. *Plant, Cell & Environment*, 7(6), 457–466.

Borowitzka, M.A. and Larkum, A.W.D. (1976a). Calcification in the green alga *Halimeda*: III. The sources of inorganic carbon for photosynthesis and calcification and a model of the mechanism of calcification. *Journal of Experimental Botany*, 27(5), 879–893.

Borowitzka, M.A. and Larkum, A.W.D. (1976b). Calcification in the green alga *Halimeda*: IV. The action of metabolic inhibitors on photosynthesis and calcification. *Journal of Experimental Botany*, 27(5), 894–907.

Borum, J., Pedersen, O., Kotula, L., Fraser, M.W., Statton, J., Colmer, T.D., Kendrick, G.A. (2016). Photosynthetic response to globally increasing CO_2 of co-occurring temperate seagrass species. *Plant, Cell & Environment*, 39(6), 1240–1250.

Boudreau, B.P. (1997). *Diagenetic Models and their Implementation Modelling Transport and Reactions in Aquatic Sediments*. Springer, Berlin.

Brandenburg, K.M., Velthuis, M., Van De Waal, D.B. (2019). Meta-analysis reveals enhanced growth of marine harmful algae from temperate regions with warming and elevated CO_2 levels. *Global Change Biology*, 25(8), 2607–2618.

Buitenhuis, E.T., De Baar, H.J.W., Veldhuis, M.J.W. (1999). Photosynthesis and calcification by *Emiliania huxleyi* (Prymnesiophyceae) as a function of inorganic carbon species. *Journal of Phycology*, 35(5), 949–959.

Burkhardt, S., Amoroso, G., Riebesell, U., Sültemeyer, D. (2001). CO_2 and HCO_3^- uptake in marine diatoms acclimated to different CO_2 concentrations. *Limnology and Oceanography*, 46(6), 1378–1391.

Burow, M.D., Chen, Z.Y., Mouton, T.M., Moroney, J.V. (1996). Isolation of cDNA clones of genes induced upon transfer of *Chlamydomonas reinhardtii* cells to low CO_2. *Plant Molecular Biology*, 31(2), 443–448.

Clement, R., Dimnet, L., Maberly, S.C., Gontero, B. (2016). The nature of the CO_2-concentrating mechanisms in a marine diatom, *Thalassiosira pseudonana*. *New Phytologist*, 209(4), 1417–1427.

Coleman, J.R. (1991). The molecular and biochemical analyses of CO_2-concentrating mechanisms in cyanobacteria and microalgae. *Plant Cell and Environment*, 14(8), 861–867.

Colman, B., Huertas, I.E., Bhatti, S., Dason, J.S. (2002). The diversity of inorganic carbon acquisition mechanisms in eukaryotic microalgae. *Functional Plant Biology*, 29(2–3), 261–270.

Cordat, E. and Casey, J.R. (2009). Bicarbonate transport in cell physiology and disease. *Biochemical Journal*, 417, 423–439.

Cornwall, C.E., Hepburn, C.D., Pilditch, C.A., Hurd, C.L. (2013). Concentration boundary layers around complex assemblages of macroalgae: Implications for the effects of ocean acidification on understory coralline algae. *Limnology and Oceanography*, 58(1), 121–130.

Cornwall, C.E., Revill, A.T., Hall-Spencer, J.M., Milazzo, M., Raven, J.A., Hurd, C.L. (2017). Inorganic carbon physiology underpins macroalgal responses to elevated CO_2. *Scientific Reports*, 7(1), 46297.

Dason, J.S. and Colman, B. (2004). Inhibition of growth in two dinoflagellates by rapid changes in external pH. *Canadian Journal of Botany*, 82(4), 515–520.

Dason, J.S., Huertas, I.E., Colman, B. (2004). Source of inorganic carbon for photosynthesis in two marine dinoflagellates. *Journal of Phycology*, 40(2), 285–292.

Dickson, A.G. (1981). An exact definition of total alkalinity and a procedure for the estimation of alkalinity and total inorganic carbon from titration data. *Deep-Sea Research Part A – Oceanographic Research Papers*, 28(6), 609–623.

Dickson, A.G. and Millero, F.J. (1987). A comparison of the equilibrium-constants for the dissociation of carbonic-acid in seawater media. *Deep-Sea Research Part A – Oceanographic Research Papers*, 34(10), 1733–1743.

Doney, S.C., Busch, D.S., Cooley, S.R., Kroeker, K.J. (2020). The impacts of ocean acidification on marine ecosystems and reliant human communities. *Annual Review of Environment and Resources*, 45(1), 83–112.

Drechsler, Z., Sharkia, R., Cabantchik, Z.I., Beer, S. (1993). Bicarbonate uptake in the marine macroalga *Ulva* sp. is inhibited by classical probes of anion exchange by red blood cells. *Planta*, 191(1), 34–40.

Du, J.H., Forster, B., Rourke, L., Howitt, S.M., Price, G.D. (2014). Characterisation of cyanobacterial bicarbonate transporters in *E. coli* shows that SbtA homologs are functional in this heterologous expression system. *PLOS ONE*, 9(12).

Duanmu, D., Miller, A.R., Horken, K.M., Weeks, D.P., Spalding, M.H. (2009a). Knockdown of limiting-CO_2-induced gene HLA3 decreases HCO_3^- transport and photosynthetic C_i affinity in *Chlamydomonas reinhardtii*. *Proceedings of the National Academy of Sciences of the United States of America*, 106(14), 5990–5995.

Duanmu, D., Wang, Y.J., Spalding, M.H. (2009b). Thylakoid lumen carbonic anhydrase (CAH3) mutation suppresses air-dier phenotype of LCIB mutant in *Chlamydomonas reinhardtii*. *Plant Physiology*, 149(2), 929–937.

Egleston, E.S., Sabine, C.L., Morel, F.M.M. (2010). Revelle revisited: Buffer factors that quantify the response of ocean chemistry to changes in DIC and alkalinity. *Global Biogeochemical Cycles*, 24, 1–9.

Eichner, M., Kranz, S.A., Rost, B. (2014). Combined effects of different CO_2 levels and N sources on the diazotrophic cyanobacterium *Trichodesmium*. *Physiologia Plantarum*, 152(2), 316–330.

Eigen, M. (1964). Proton transfer, acid-base catalysis, and enzymatic hydrolysis. Part I: Elementary processes. *Angewandte Chemie International Edition*, 3(1), 1–19.

Endeward, V., Musa-Aziz, R., Cooper, G.J., Chen, L.M., Pelletier, M.F., Virkki, L.V., Supuran, C.T., King, L.S., Boron, W.F., Gros, G. (2006). Evidence that aquaporin 1 is a major pathway for CO_2 transport across the human erythrocyte membrane. *FASEB Journal*, 20(12), 1974–1981.

Endeward, V., Cartron, J.P., Ripoche, P., Gros, G. (2008). RhaG protein of the rhesus complex is a CO_2 channel in the human red cell membrane. *FASEB Journal*, 22(1), 64–73.

Falkowski, P.G., Schofield, O., Katz, M.E., Van De Schootbrugge, B., Knoll, A.H. (2004). Why is the land green and the ocean red? In *Coccolithophores: From Molecular Processes to Global Impact*, Thierstein, H.R. and Young, J.R. (eds). Springer, Berlin.

Fang, W., Si, Y.Q., Douglass, S., Casero, D., Merchant, S.S., Pellegrini, M., Ladunga, I., Liu, P., Spalding, M.H. (2012). Transcriptome-wide changes in *Chlamydomonas reinhardtii* gene expression regulated by carbon dioxide and the CO_2-concentrating mechanism regulator CIA5/CCM1. *Plant Cell*, 24(5), 1876–1893.

Fernández, P.A., Hurd, C.L., Roleda, M.Y. (2014). Bicarbonate uptake via an anion exchange protein is the main mechanism of inorganic carbon acquisition by the giant kelp *Macrocystis pyrifera* (Laminariales, Phaeophyceae) under variable pH. *Journal of Phycology*, 50(6), 998–1008.

Fick, A. (1855). Über Diffusion. *Annalen der Physik*, 170(1), 59–86.

Field, C.B., Behrenfeld, M.J., Randerson, J.T., Falkowski, P. (1998). Primary production of the biosphere: Integrating terrestrial and oceanic components. *Science*, 281(5374), 237–240.

Flynn, K.J., Blackford, J.C., Baird, M.E., Raven, J.A., Clark, D.R., Beardall, J., Brownlee, C., Fabian, H., Wheeler, G.L. (2012). Changes in pH at the exterior surface of plankton with ocean acidification. *Nature Climate Change*, 2(7), 510–513.

Gafar, N.A., Eyre, B.D., Schulz, K.G. (2019). Particulate inorganic to organic carbon production as a predictor for coccolithophorid sensitivity to ongoing ocean acidification. *Limnology and Oceanography Letters*, 4(3), 62–70.

Galmés, J., Hermida-Carrera, C., Laanisto, L., Niinemets, Ü. (2016). A compendium of temperature responses of Rubisco kinetic traits: Variability among and within photosynthetic groups and impacts on photosynthesis modeling. *Journal of Experimental Botany*, 67(17), 5067–5091.

Gao, H., Wang, Y.J., Fei, X.W., Wright, D.A., Spalding, M.H. (2015). Expression activation and functional analysis of HLA3, a putative inorganic carbon transporter in *Chlamydomonas reinhardtii*. *Plant Journal*, 82(1), 1–11.

Giordano, M. and Bowes, G. (1997). Gas exchanges, metabolism, and morphology of *Dunaliella salina* in response to the CO_2 concentration and nitrogen source used for growth. *Plant Physiology*, 115, 1049.

Goyal, A. and Tolbert, N.E. (1990). Salicylhydroxamic acid (SHAM) inhibition of the dissolved inorganic carbon concentrating process in unicellular green algae. *Plant Physiology*, 92(3), 630–636.

Gutknecht, J., Bisson, M.A., Tosteson, F.C. (1977). Diffusion of carbon-dioxide through lipid bilayer membranes – Effects of carbonic-anhydrase, bicarbonate, and unstirred layers. *Journal of General Physiology*, 69(6), 779–794.

Harada, H. and Matsuda, Y. (2005). Promoter analysis of two CO_2-inducible carbonic anhydrases in the marine diatom *Phaeodactylum tricornutum*. *Plant and Cell Physiology*, 46, S88.

Harada, H., Nakajima, K., Sakaue, K., Matsuda, Y. (2006a). CO_2 sensing at ocean surface mediated by cAMP in a marine diatom. *Plant Physiology*, 142(3), 1318–1328.

Harada, H., Sakaue, K., Matsuda, Y. (2006b). CO_2 sensing mechanism mediated by cAMP in the marine diatom *Phaeodactylum tricornutum*. *Plant and Cell Physiology*, 47, S52.

Haslam, R.P., Keys, A.J., Andralojc, P.J., Madgwick, P.J., Andersson, I., Grimsrud, A., Eilertsen, H.C., Parry, M.A.J. (2005). Specificity of diatom Rubisco. In *Plant Responses to Air Pollution and Global Change*, Omasa, K., Nouchi, I., De Kok, L.J. (eds). Springer, Tokyo.

Hennacy, J.H. and Jonikas, M.C. (2020). Prospects for engineering biophysical CO_2 concentrating mechanisms into land plants to enhance yields. *Annual Review of Plant Biology*, 71(1), 461–485.

Henry, W. (1803). Experiments on the quantity of gases absorbed by water, at different temperatures, and under different pressures. *Philosophical Transactions of the Royal Society*, 93, 29–274.

Herfort, L., Thake, B., Roberts, J. (2002). Acquisition and use of bicarbonate by *Emiliania huxleyi*. *New Phytologist*, 156(3), 427–436.

Hoins, M., Eberlein, T., Groβmann, C.H., Brandenburg, K., Reichart, G.-J., Rost, B., Sluijs, A., Van De Waal, D.B. (2016a). Combined effects of ocean acidification and light or nitrogen availabilities on ^{13}C fractionation in marine dinoflagellates. *PLOS ONE*, 11(5), e0154370.

Hoins, M., Eberlein, T., Van De Waal, D.B., Sluijs, A., Reichart, G.J., Rost, B. (2016b). CO_2-dependent carbon isotope fractionation in dinoflagellates relates to their inorganic carbon fluxes. *Journal of Experimental Marine Biology and Ecology*, 481, 9–14.

Holtz, L.M., Wolf-Gladrow, D., Thoms, S. (2015). Simulating the effects of light intensity and carbonate system composition on particulate organic and inorganic carbon production in *Emiliania huxleyi*. *Journal of Theoretical Biology*, 372, 192–204.

Hopkins, J. and Balch, W.M. (2018). A new approach to estimating coccolithophore calcification rates from space. *Journal of Geophysical Research: Biogeosciences*, 123(5), 1447–1459.

Hopkinson, B.M. (2014). A chloroplast pump model for the CO_2 concentrating mechanism in the diatom *Phaeodactylum tricornutum*. *Photosynthesis Research*, 121(2–3), 223–233.

Hopkinson, B.M., Dupont, C.L., Allen, A.E., Morel, F.M.M. (2011). Efficiency of the CO_2-concentrating mechanism of diatoms. *Proceedings of the National Academy of Sciences of the United States of America*, 108(10), 3830–3837.

Hopkinson, B.M., Meile, C., Shen, C. (2013). Quantification of extracellular carbonic anhydrase activity in two marine diatoms and investigation of its role. *Plant Physiology*, 162(2), 1142–1152.

Hopkinson, B.M., Dupont, C.L., Matsuda, Y. (2016). The physiology and genetics of CO_2 concentrating mechanisms in model diatoms. *Current Opinion in Plant Biology*, 31, 51–57.

Hoppe, C.J.M., Wolf, K.K.E., Schuback, N., Tortell, P.D., Rost, B. (2018). Compensation of ocean acidification effects in arctic phytoplankton assemblages. *Nature Climate Change*, 8(6), 529–533.

Huang, W., Han, S., Jiang, H., Gu, S., Li, W., Gontero, B., Maberly, S.C. (2020). External α-carbonic anhydrase and solute carrier 4 are required for bicarbonate uptake in a freshwater angiosperm. *Journal of Experimental Botany*, 71(19), 6004–6014.

Huertas, I.E., Espie, G.S., Colman, B., Lubian, L.L. (2000). Light dependent bicarbonate transport and CO_2 efflux in the marine microalga *Nannochloropsis gaditana*. *Planta*, 211, 43.

Huertas, I.E., Colman, B., Espie, G.S. (2002). Inorganic carbon acquisition and its energization in eustigmatophyte algae. *Functional Plant Biology*, 29, 271.

Iglesias-Rodriguez, D., Nimer, N., Merret, M.J. (1998). Carbon dioxide-concentrating mechanism and the development of extracellular carbonic anhydrase in the marine picoeukaryote *Micromonas pusilla*. *New Phytologist*, 140(4), 685–690.

Iñiguez, C., Galmés, J., Gordillo, F.J.L. (2018). Rubisco carboxylation kinetics and inorganic carbon utilization in polar versus cold-temperate seaweeds. *Journal of Experimental Botany*, 70(4), 1283–1297.

Iñiguez, C., Capó-Bauçà, S., Niinemets, Ü., Stoll, H., Aguiló-Nicolau, P., Galmés, J. (2020). Evolutionary trends in Rubisco kinetics and their co-evolution with CO_2 concentrating mechanisms. *The Plant Journal*, 101(4), 897–918.

Invers, O., Zimmerman, R.C., Alberte, R.S., Pérez, M., Romero, J. (2001). Inorganic carbon sources for seagrass photosynthesis: An experimental evaluation of bicarbonate use in species inhabiting temperate waters. *Journal of Experimental Marine Biology and Ecology*, 265(2), 203–217.

Israel, A. and Hophy, M. (2002). Growth, photosynthetic properties and Rubisco activities and amounts of marine macroalgae grown under current and elevated seawater CO_2 concentrations. *Global Change Biology*, 8(9), 831–840.

Itel, F., Al-Samir, S., Oberg, F., Chami, M., Kumar, M., Supuran, C.T., Deen, P.M.T., Meier, W., Hedfalk, K., Gros, G., Endeward, V. (2012). CO_2 permeability of cell membranes is regulated by membrane cholesterol and protein gas channels. *FASEB Journal*, 26(12), 5182–5191.

Iversen, L.L., Winkel, A., Baastrup-Spohr, L., Hinke, A.B., Alahuhta, J., Baattrup-Pedersen, A., Birk, S., Brodersen, P., Chambers, P.A., Ecke, F., Feldmann, T., Gebler, D., Heino, J., Jespersen, T.S., Moe, S.J., Riis, T., Sass, L., Vestergaard, O., Maberly, S.C., Sand-Jensen, K., Pedersen, O. (2019). Catchment properties and the photosynthetic trait composition of freshwater plant communities. *Science*, 366(6467), 878–881.

James, P.L. and Larkum, A.W.D. (1996). Photosynthetic inorganic carbon acquisition of *Posidonia australis*. *Aquatic Botany*, 55(3), 149–157.

Ji, X., Verspagen, J.M.H., Waal, D.B.V.D., Rost, B., Huisman, J. (2020). Phenotypic plasticity of carbon fixation stimulates cyanobacterial blooms at elevated CO_2. *Science Advances*, 6(8), eaax2926.

Jin, S.Y., Sun, J., Wunder, T., Tang, D.S., Cousins, A.B., Sze, S.K., Mueller-Cajar, O., Gao, Y.G. (2016). Structural insights into the LCIB protein family reveals a new group of beta-carbonic anhydrases. *Proceedings of the National Academy of Sciences of the United States of America*, 113(51), 14716–14721.

Johnson, K.S. (1982). Carbon-dioxide hydration and dehydration kinetics in sea-water. *Limnology and Oceanography*, 27(5), 849–855.

Johnston, A.M. (1991). The acquisition of inorganic carbon by marine macroalgae. *Canadian Journal of Botany*, 69, 1123.

Johnston, A.M. and Raven, J.A. (1996). Inorganic carbon accumulation by the marine diatom *Phaeodactylum tricornutum. European Journal of Phycology*, 31(3), 285–290.

Johnston, A.M., Maberly, S.C., Raven, J.A. (1992). The acquisition of inorganic carbon by four red macroalgae. *Oecologia*, 92(3), 317–326.

Kaplan, A. and Reinhold, L. (1999). CO_2-concentrating mechanisms in photosynthetic microorganisms. *Annual Review of Plant Physiology and Plant Molecular Biology*, 50(1), 539–570.

Keeling, P.J. (2013). The number, speed, and impact of plastid endosymbioses in eukaryotic evolution. *Annual Review of Plant Biology*, 64, 583–607.

Kikutani, S., Nakajima, K., Nagasato, C., Tsuji, Y., Miyatake, A., Matsuda, Y. (2016). Thylakoid luminal theta-carbonic anhydrase critical for growth and photosynthesis in the marine diatom *Phaeodactylum tricornutum. Proceedings of the National Academy of Sciences of the United States of America*, 113(35), 9828–9833.

Klaas, C. and Archer, D.E. (2002). Association of sinking organic matter with various types of mineral ballast in the deep sea: Implications for the rain ratio. *Global Biogeochemical Cycles*, 16(4), 1116–1130.

Koch, M., Bowes, G., Ross, C., Zhang, X.-H. (2013). Climate change and ocean acidification effects on seagrasses and marine macroalgae. *Global Change Biology*, 19(1), 103–132.

Kono, A. and Spalding, M.H. (2020). LCI1, a *Chlamydomonas reinhardtii* plasma membrane protein, functions in active CO_2 uptake under low CO_2. *Plant Journal*, 102(6), 1127–1141.

Kono, A., Chou, T.H., Radhakrishnan, A., Bolla, J.R., Sankar, K., Shome, S., Su, C.C., Jernigan, R.L., Robinson, C.V., Yu, E.W.W., Spalding, M.H. (2020). Structure and function of LCI1: A plasma membrane CO_2 channel in the *Chlamydomonas* CO_2-concentrating mechanism. *Plant Journal*, 102(6), 1107–1126.

Kottmeier, D.M., Rokitta, S.D., Tortell, P.D., Rost, B. (2014). Strong shift from HCO_3^- to CO_2 uptake in *Emiliania huxleyi* with acidification: New approach unravels acclimation versus short-term pH effects. *Photosynthesis Research*, 121(2–3), 265–275.

Kottmeier, D.M., Rokitta, S.D., Rost, B. (2016a). Acidification, not carbonation, is the major regulator of carbon fluxes in the coccolithophore *Emiliania huxleyi. New Phytologist*, 211(1), 126–137.

Kottmeier, D.M., Rokitta, S.D., Rost, B. (2016b). H^+-driven increase in CO_2 uptake and decrease in HCO_3^- uptake explain coccolithophores' acclimation responses to ocean acidification. *Limnology and Oceanography*, (61), 2045–2057.

Kranz, S.A., Sültemeyer, D., Richter, K.U., Rost, B. (2009). Carbon acquisition by *Trichodesmium*: The effect of pCO_2 and diurnal changes. *Limnology and Oceanography*, 54(2), 548–559.

Kranz, S.A., Wolf-Gladrow, D., Nehrke, G., Langer, G., Rost, B. (2010). Calcium carbonate precipitation induced by the growth of the marine cyanobacterium *Trichodesmium. Limnology and Oceanography*, 55(6), 2563–2569.

Kranz, S.A., Eichner, M., Rost, B. (2011). Interactions between CCM and N_2 fixation in *Trichodesmium. Photosynthesis Research*, 109(1), 73–84.

Kranz, S.A., Young, J.N., Hopkinson, B.M., Goldman, J.A.L., Tortell, P.D., Morel, F.M.M. (2015). Low temperature reduces the energetic requirement for the CO_2 concentrating mechanism in diatoms. *New Phytologist*, 205(1), 192–201.

Kroeker, K.J., Kordas, R.L., Crim, R., Hendriks, I.E., Ramajo, L., Singh, G.S., Duarte, C.M., Gattuso, J.P. (2013). Impacts of ocean acidification on marine organisms: Quantifying sensitivities and interaction with warming. *Global Change Biology*, 19(6), 1884–1896.

Lane, T.W. and Morel, F.M.M. (2000). Regulation of carbonic anhydrase expression by zinc, cobalt, and carbon dioxide in the marine diatom Thalassiosira weissflogii. *Plant Physiology*, 123(1), 345–352.

Lane, T.W., Saito, M.A., George, G.N., Pickering, I.J., Prince, R.C., Morel, F.M.M. (2005). A cadmium enzyme from a marine diatom. *Nature*, 435(7038), 42.

Larkum, A.W.D., Mccomb, A.J., Shephard, S.A. (1989). *Biology of Seagrasses: A Treatise on the Biology of Seagrasses with Special Reference to the Australian Region*. Elsevier Science Publications, Amsterdam.

Larkum, A.W.D., Davey, P.A., Kuo, J., Ralph, P.J., Raven, J.A. (2017). Carbon-concentrating mechanisms in seagrasses. *Journal of Experimental Botany*, 68(14), 3773–3784.

Leggat, W., Badger, M.R., Yellowlees, D. (1999). Evidence for an inorganic carbon-concentrating mechanism in the symbiotic dinoflagellate *Symbiodinium* sp. *Plant Physiology*, 121(4), 1247–1255.

Leonardos, N., Read, B., Thake, B., Young, J.R. (2009). No mechanistic dependence of photosynthesis on calcification in the coccolithophorid *Emiliania huxleyi* (Haptophyta). *Journal of Phycology*, 45(5), 1046–1051.

Lucas, W.J. and Smith, F.A. (1973). The formation of alkaline and acid regions at the surface of *Chara corallina* cells. *Journal of Experimental Botany*, 24(1), 1–14.

Luo, Y.-W., Shi, D., Kranz, S.A., Hopkinson, B.M., Hong, H., Shen, R., Zhang, F. (2019). Reduced nitrogenase efficiency dominates response of the globally important nitrogen fixer *Trichodesmium* to ocean acidification. *Nature Communications*, 10(1), 1521.

Maberly, S.C. (1990). Exogenous sources of inorganic carbon for photosynthesis by marine macroalgae. *Journal of Phycology*, 26(3), 439–449.

Maberly, S.C. (1992). Carbonate ions appear to neither inhibit nor stimulate use of bicarbonate ions in photosynthesis by *Ulva lactuca*. *Plant, Cell & Environment*, 15(2), 255–260.

Maberly, S.C. and Madsen, T.V. (1998). Affinity for CO_2 in relation to the ability of freshwater macrophytes to use HCO_3^-. *Functional Ecology*, 12(1), 99–106.

Maberly, S.C. and Madsen, T.V. (2002). Freshwater angiosperm carbon concentrating mechanisms: Processes and patterns. *Functional Plant Biology*, 29(3), 393–405.

Maberly, S.C., Raven, J.A., Johnston, A.M. (1992). Discrimination between ^{12}C and ^{13}C by marine plants. *Oecologia*, 91(4), 481–492.

Maberly, S.C., Gontero, B., Puppo, C., Villain, A., Severi, I., Giordano, M. (2021). Inorganic carbon uptake in a freshwater diatom, *Asterionella formosa* (Bacillariophyceae): From ecology to genomics. *Phycologia*, 1–12.

Mackinder, L.C.M. (2018). The *Chlamydomonas* CO_2-concentrating mechanism and its potential for engineering photosynthesis in plants. *New Phytologist*, 217(1), 54–61.

Mackinder, L.C.M., Bach, L., Schulz, K., Wheeler, G., Schroeder, D., Riebesell, U., Brownlee, C. (2011). The molecular basis of inorganic carbon uptake mechanisms in the coccolithophore *Emiliania huxleyi*. *European Journal of Phycology*, 46, 142–143.

Mackinder, L.C.M., Meyer, M.T., Mettler-Altmann, T., Chen, V.K., Mitchell, M.C., Caspari, O., Rosenzweig, E.S.F., Pallesen, L., Reeves, G., Itakura, A., Roth, R., Sommer, F., Geimer, S., Muhlhaus, T., Schroda, M., Goodenough, U., Stitt, M., Griffiths, H., Jonikas, M.C. (2016). A repeat protein links Rubisco to form the eukaryotic carbon-concentrating organelle. *Proceedings of the National Academy of Sciences of the United States of America*, 113(21), 5958–5963.

Maeda, S., Badger, M.R., Price, G.D. (2002). Novel gene products associated with NdhD3/D4-containing NDH-1 complexes are involved in photosynthetic CO_2 hydration in the cyanobacterium *Synechococcus* sp. PCC7942. *Molecular Microbiology*, 43, 425.

Martin, C.L. and Tortell., P.D. (2006). Bicarbonate transport and extracellular carbonate anhydrase in Bering Sea phytoplankton assemblages: Results from isotopic disequilibrium experiments. *Limnology and Oceanography*, 51(5), 2111–2121.

McConnaughey, T.A. (1991). Calcification in *Chara corallina*: CO_2 hydroxylation generates protons for bicarbonate assimilation. *Limnology and Oceanography*, 36(4), 619–628.

McConnaughey, T.A. and Falk, R.H. (1991). Calcium-proton exchange during algal calcification. *The Biological Bulletin*, 180(1), 185–195.

McConnaughey, T.A. and Whelan, J.F. (1997). Calcification generates protons for nutrient and bicarbonate uptake. *Earth-Science Reviews*, 42(1), 95–117.

Mehrbach, C., Culberson, C.H., Hawley, J.E., Pytkowicz, R.M. (1973). Measurement of apparent dissociation – Constants of carbonic-acid in seawater at atmospheric pressure. *Limnology and Oceanography*, 18(6), 897–907.

Middelboe, A.L. and Hansen, P.J. (2007). High pH in shallow-water macroalgal habitats. *Marine Ecology Progress Series*, 338, 107–117.

Middelburg, J.J., Soetaert, K., Hagens, M. (2020). Ocean alkalinity, buffering and biogeochemical processes. *Reviews of Geophysics*, 58(3), e2019RG000681.

Missner, A., Kugler, P., Antonenko, Y.N., Pohl, P. (2008). Passive transport across bilayer lipid membranes: Overton continues to rule. *Proceedings of the National Academy of Sciences of the United States of America*, 105(52), E123–E123.

Moore, T.S., Dowell, M.D., Franz, B.A. (2012). Detection of coccolithophore blooms in ocean color satellite imagery: A generalized approach for use with multiple sensors. *Remote Sensing of Environment*, 117, 249–263.

Moroney, J.V., Ma, Y.B., Frey, W.D., Fusilier, K.A., Pham, T.T., Simms, T.A., Dimario, R.J., Yang, J., Mukherjee, B. (2011). The carbonic anhydrase isoforms of *Chlamydomonas reinhardtii*: Intracellular location, expression, and physiological roles. *Photosynthesis Research*, 109(1–3), 133–149.

Morse, D., Salois, P., Markovic, P., Hastings, J.W. (1995). A nuclear-encoded form II Rubisco in dinoflagellates. *Science*, 268(5217), 1622–1624.

Mukherjee, A., Lau, C.S., Walker, C.E., Rai, A.K., Prejean, C.I., Yates, G., Emrich-Mills, T., Lemoine, S.G., Vinyard, D.J., Mackinder, L.C.M., Moroney, J.V. (2019). Thylakoid localized bestrophin-like proteins are essential for the CO_2 concentrating mechanism of *Chlamydomonas reinhardtii. Proceedings of the National Academy of Sciences of the United States of America*, 116(34), 16915–16920.

Müller, M.N., Trull, T.W., Hallegraeff, G.M. (2015). Differing responses of three Southern Ocean *Emiliania huxleyi* ecotypes to changing seawater carbonate chemistry. *Marine Ecology Progress Series*, 531, 81–90.

Nakajima, K., Tanaka, A., Matsuda, Y. (2013). SLC4 family transporters in a marine diatom directly pump bicarbonate from seawater. *Proceedings of the National Academy of Sciences of the United States of America*, 110(5), 1767–1772.

Nielsen, M.V. (1995). Photosynthetic characteristics of the coccolithophorid *Emiliania huxleyi* (Prymnesiophyceae) exposed to elevated concentrations of dissolved inorganic carbon. *Journal of Phycology*, 31(5), 715–719.

Nimer, N.A. and Merrett, M.J. (1993). Calcification rate in *Emiliania Huxleyi* Lohmann in response to light, nitrate and availability of inorganic carbon. *New Phytologist*, 123(4), 673–677.

Nimer, N.A. and Merrett, M.J. (1996). The development of a CO_2-concentrating mechanism in *Emiliania huxleyi. New Phytologist*, 133(3), 383–389.

Nimer, N.A., Guan, Q., Merret, M.J. (1994). Extra- and intra-cellular carbonic anhydrase in relation to culture age in a high-calcifying strain of *Emiliania huxleyi* Lohmann. *New Phytologist*, 126(4), 601–607.

Ohkawa, H., Pakrasi, H.B., Ogawa, T. (2000a). Two types of functionally distinct NAD(P)H dehydrogenases in *Synechocystis* sp. strain PCC6803. *Journal of Biological Chemistry*, 275, 31630.

Ohkawa, H., Price, G.D., Badger, M.R., Ogawa, T. (2000b). Mutation of ndh genes leads to inhibition of CO_2 uptake rather than HCO_3^- uptake in *Synechocystis* sp. strain PCC 6803. *Journal of Bacteriology*, 182, 2591.

Ohkawa, H., Sonoda, M., Shibata, M., Ogawa, T. (2001). Localization of NAD(P)H dehydrogenase in the cyanobacterium *Synechocystis* sp. strain PCC 6803. *Journal of Bacteriology*, 183, 4938–4939.

Ohnishi, N., Mukherjee, B., Tsujikawa, T., Yanase, M., Nakano, H., Moroney, J.V., Fukuzawa, H. (2010). Expression of a low CO_2-inducible protein, LCI1, increases inorganic carbon uptake in the green alga *Chlamydomonas reinhardtii*. *Plant Cell*, 22(9), 3105–3117.

Omata, T., Price, G.D., Badger, M.R., Okamura, M., Gohta, S., Ogawa, T. (1999). Identification of an ATP-binding cassette transporter involved in bicarbonate uptake in the cyanobacterium *Synechococcus* sp. strain PCC 7942. *Proceedings of the National Academy of Sciences of the United States of America*, 96(23), 13571–13576.

Orr, D.J., Alcântara, A., Kapralov, M.V., Andralojc, P.J., Carmo-Silva, E., Parry, M.A.J. (2016). Surveying Rubisco diversity and temperature response to improve crop photosynthetic efficiency. *Plant Physiology*, 172(2), 707–717.

Petrou, K., Baker, K.G., Nielsen, D.A., Hancock, A.M., Schulz, K.G., Davidson, A.T. (2019). Acidification diminishes diatom silica production in the Southern Ocean. *Nature Climate Change*, 9(10), 781.

Prasad, G.V.R., Coury, L.A., Finn, F., Zeidel, M.L. (1998). Reconstituted aquaporin 1 water channels transport CO_2 across membranes. *Journal of Biological Chemistry*, 273(50), 33123–33126.

Price, G.D. and Badger, M.R. (1989a). Expression of human carbonic-anhydrase in the cyanobacterium *Synechococcus* Pcc7942 creates a high CO_2-requiring phenotype – Evidence for a central role for carboxysomes in the CO_2-concentrating mechanism. *Plant Physiology*, 91(2), 505–513.

Price, G.D. and Badger, M.R. (1989b). Isolation and characterization of high CO_2-requiring-mutants of the cyanobacterium *Synechococcus* Pcc7942 – 2 phenotypes that accumulate inorganic carbon but are apparently unable to generate CO_2 within the carboxysome. *Plant Physiology*, 91(2), 514–525.

Price, G.D. and Howitt, S.M. (2011). The cyanobacterial bicarbonate transporter BicA: Its physiological role and the implications of structural similarities with human SLC26 transporters. *Biochemistry and Cell Biology*, 89(2), 178–188.

Price, G.D., Maeda, S., Omata, T., Badger, M.R. (2002). Modes of active inorganic carbon uptake in the cyanobacterium *Synechococcus* sp. PCC7942. *Functional Plant Biology*, 29(2–3), 131–149.

Price, G.D., Woodger, F.J., Badger, M.R., Howitt, S.M., Tucker, L. (2004). Identification of a SulP-type bicarbonate transporter in marine cyanobacteria. *Proceedings of the National Academy of Sciences of the United States of America*, 101(52), 18228–18233.

Price, G.D., Badger, M.R., Woodger, F.J., Long, B.M. (2008). Advances in understanding the cyanobacterial CO_2-concentrating-mechanism (CCM): Functional components, C_i transporters, diversity, genetic regulation and prospects for engineering into plants. *Journal of Experimental Botany*, 59(7), 1441–1461.

Price, G.D., Shelden, M.C., Howitt, S.M. (2011). Membrane topology of the cyanobacterial bicarbonate transporter, SbtA, and identification of potential regulatory loops. *Molecular Membrane Biology*, 28(5), 265–275.

Prins, H.B.A., Snel, J.F.H., Helder, R.J., Zanstra, P.E. (1980). Photosynthetic HCO_3^- utilization and OH^- excretion in aquatic angiosperms: Light-induced pH changes at the leaf surface. *Plant Physiology*, 66(5), 818–822.

Rae, B.D., Long, B.M., Forster, B., Nguyen, N.D., Velanis, C.N., Atkinson, N., Hee, W.Y., Mukherjee, B., Price, G.D., Mccormick, A.J. (2017). Progress and challenges of engineering a biophysical CO_2-concentrating mechanism into higher plants. *Journal of Experimental Botany*, 68(14), 3717–3737.

Raven, J.A. (1986). Biochemical disposal of excess H^+ in growing plants. *New Phytologist*, 104(2), 175–206.

Raven, J.A. (2011a). Acidification and warming in the response of marine phytoplankton to environmental change. *Journal of Phycology*, 47, S2–S3.

Raven, J.A. (2011b). Effects on marine algae of changed seawater chemistry with increasing atmospheric CO_2. *Biology and Environment: Proceedings of the Royal Irish Academy*, 111b(1), 1–17.

Raven, J.A. and Beardall, J. (2016). The ins and outs of CO_2. *Journal of Experimental Botany*, 67(1), 1–13.

Raven, J.A. and Hurd, C.L. (2012). Ecophysiology of photosynthesis in macroalgae. *Photosynthesis Research*, 113(1), 105–125.

Raven, J.A. and Lucas, W.J. (1985). Energy costs of carbon acquisition. In *Inorganic Carbon Uptake by Aquatic Photosynthetic Organisms*, Lucas, W.J. and Berry, J.A. (eds). Americam Society of Plant Physiologists, Rockville, MD.

Raven, J.A., Johnston, A.M., Kübler, J.E., Korb, R., McInroy, S.G., Handley, L.L., Scrimgeour, C.M., Walker, D.I., Beardall, J., Vanderklift, M., Fredriksen, S., Dunton, K.H. (2002). Mechanistic interpretation of carbon isotope discrimination by marine macroalgae and seagrasses. *Functional Plant Biology*, 29(3), 355–378.

Raven, J.A., Ball, L.A., Beardall, J., Giordano, M., Maberly, S.C. (2005). Algae lacking carbon-concentrating mechanisms. *Canadian Journal of Botany*, 83(7), 879–890.

Raven, J.A., Beardall, J., Giordano, M. (2014). Energy costs of carbon dioxide concentrating mechanisms in aquatic organisms. *Photosynthesis Research*, 121(2–3), 111–124.

Read, B.A. and Tabita, F.R. (1994). High substrate-specificity factor ribulose-bisphosphate carboxylase oxygenase from eukaryotic marine-algae and properties of recombinant cyanobacterial Rubisco containing algal residue modifications. *Archives of Biochemistry and Biophysics*, 312(1), 210–218.

Roberts, S.B., Lane, T.W., Morel, F.M.M. (1997). Carbonic anhydrase in the marine diatom *Thalassiosira weissflogii* (Bacillariophyceae). *Journal of Phycology*, 33(5), 845–850.

Rokitta, S.D. and Rost, B. (2012). Effects of CO_2 and their modulation by light in the life-cycle stages of the coccolithophore *Emiliania huxleyi*. *Limnology and Oceanography*, 57(2), 607–618.

Rokitta, S.D., De Nooijer, L.J., Trimborn, S., De Vargas, C., Rost, B., John, U. (2011). Transcriptome analyses reveal differential gene expression patterns between the life-cycle stages of *Emiliania Huxleyi* (Haptophyta) and reflect specialization to different ecological niches. *Journal of Phycology*, 47(4), 829–838.

Rost, B., Riebesell, U., Burkhardt, S., Sültemeyer, D. (2003). Carbon acquisition of bloom-forming marine phytoplankton. *Limnology and Oceanography*, 48, 55.

Rost, B., Richter, K.U., Riebesell, U., Hansen, P.J. (2006a). Inorganic carbon acquisition in red tide dinoflagellates. *Plant Cell and Environment*, 29(5), 810–822.

Rost, B., Riebesell, U., Sültemeyer, D. (2006b). Carbon acquisition of marine phytoplankton: Effect of photoperiod length. *Limnology and Oceanography*, 51(1), 12–20.

Rost, B., Zondervan, I., Wolf-Gladrow, D. (2008). Sensitivity of phytoplankton to future changes in ocean carbonate chemistry: Current knowledge, contradictions and research directions. *Marine Ecology Progress Series*, 373, 227–237.

Rotatore, C., Colman, B., Kuzma, M. (1995). The active uptake of carbon-dioxide by the marine diatoms *Phaeodactylum Tricornutum* and *Cyclotella* sp. *Plant Cell and Environment*, 18(8), 913–918.

Roy, R.N., Roy, L.N., Vogel, K.M., Portermoore, C., Pearson, T., Good, C.E., Millero, F.J., Campbell, D.M. (1993). The dissociation-constants of carbonic-acid in seawater at salinities 5 to 45 and temperatures 0°C to 45°C. *Marine Chemistry*, 44(2–4), 249–267.

Samukawa, M., Shen, C., Hopkinson, B.M., Matsuda, Y. (2014). Localization of putative carbonic anhydrases in the marine diatom, *Thalassiosira pseudonana*. *Photosynthesis Research*, 121(2–3), 235–249.

Sasso, S., Stibor, H., Mittag, M., Grossman, A.R. (2018). From molecular manipulation of domesticated *Chlamydomonas reinhardtii* to survival in nature. *eLife*, 7, 1–14.

Schulz, K.G., Riebesell, U., Rost, B., Thoms, S., Zeebe, R.E. (2006). Determination of the rate constants for the carbon dioxide to bicarbonate inter-conversion in pH-buffered seawater systems. *Marine Chemistry*, 100(1–2), 53–65.

Schulz, K.G., Rost, B., Burkhardt, S., Riebesell, U., Thoms, S., Wolf-Gladrow, D.A. (2007a). The effect of iron availability on the regulation of inorganic carbon acquisition in the coccolithophore emiliania huxleyi and the significance of cellular compartmentation for stable carbon isotope fractionation. *Geochimica et Cosmochimica Acta*, 71(22), 5301–5312.

Schulz, K.G., Rost, B., Burkhardt, S., Riebesell, U., Thoms, S., Wolf-Gladrow, D.A. (2007b). The effect of iron availability on the regulation of inorganic carbon acquisition in the coccolithophore *Emiliania huxleyi* and the significance of cellular compartmentation for stable carbon isotope fractionation. *Geochimica et Cosmochimica Acta*, 71, 5301–5312.

Schwarz, D., Nodop, A., Huge, J., Purfurst, S., Forchhammer, K., Michel, K.P., Bauwe, H., Kopka, J., Hagemann, M. (2011). Metabolic and transcriptomic phenotyping of inorganic carbon acclimation in the cyanobacterium *Synechococcus elongatus* PCC 7942. *Plant Physiology*, 155(4), 1640–1655.

Seifert, M., Rost, B., Trimborn, S., Hauck, J. (2020). Meta-analysis of multiple driver effects on marine phytoplankton highlights modulating role of pCO_2. *Global Change Biology*, 26(12), 6787–6804.

Sekino, K. and Shiraiwa, Y. (1994). Accumulation and utilization of dissolved inorganic carbon by a marine unicellular coccolithophorid, *Emiliania huxleyi*. *Plant and Cell Physiology*, 35(3), 353–361.

Selim, K.A., Haase, F., Hartmann, M.D., Hagemann, M., Forchhammer, K. (2018). P-II-like signaling protein SbtB links cAMP sensing with cyanobacterial inorganic carbon response. *Proceedings of the National Academy of Sciences of the United States of America*, 115(21), E4861–E4869.

Sethi, D., Butler, T.O., Shuhaili, F., Vaidyanathan, S. (2020). Diatoms for carbon sequestration and bio-based manufacturing. *Biology*, 9(8), 217.

Shao, H., Gontero, B., Maberly, S.C., Jiang, H.S., Cao, Y., Li, W., Huang, W.M. (2017). Responses of *Ottelia alismoides*, an aquatic plant with three CCMs, to variable CO_2 and light. *Journal of Experimental Botany*, 68(14), 3985–3995.

Sharkey, T.D. and Berry, J.A. (1985). Carbon isotope fractionation of algae influenced by an inducible CO_2-concentrating mechanism. In *Inorganic Carbon Uptake by Aquatic Photosynthetic Organisms*, Lucas, W.J. and Berry, J.A. (eds). American Society of Plant Physiologists, Rockville.

Shibata, M., Ohkawa, H., Kaneko, T., Fukuzawa, H., Tabata, S., Kaplan, A., Ogawa, T. (2001). Distinct constitutive and low-CO_2-induced CO_2 uptake systems in cyanobacteria: Genes involved and their phylogenetic relationship with homologous genes in other organisms. *Proceedings of the National Academy of Sciences of the United States of America*, 98(20), 11789–11794.

Sikes, C.S. and Wheeler, A.P. (1982). Carbonic anhydrase and carbon fixation in coccolithophorids. *Journal of Phycology*, 18(3), 423–426.

Sikes, C.S., Roer, R.D., Wilbur, K.M. (1980). Photosynthesis and coccolith formation – Inorganic carbon-sources and net inorganic reaction of deposition. *Limnology and Oceanography*, 25(2), 248–261.

Sinetova, M.A., Kupriyanova, E.V., Markelova, A.G., Allakhverdiev, S.I., Pronina, N.A. (2012). Identification and functional role of the carbonic anhydrase Cah3 in thylakoid membranes of pyrenoid of *Chlamydomonas reinhardtii*. *Biochimica et Biophysica Acta (BBA) – Bioenergetics*, 1817(8), 1248–1255.

Spalding, M.H. (2008). Microalgal carbon-dioxide-concentrating mechanisms: *Chlamydomonas* inorganic carbon transporters. *Journal of Experimental Botany*, 59(7), 1463–1473.

Stepien, C.C. (2015). Impacts of geography, taxonomy and functional group on inorganic carbon use patterns in marine macrophytes. *Journal of Ecology*, 103(6), 1372–1383.

Stepien, C.C., Pfister, C.A., Wootton, J.T. (2016). Functional traits for carbon access in macrophytes. *PLOS ONE*, 11(7), e0159062.

Stojkovic, S., Beardall, J., Matear, R. (2013). CO_2-concentrating mechanisms in three southern hemisphere strains of *Emiliania huxleyi*. *Journal of Phycology*, 49(4), 670–679.

Sültemeyer, D. and Rinast, K.A. (1996). The CO_2 permeability of the plasma membrane of *Chlamydomonas reinhardtii*: Mass-spectrometric ^{18}O-exchange measurements from $^{13}C^{18}O_2$ in suspensions of carbonic anhydrase-loaded plasma-membrane vesicles. *Planta*, 200(3), 358–368.

Sültemeyer, D.F., Miller, A.G., Espie, G.S., Fock, H.P., Canvin, D.T. (1989). Active CO_2 transport by the green alga *Chlamydomonas Reinhardtii*. *Plant Physiology*, 89(4), 1213–1219.

Sültemeyer, D.F., Fock, H.P., Canvin, D.T. (1991). Active uptake of inorganic carbon by *Chlamydomonas Reinhardtii* – Evidence for simultaneous transport of HCO_3^- and CO_2 and characterization of active CO_2 transport. *Canadian Journal of Botany*, 69(5), 995–1002.

Sweeney, C. (2003). The annual cycle of surface water CO_2 and O_2 in the Ross Sea. In *Biogeochemistry of the Ross Sea*, Dunbar, R.B. and DiTullio, G.R. (eds). AGU, Washington, D.C.

Tachibana, M., Allen, A.E., Kikutani, S., Endo, Y., Bowler, C., Matsuda, Y. (2011). Localization of putative carbonic anhydrases in two marine diatoms, *Phaeodactylum tricornutum* and *Thalassiosira pseudonana*. *Photosynthesis Research*, 109(1–3), 205–221.

Tanaka, A., Ohno, N., Nakajima, K., Matsuda, Y. (2016). Light and CO_2/cAMP signal cross talk on the promoter elements of chloroplastic beta-carbonic anhydrase genes in the marine diatom *Phaeodactylum tricornutum*. *Plant Physiology*, 170(2), 1105–1116.

Tcherkez, G.G.B., Farquhar, G.D., Andrews, T.J. (2006). Despite slow catalysis and confused substrate specificity, all ribulose bisphosphate carboxylases may be nearly perfectly optimized. *Proceedings of the National Academy of Sciences of the United States of America*, 103(19), 7246–7251.

Tchernov, D., Silverman, J., Luz, B., Reinhold, L., Kaplan, A. (2003). Massive light-dependent cycling of inorganic carbon between oxygenic photosynthetic microorganisms and their surroundings. *Photosynthesis Research*, 77(2), 95–103.

Tortell, P.D. (2000). Evolutionary and ecological perspectives on carbon acquisition in phytoplankton. *Limnology and Oceanography*, 45(3), 744–750.

Tortell, P.D., Payne, C.D., Li, Y., Trimborn, S., Rost, B., Smith, W.O., Riesselman, C., Dunbar, R., Sedwick, P., Ditullio, G. (2008a). The CO_2 response of Southern Ocean phytoplankton. *Geophysical Research Letters*, 35, L04605.

Tortell, P.D., Payne, C.D., Li, Y.Y., Trimborn, S., Rost, B., Smith, W.O., Riesselman, C., Dunbar, R.B., Sedwick, P., Ditullio, G.R. (2008b). CO_2 sensitivity of southern ocean phytoplankton. *Geophysical Research Letters*, 35(4), 1–5.

Trimborn, S., Langer, G., Rost, B. (2007). Effect of varying calcium concentrations and light intensities on calcification and photosynthesis in *Emiliania huxleyi*. *Limnology and Oceanography*, 52(5), 2285–2293.

Trimborn, S., Lundholm, N., Thoms, S.A., Richter, K.-U., Krock, B., Hansen, P.J., Rost, B. (2008). Inorganic carbon acquisition in potentially toxic and non-toxic diatoms: The effect of pH-induced changes in seawater carbonate chemistry. *Physiologia Plantarum*, 133, 92–105.

Trimborn, S., Brenneis, T., Sweet, E., Rost, B. (2013). Sensitivity of Antarctic phytoplankton species to ocean acidification: Growth, carbon acquisition, and species interaction. *Limnology and Oceanography*, 58(3), 997–1007.

Tsuji, Y., Mahardika, A., Matsuda, Y. (2017a). Evolutionarily distinct strategies for the acquisition of inorganic carbon from seawater in marine diatoms. *Journal of Experimental Botany*, 68(14), 3949–3958.

Tsuji, Y., Nakajima, K., Matsuda, Y. (2017b). Molecular aspects of the biophysical CO_2-concentrating mechanism and its regulation in marine diatoms. *Journal of Experimental Botany*, 68(14), 3763–3772.

Tu, C., Wynns, G.C., Mcmurray, R.E., Silverman, D.N. (1978). CO_2 kinetics in red-cell suspensions measured by ^{18}O exchange. *Journal of Biological Chemistry*, 253(22), 8178–8184.

Uehlein, N., Lovisolo, C., Siefritz, F., Kaldenhoff, R. (2003). The tobacco aquaporin NtAQP1 is a membrane CO_2 pore with physiological functions. *Nature*, 425(6959), 734–737.

Van De Waal, D.B., Brandenburg, K.M., Keuskamp, J., Trimborn, S., Rokitta, S., Kranz, S.A., Rost, B. (2019). Highest plasticity of carbon-concentrating mechanisms in earliest evolved phytoplankton. *Limnology and Oceanography Letters*, 4(2), 37–43.

Von Dassow, P., Ogata, H., Probert, I., Wincker, P., Da Silva, C., Audic, S., Claverie, J.M., De Vargas, C. (2009). Transcriptome analysis of functional differentiation between haploid and diploid cells of *Emiliania huxleyi*, a globally significant photosynthetic calcifying cell. *Genome Biology*, 10(10), 1–33.

Wang, Y.J. and Spalding, M.H. (2014a). Acclimation to very low CO_2: Contribution of limiting CO_2 inducible proteins, LCIB and LCIA, to inorganic carbon uptake in *Chlamydomonas reinhardtii*. *Plant Physiology*, 166(4), 2040–2050.

Wang, Y.J. and Spalding, M.H. (2014b). LCIB in the *Chlamydomonas* CO_2-concentrating mechanism. *Photosynthesis Research*, 121(2–3), 185–192.

Wang, L.Y., Yamano, T., Takane, S., Niikawa, Y., Toyokawa, C., Ozawa, S., Tokutsu, R., Takahashi, Y., Minagawa, J., Kanesaki, Y., Yoshikawa, H., Fukuzawa, H. (2016). Chloroplast-mediated regulation of CO_2-concentrating mechanism by Ca^{2+}-binding protein CAS in the green alga *Chlamydomonas reinhardtii*. *Proceedings of the National Academy of Sciences of the United States of America*, 113(44), 12586–12591.

Watson, G.M.F. and Tabita, F.R. (1997). Microbial ribulose 1,5-bisphosphate carboxylase/oxygenase: A molecule for phylogenetic and enzymological investigation. *FEMS Microbiology Letters*, 146(1), 13–22.

Whitney, S.M. and Andrews, T.J. (2001). Plastome-encoded bacterial ribulose-1,5-bisphosphate carboxylase/oxygenase (Rubisco) supports photosynthesis and growth in tobacco. *Proceedings of the National Academy of Sciences of the United States of America*, 98(25), 14738–14743.

Whitney, S.M., Houtz, R.L., Alonso, H. (2011a). Advancing our understanding and capacity to engineer nature's CO_2-sequestering enzyme, Rubisco. *Plant Physiology*, 155(1), 27–35.

Whitney, S.M., Sharwood, R.E., Orr, D., White, S.J., Alonso, H., Galmes, J. (2011b). Isoleucine 309 acts as a C-4 catalytic switch that increases ribulose-1,5-bisphosphate carboxylase/oxygenase (Rubisco) carboxylation rate in *Flaveria*. *Proceedings of the National Academy of Sciences of the United States of America*, 108(35), 14688–14693.

Wolf-Gladrow, D.A., Bijma, J., Zeebe, R.E. (1999). Model simulation of the carbonate chemistry in the microenvironment of symbiont bearing foraminifera. *Marine Chemistry*, 64(3), 181–198.

Wolf-Gladrow, D.A., Zeebe, R.E., Klaas, C., Körtzinger, A., Dickson, A.G. (2007). Total alkalinity: The explicit conservative expression and its application to biogeochemical processes. *Marine Chemistry*, 106(1–2), 287–300.

Xu, Y., Feng, L., Jeffrey, P.D., Shi, Y.G., Morel, F.M.M. (2008). Structure and metal exchange in the cadmium carbonic anhydrase of marine diatoms. *Nature*, 452(7183), 56–61.

Yamano, T., Tsujikawa, T., Hatano, K., Ozawa, S., Takahashi, Y., Fukuzawa, H. (2010). Light and low-CO2-dependent LCIB-LCIC complex localization in the chloroplast supports the carbon-concentrating mechanism in *Chlamydomonas reinhardtii*. *Plant and Cell Physiology*, 51(9), 1453–1468.

Yamano, T., Sato, E., Iguchi, H., Fukuda, Y., Fukuzawa, H. (2015). Characterization of cooperative bicarbonate uptake into chloroplast stroma in the green alga *Chlamydomonas reinhardtii*. *Proceedings of the National Academy of Sciences of the United States of America*, 112(23), 7315–7320.

Yang, B.X., Fukuda, N., Van Hoek, A., Matthay, M.A., Ma, T.H., Verkman, A.S. (2000). Carbon dioxide permeability of aquaporin-1 measured in erythrocytes and lung of aquaporin-1 null mice and in reconstituted proteoliposomes. *Journal of Biological Chemistry*, 275(4), 2686–2692.

Yoon, H.S., Hackett, J.D., Bhattacharya, D. (2002). A single origin of the peridinin- and fucoxanthin-containing plastids in dinoflagellates through tertiary endosymbiosis. *Proceedings of the National Academy of Sciences of the United States of America*, 99(18), 11724–11729.

Yoon, H.S., Hackett, J.D., Van Dolah, F.M., Nosenko, T., Lidie, L., Bhattacharya, D. (2005). Tertiary endosymbiosis driven genome evolution in dinoflagellate algae. *Molecular Biology and Evolution*, 22(5), 1299–1308.

Young, J.N., Rickaby, R.E.M., Kapralov, M.V., Filatov, D.A. (2012). Adaptive signals in algal Rubisco reveal a history of ancient atmospheric carbon dioxide. *Philosophical Transactions of the Royal Society B – Biological Sciences*, 367(1588), 483–492.

Young, J.N., Heureux, A.M.C., Sharwood, R.E., Rickaby, R.E.M., Morel, F.M.M., Whitney, S.M. (2016). Large variation in the Rubisco kinetics of diatoms reveals diversity among their carbon-concentrating mechanisms. *Journal of Experimental Botany*, 67(11), 3445–3456.

Zeebe, R.E. and Wolf-Gladrow, D.A. (2001). *CO_2 in Seawater: Equilibrium, Kinetics, Isotopes*, Elsevier Science B.V., Amsterdam.

Zhang, Y., Yin, L., Jiang, H.-S., Li, W., Gontero, B., Maberly, S.C. (2014). Biochemical and biophysical CO_2 concentrating mechanisms in two species of freshwater macrophyte within the genus *Ottelia* (Hydrocharitaceae). *Photosynthesis Research*, 121(2), 285–297.

5

Biochemical Carbon Dioxide Concentrating Mechanisms

Brigitte GONTERO[1] and Stephen Christopher MABERLY[2]

[1] *Aix-Marseille Université, CNRS BIP UMR 7281, Marseille, France*

[2] *Lake Ecosystems Group, UK Centre for Ecology & Hydrology, Lancaster, United Kingdom*

5.1. Introduction

The productivity of terrestrial plants can be limited by water supply and high temperature (Hsu *et al.* 2012; Sage *et al.* 2018), but these factors are often replaced in aquatic plants by low light and restricted inorganic carbon supply (Maberly and Gontero 2017, 2018). Furthermore, oxygen concentrations in water can be over two times greater than air-equilibrium and low diffusion will restrict losses from the chloroplast to the environment (Maberly and Gontero 2018). These conditions can cause underwater photosynthesis to be carbon-limited, promoting photorespiration and favoring the possession of a CO_2 concentrating mechanism (CCM). Not all aquatic species possess a CCM, but the most frequent strategy for enhancing carbon acquisition in aquatic photoautotrophs is a biophysical CCM based on the use of HCO_3^- (Giordano *et al.* 2005; Meyer and Griffiths 2013; Raven *et al.* 2011; Raven and Hurd 2012) (see Chapters 4 and 6). A biophysical CCM occurs in about 50% of freshwater plants (Maberly and Gontero 2017; Iversen *et al.* 2019), about 80% of freshwater filamentous algae (Maberly and Gontero 2017) and is widespread, but not ubiquitous, in marine macroalgae (Maberly 1990; Koch *et al.* 2013; Stepien 2015), freshwater and marine microalgae (Colman *et al.* 2002; Maberly *et al.* 2009) and seagrasses (Koch *et al.* 2013; Borum *et al.* 2016; Larkum *et al.* 2017). Use of HCO_3^- is, however, not the only strategy used to counter underwater carbon limitation. Some aquatic plants also have carbon-fixation enzymes and pathways, analogous to those in some terrestrial plants, that reduce carbon-limitation by increasing the concentration of CO_2 around the primary carboxylation enzyme.

5.2. Carbon-fixation by Rubisco in the C_3 pathway

Ribulose-1,5-bisphosphate carboxylase/oxygenase (Rubisco) is responsible for more than 99.5% of the CO_2 fixation on Earth (Raven 2009); the process uses ATP and NADPH from the primary phase of photosynthesis. Rubisco has a relatively low specific activity and can fix O_2 as well as CO_2, leading to photorespiration that results in a loss of about 30% of the photosynthetic output of terrestrial C_3 plants (Erb and Zarzycki 2016) and incurs an additional energy and reducing-power cost.

Rubisco is present in Bacteria, Archaea and Eukarya and exists in four different forms reflecting the evolutionary diversity of photoautotrophic organisms (Tabita *et al.* 2008; Flamholz *et al.* 2019; Iniguez *et al.* 2020) (see Chapter 3; Figure 5.1). The form III protein occurs in Archaea and is believed to be the ancestral Rubisco that evolved about 3500 million years ago (Tabita *et al.* 2007) or possibly as early as 4 Ga when the Earth was anoxic (Knoll and Nowak 2017) (see Chapter 2) (Tabita *et al.* 2008). Forms I, II and III are involved in autotrophic CO_2 fixation (Iniguez *et al.* 2020) while form IV, known as a Rubisco-like protein, is largely found in some bacteria but does not have carboxylation or oxygenation activity (Tabita *et al.* 2008). Form II is found in proteobacteria and the red-lineage eukaryote dinophytes (Table 5.1, Figure 5.1), although some dinophytes that evolved after a tertiary endosymbiotic event with a green alga contain form IB and ID (Minge *et al.* 2010). Form IA is found in the prokaryotic proteobacteria and cyanobacteria, and can be subdivided based on the amino acid sequence of their small subunit. Genes for IAc are clustered with genes for α-carboxysomes in an operon and are found in cyanobacteria and proteobacteria (Badger and Bek 2008) while genes for the other form, IAq, are only found in proteobacteria. Form IB is found in cyanobacteria and also in eukaryotes from the green lineage. Like form IA, form IB is subdivided into two forms, IB and IBc; the former is found in green-lineage eukaryotes and the latter is associated with β-carboxysomes and is restricted to cyanobacteria (Badger and Bek 2008). The relationship between the forms of Rubisco and the presence or absence of carboxysomes is also found for other enzymes such as ATP sulfurylase (Prioretti *et al.* 2016). Form IC is restricted to proteobacteria and Chloroflexi, and

ID is found in proteobacteria and also in ochrophytes, rhodophytes and haptophytes from the eukaryote red-lineage.

Two types of subunit can form the active enzyme, a large (L) subunit of 50 to 55 kDa and a small (S) subunit of 10 to 15 kDa. Two active sites of Rubisco are formed at the interface of a head-to-tail dimer of the large subunit, and this is the minimal active unit (Tabita *et al.* 2008). The composition and nature of subunits that make up the active Rubisco also vary among clades (Table 5.1 and Figure 5.1). Forms II and III only contain the large subunits but form I Rubisco also contains small subunits that provide structural stability and maximize catalytic activity (Spreitzer 2003). In the green microalga *Chlamydomonas reinhardtii*, the small subunits also control the formation of the pyrenoid, a protein-rich sub-cellular microcompartment within the chloroplast where Rubisco is concentrated (Genkov *et al.* 2010; Meyer *et al.* 2012) (see Chapter 7). The assembly of Rubisco in red and green lineage involves the transfer of the small subunit into the chloroplast since it is encoded in the nucleus, while the large subunit is encoded in the chloroplast. In contrast, in diatoms, both subunits are encoded by the chloroplast genome (Hwang and Tabita 1989, 1991).

Form	IA_q	IA_c	IB	IB_c	IC	ID	II	III
Large subunits	8	8	8	8	8	8	2/n	1/2
Small subunits	8	8	8	8	8	8	0	0
Archaea								
Proteobacteria								
Chloroflexi								
Cyanobacteria								
Green-lineage								
Red-lineage								
Red-lineage dinophytes								

Table 5.1. *Distribution of forms of Rubisco among different lineages of autotrophs, the number of large and small subunits in the mature enzyme and the structure of the Forms I, II and III (Tabita et al. 2008). IAc is found in cyanobacteria with α-carboxysomes and in proteobacteria while IAq is only found in proteobacteria; IBc is found in cyanobacteria with β-carboxysomes. n represents a variable number. For a color version of this table, see www.iste.co.uk/maberly/photosynthesis.zip*

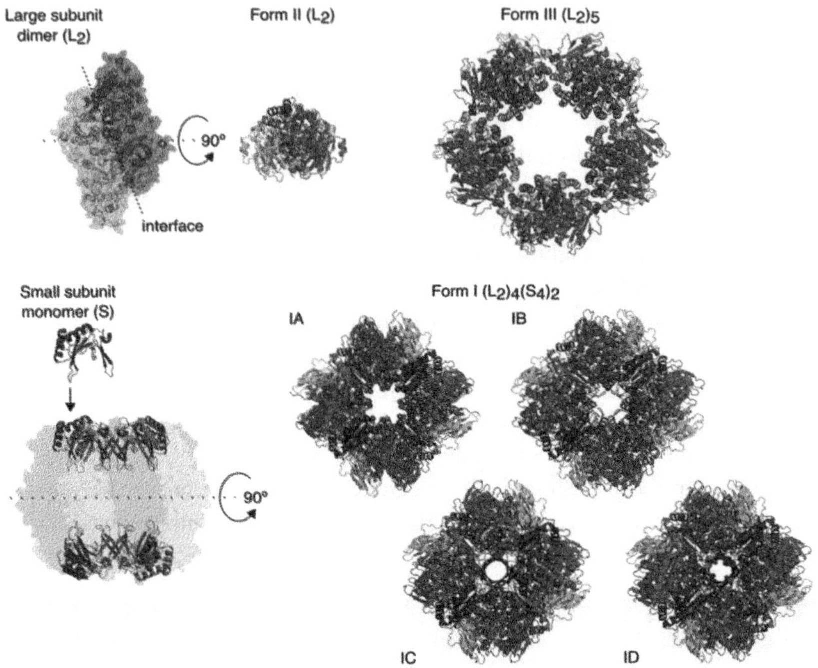

Figure 5.1. *Crystal structure of the main forms of Rubisco from the protein database (pdb with accession number): IA* (Halothiobacillus neapolitanus, *1svd.pdb); IB* (Spinacia oleracea *1rcx.pdb); IC* (Rhodobacter sphaeroides, *5nv3.pdb); ID* (Thalassiosira antarctica, *5mz2.pdb); II;* Rhodospirillum rubrum, *(9rub.pdb) III* (Thermococcus kodakarensis, *1geh.pdb). For a color version of this figure, see www.iste.co.uk/maberly/photosynthesis.zip*

Different forms of Rubisco have different kinetic properties. This relates to their phylogenetic/evolutionary history and also to the concentrations of CO_2 and O_2 at the active site depending on external conditions and the presence of a CCM (Flamholz *et al.* 2019; Iniguez *et al.* 2020). Thus, form III Rubisco from Archaea that grows in anoxic environments, and probably at relatively high concentrations of CO_2, has a low selectivity (S_{co}, sometimes called tau, τ) for CO_2 vs. O_2 (Figure 5.2). In contrast, the selectivity for CO_2 vs. O_2 in oxygenic green- and red-lineage plants and algae is much higher, which corresponds to their growth in the presence of oxygen and concentrations of CO_2 that are often rate-limiting. The average half-saturation constant for CO_2 (K_c) in the ancestral form III Rubisco is about 50 µmol L^{-1}, but it is much lower (i.e. a higher affinity) in the oxygenic green- and red-lineage plants and algae. The cyanobacteria have an intermediate S_{co} and high K_c (i.e. low affinity), but they have the highest median catalytic constant (k_{cat}) of any group. These properties correspond to a very effective biophysical CCM, producing high CO_2 concentrations at the active site of Rubisco (see Chapter 4). There is also a correlation between Rubisco kinetics and the presence of a CCM within the oxygenic green- and red-lineages (Figure 5.3). Species with a CCM have a higher K_c, a lower S_{co} and a higher k_{cat} than those without a CCM, as this increases the concentration of CO_2 at the active site of Rubisco. In the red-lineage, the catalytic efficiency (k_{cat}/K_c) is greater in species without a CCM, but in the green-lineage, dominated by terrestrial plants in the present dataset (Iniguez *et al.* 2020), k_{cat}/K_c is high and not statistically different between species with and without a CCM. There are evolutionary trends associated with the properties of Rubisco. Tortell (2000) and Shih *et al.* (2016) showed that the specificity of Rubisco for CO_2 vs. O_2 generally increased over geological time as the atmospheric concentrations of CO_2 declined and the concentrations of O_2 increased (Raven *et al.* 2012) (see Chapter 2).

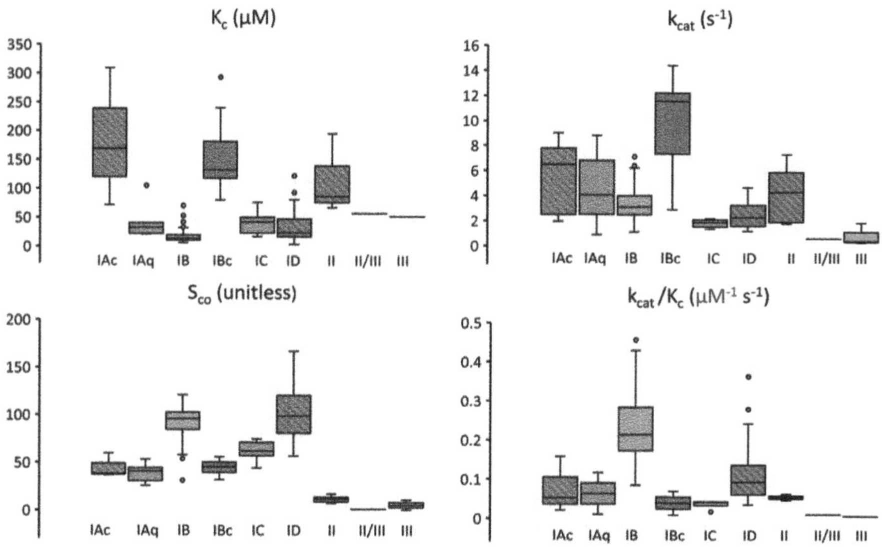

Figure 5.2. *Kinetic characteristics of different forms of Rubisco for CO_2 based on data in Iniguez et al. (2020). See Figure 5.1 for the phylogenetic distribution of the Rubisco forms. The box and whisker plots show the median (central line) interquartile range (box) and the error bars are the smallest and largest value up to 1.5 times the interquartile range. Outliers, values that exceed 1.5 times the interquartile range, are shown as individual points. Form II/III is intermediate between II and III and currently known from one species. For a color version of this figure, see www.iste.co.uk/maberly/photosynthesis.zip*

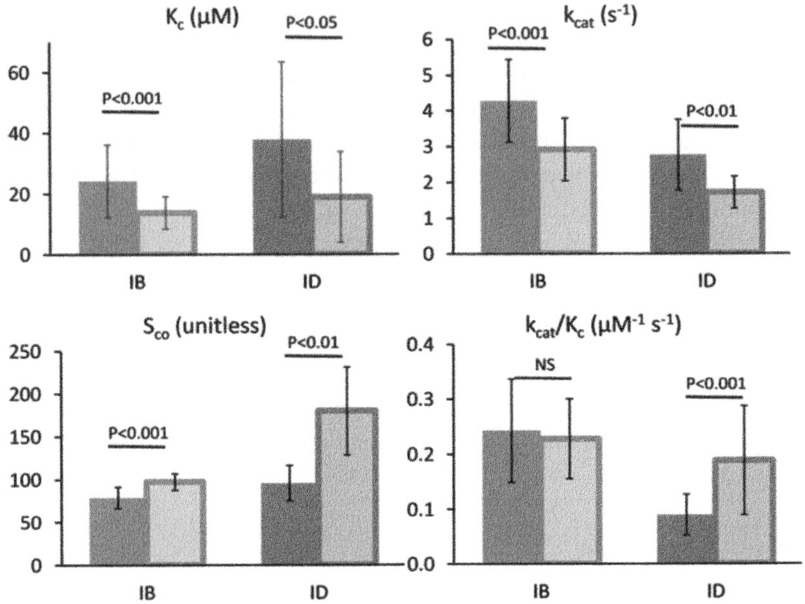

Figure 5.3. *Comparison of Rubisco kinetics for species of the green- and red-lineage with (solid fill) or without (hatched fill) a CCM for IB and ID forms. Error bars show one standard deviation. The significance of a difference for the presence or absence of a CCM is shown (ANOVA). NS, not significant. Data from Iniguez et al. (2020)*

5.3. The C_4 CO_2 concentrating mechanism

5.3.1. *C_4 in terrestrial plants*

It was widely believed that the Calvin–Benson–Bassham (CBB) cycle was the only pathway involved in carbon fixation in plants until the latter half of the 1960s, when Kortschak *et al.* (1965) found a C_4 acid, and not a three-carbon compound, as the first product of carbon fixation in leaves of sugarcane (*Saccharum officinarum*). Hatch and Slack (1966) elucidated the pathway responsible: the C_4 dicarboxylic pathway of photosynthesis, abbreviated later to the C_4 pathway or C_4 photosynthesis (Edwards 2019; Edwards *et al.* 2004; Sage *et al.* 2012). This initial carbon fixation is performed by the enzyme phosphoenolpyruvate carboxylase (PEPC), which is oxygen insensitive and uses HCO_3^- (in a pH-dependent equilibrium with dissolved CO_2) to carboxylate phosphoenolpyruvate, yielding a four-carbon molecule, oxaloacetate, that is subsequently converted to malate or aspartate (Edwards *et al.* 2004; Sage *et al.* 2012; Bräutigam *et al.* 2014) (Figure 5.4)). The decarboxylation of the C_4 compound around Rubisco increases the CO_2 concentration at its active site and minimizes photorespiration. The enzyme pyruvate phosphate dikinase (PPDK) is also required in C_4 fixation as it regenerates PEP, the initial acceptor of HCO_3^-. Three alternative decarboxylating enzymes occur in C_4 plants. For example, the C_4 compound is decarboxylated by the NADP-malic enzyme in sugarcane, by NAD-malic enzyme (NAD-ME) in *Pennisetum glaucum* and by PEP carboxykinase (PEPCK) in *Panicum maximum*. These three decarboxylation enzymes are used to classify terrestrial C_4 plants into sub-types that have slightly different physiological properties (Blätke and Bräutigam 2019), although there is some flexibility in the operation of the different pathways (Furbank 2011).

C_4 photosynthesis is a dramatic example of convergent evolution, is believed to have evolved on at least 63 occasions and is found in about 7500 species (Sage *et al.* 2011, 2012, 2018), and C_4 plants are estimated to contribute 23% to terrestrial primary productivity (Still *et al.* 2003). Molecular clock approaches suggest C_4 photosynthesis first evolved around 30 Ma (Christin *et al.* 2011; Sage *et al.* 2012), at a time when atmospheric concentrations of CO_2 were once more falling rapidly from above 1000 ppm to around present day levels. However, the expansion of species with the C_4 syndrome, and the first fossil evidence, occurred later, around 20 Ma (Edwards *et al.* 2010; Sage *et al.* 2018). Low atmospheric CO_2 *per se* is not a sufficient pressure to promote C_4 photosynthesis: C_3 plants dominate the flora in temperate and boreal regions today, for example. However, C_4 photosynthesis is favored in combination with other factors, especially those linked to water stress and low intracellular concentrations of CO_2, such as high temperature and high light, salinity and aridity (Sage *et al.* 2012). While the C_4 cycle via malic enzymes adds an additional energy cost of 2 ATP per carbon fixed, Berry and Farquhar (1978) unrestrained photorespiration comes with an energy cost of 3.5 ATP and 2 NADPH (Farquhar *et al.* 1980) and so C_4 will be beneficial when rates of photorespiration are high but a disadvantage when photorespiration is low.

To prevent futile cycling between carbon fixation by PEPC and CO_2 production by one of the decarboxylation enzymes, C_4 photosynthesis is usually associated with a characteristic structure called Kranz anatomy. This involves two different types of cell, mesophyll and bundle sheath, and around twenty-two different types of Kranz anatomy are recognized (Edwards and Voznesenskya 2011). PEPC and PPDK are located in mesophyll cells and produce a C_4 compound. The mesophyll cells surround bundle sheath cells and the C_4 compound is transported there where it is decarboxylated producing CO_2 around Rubisco and a C_3 compound that moves back to the mesophyll cells (Figure 5.4). Chloroplast ultrastructure can differ in mesophyll and bundle sheath cells, producing so-called dimorphic chloroplasts that are linked to the energy requirements of the two cell types and the production of starch (Edwards *et al.* 2004). In the NADP-ME subtype, chloroplasts in bundle sheath cells are often depleted in grana and photosystem II activity while grana and photosystem II activity are present in bundle sheath chloroplasts in the NAD-ME and PEPCK subtypes. Starch is predominantly found in bundle sheath cells where Rubisco is located.

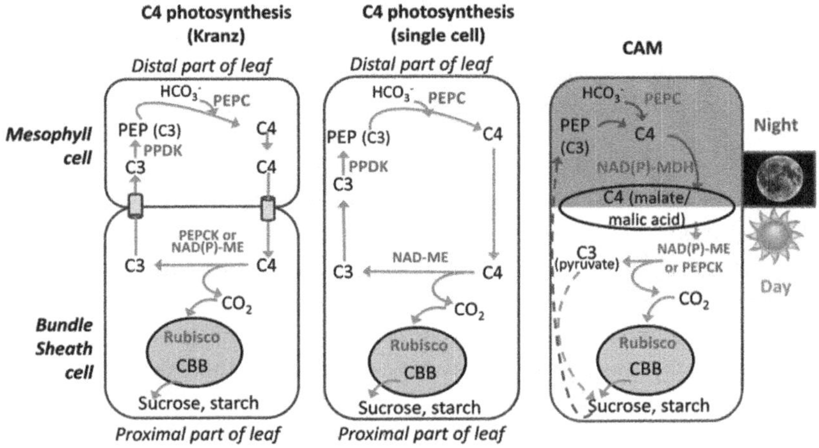

Figure 5.4. *Scheme of C_4 photosynthesis and Crassulacean Acid Metabolism (CAM). Orange and blue arrows show fluxes in the day and night respectively. C3 and C4 stand for three carbon and four carbon compounds respectively; the C4 compound is variable depending on the species; CBB, Calvin–Benson–Bassham cycle. The blue cylinders represent plasmodesmata. Enzyme names in green: NAD(P) MDH, NAD(P) malate dehydrogenase; NAD(P)ME, NAD(P) malic enzyme; PEPC, phosphoenolpyruvate carboxylase; PEPCK, phosphoenolpyruvate carboxykinase; PPDK, pyruvate phosphate dikinase; Rubisco, ribulose-1,5-bisphosphate carboxylase/ oxygenase. For a color version of this figure, see www.iste.co.uk/maberly/ photosynthesis.zip*

It was thought for many years that Kranz anatomy was necessary for C_4 photosynthesis. However, in 2000, it was shown that two succulent plants from the Chenopodiaceae, *Suaeda aralocaspica* (formerly called *Borszczowia aralocaspica*) and *Bienertia cycloptera*, lack Kranz anatomy but perform C_4 photosynthesis in a single cell (Voznesenskaya *et al.* 2001; Offermann *et al.* 2011; Koteyeva *et al.* 2016). Subsequently, two more species from the genus *Bienertia*, *B. sinuspersici* and *B. kavirense*, have been shown also to possess single-cell C_4 photosynthesis (Akhani *et al.* 2005; Akhani *et al.* 2012). These plants grow south of the Caspian Sea in the central Asian and Iranian salt deserts and around the Persian Gulf (Akhani *et al.*

2012). The occurrence of C_4 in single cells relies on spatial separation of carboxylation and decarboxylation within elongated cells in the case of *S. aralocaspica* (Figure 5.4) and clusters of chloroplasts with Rubisco being located in the center in the case of *Bienertia*. In these species, NAD-ME appears to be the decarboxylase (Offermann *et al.* 2011).

5.3.2. *C_4 in aquatic plants and algae*

5.3.2.1. *C_4 in aquatic angiosperms*

No known aquatic plants have Kranz anatomy apart from the partially aquatic C_4 grasses *Neostapfia colusana* and *Tuctoria greenei* that grow in vernal pools and flower when the pools have dried (Keeley 1998a). Nevertheless, several fully aquatic species can perform C_4 photosynthesis, of which the submerged freshwater monocotyledon *Hydrilla verticillata* (Hydrocharitaceae) is one of the best-characterized and the first plant shown to have single-cell C_4 photosynthesis (Holaday and Bowes 1980; Salvucci and Bowes 1981; Reiskind *et al.* 1989; Bowes *et al.* 2002; Bowes 2011). In *H. verticillata,* C_4 is facultative and induced by a decrease in the availability of CO_2. For example, plants growing in dense canopies deplete concentrations of inorganic carbon substantially (e.g. CO_2 <0.03 and HCO_3^- 115 μmol L^{-1}) at pH 9.8 and have the C_4 phenotype, while nearby plants in the open water with the C_3 phenotype experience much higher concentrations (e.g. CO_2 20 and HCO_3^- 2050 μmol L^{-1}) at pH 8.0 (Reiskind *et al.* 1997). Plants grown at high light and high temperature also develop C_4 photosynthesis as this indirectly imposes inorganic carbon limitation (Salvucci and Bowes 1981). Detailed research by Magnin *et al.* (1997) showed that when the C_3 phenotype of *H. verticillata* was transferred to low concentrations of CO_2, the compensation point fell from about 60 to 20 μL L^{-1} and in parallel, % oxygen inhibition fell from about 45 to 12% over 12 days (Figure 5.5). These physiological changes were underpinned by increased amounts of key C_4 enzymes. There was a 11-fold increase in PEPC activity and little change in Rubisco activity so that the PEPC to Rubisco ratio increased about 15-fold from 0.3 to 4.9 (Figure 5.5) and PPDK and NADP-ME increased 8- and 6-fold respectively over 12 days. In an earlier experiment Salvucci and Bowes (1981) showed that a different decarboxylase, NAD-ME, increased by about 3-fold in the C_4 versus the C_3 phenotype. However, 95% of the activity of NAD-ME was located in the mitochondria (Magnin *et al.* 1997) that could suggest this enzyme was the decarboxylase. However, since 91% of the NADP-ME was present in the chloroplast, Magnin *et al.* (1997) suggested that this enzyme was the decarboxylase in C_4 photosynthesis in *H. verticillata.*

All C_4 enzymes are also present in C_3 plants, and this holds true for PEPC, which has many functions including carbon skeletons supply to the tricarboxylic acid cycle and to ammonium assimilation to cite but two examples (Aubry *et al.* 2011). The characteristics of the enzymes involved in the C_4 CCM in *H. verticillata* have been compared to those in terrestrial C_4 plants. Three isoforms of PEPC are present in *H. verticillata*, but the transcript level of only one increased in C_4 leaves and this was therefore presumed to be the key C_4 isoform (Rao *et al.* 2002). However, the typical serine residue of PEPC in terrestrial C_4 plant at position 774 (numbered from the terrestrial C_4 plant *Flaveria trinervia*) was absent and replaced by an alanine residue. A comparison of the sequence of this PEPC isoform with PEPC from terrestrial C_4 and CAM plants showed that it was most similar to the C_3 form of PEPC from the monocot maize (*Zea mays*). Nevertheless, unlike C_3 PEPC forms from *H. verticillata*, the kinetic properties of the C_4 PEPC isoform are similar to those of PEPC from terrestrial C_4 plants: the K_M for PEP is high (0.3 mM), glucose-6-phosphate lowers the K_M, and sensitivity to malate inhibition is low (Rao *et al.* 2008). Indeed, it is well-documented that the activity of the various isoforms of plant PEPC is subject to allosteric control by metabolite effectors (Doncaster and Leegood 1987). For instance, glucose-6-phosphate and triose-phosphate are positive effectors, while malate and aspartate are negative ones. The binding site of glucose-6-phosphate to PEPC has recently been elucidated using crystallography. This showed that the binding of glucose-6-phosphate stabilized the subunit/subunit interface of the dimeric unit of PEPC (Muñoz-Clares *et al.* 2020).

H. verticillata possesses three isoforms of NADP-ME. Only one is chloroplastic and upregulated in C_4 phenotypes, and this is the presumed decarboxylase involved in C_4 in this species (Estavillo *et al.* 2007). This differs from terrestrial C_4 plants, such as *Z. mays*, that have two chloroplastic isoforms of NADP-ME, with one constitutively expressed and involved in non-photosynthetic functions, while the expression and the activity of the other isoform occurs in bundle sheath cells and is involved in decarboxylation (Maurino *et al.* 1996).

In order to avoid futile cycling, carboxylation and decarboxylation need to be spatially separated, and in *H. verticillata* this occurs within a single cell. By analyzing activities in sub-cellular fractions, Magnin *et al.* (1997) showed that PEPC was located in the cytosol while NADP-ME and PPDK were in the chloroplast with Rubisco. This confirmed an earlier study using immunogold labeling (Reiskind *et al.* 1989). These enzyme localizations produced an internal concentration of inorganic carbon in leaves of the C_4 phenotype that was five times higher than in the external medium, showing that a CCM was in operation (Reiskind *et al.* 1997). Calculations indicated that the CO_2 concentration in the chloroplast was 400 μmol L^{-1} compared to 7 µmol L^{-1} in the C_3 phenotype, and this explains the observed suppression of oxygen inhibition of photosynthesis and the reduced CO_2 compensation point.

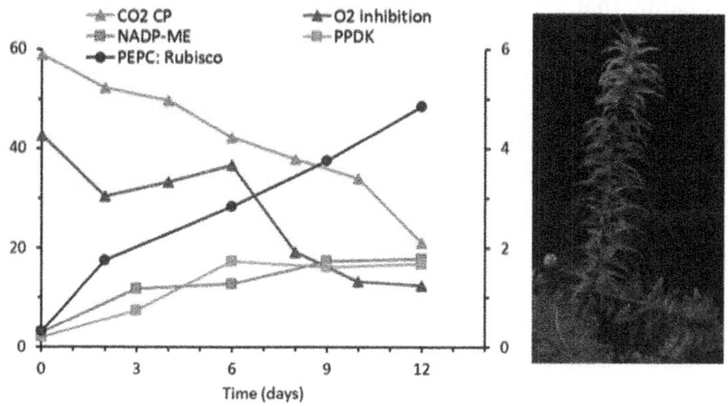

Figure 5.5. *Time course of changing CO_2 compensation point (µL L^{-1}), oxygen inhibition (%) and activities of NADP-ME and PPDK (µmol g^{-1} fresh weight h^{-1}) on the left-hand axis and the ratio of activity of PEPC: Rubisco on the right-hand axis in* H. verticillata *following transfer from high CO_2 to conditions inducing the C_4 phenotype (redrawn from Magnin* et al. *(1997)). The photograph of* H. verticillata *was taken by Fu Wenlong. For a color version of this figure, see www.iste.co.uk/maberly/photosynthesis.zip*

Another species within the Hydrocharitaceae, *Egeria densa*, is also able to perform facultative C_4 photosynthesis when grown at high light and high temperature with NADP-ME present in the chloroplasts and PEPC in the cytoplasm (Casati *et al.* 2000; Lara *et al.* 2002). The partially purified PEPC induced in the C_4 phenotype had a K_M value of about 8 µmol L^{-1} for HCO_3^-. This is lower than reported values for PEPC from terrestrial C_3 or C_4 plants. For example, the K_M for HCO_3^- was 120 and 40 µmol L^{-1} in C_3 and C_4 species of *Panicum*, 100 and 70 µmol L^{-1} in C_3 and C_4 species of *Flaveria* (Moody *et al.* 2020) and 24 to 27 µmol L^{-1} in five C_4 species with different decarboxylase enzymes (Bauwe and Chollet 1986). In *E. densa*, there is also evidence for C_4 carbon fixation in the stems in addition to the leaves (Gu *et al.* 2015) suggesting it contributes to photosynthesis as has been found for some terrestrial species (e.g. Hibberd and Quick 2002). In both leaves and stems of *E. densa*, there was about a 2.3-fold increase in PEPC activity in plants grown at low versus high CO_2.

Yet another species from the Hydrocharitaceae, *Ottelia alismoides*, also shows evidence for C_4 photosynthesis (Zhang *et al.* 2014) (Figure 5.6). While the activity of Rubisco was not significantly different in plants grown at high or low CO_2, PEPC activity increased 2-fold in leaves of *O. alismoides* grown at low CO_2 (Figure 5.6). The ratio of PEPC to Rubisco activity was higher than one: 2.8 and 5.9 at high and low CO_2, indicating that, unlike *H. verticillata* and *E. densa*, C_4 is constitutive in this species, although low CO_2 increases the PEPC:Rubisco ratio. Furthermore, in contrast to *H. verticillata* and *E. densa*, the activity of the decarboxylating enzyme NADP-ME was very low, while the high activity of NAD-ME suggested that this plant belongs to the NAD-ME type of C_4. The activity of the PEP-regenerating enzyme PPDK was similar to PEPC, and regulated to the same extent at low versus high CO_2.

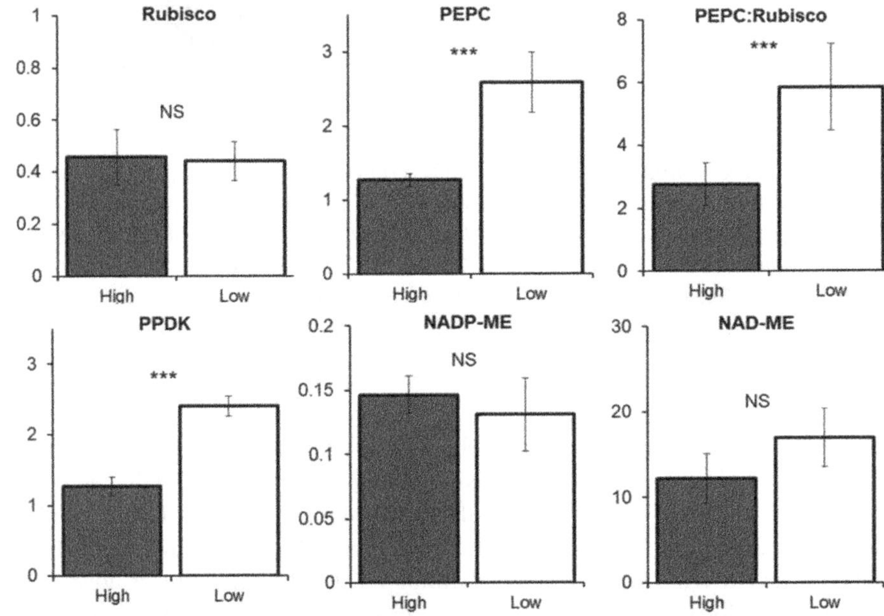

Figure 5.6. *Enzyme activities (μmol mg^{-1} protein minute^{-1}) of* Ottelia alismoides *grown at high and low concentrations of CO_2. Error bars represent one standard deviation. The statistics show the significance of the difference between plants grown at high and low CO_2: NS, not significant; ***, $P<0.001$. Data derived from Zhang* et al. *(2014)*

Mature leaves of *O. alismoides* are large, up to 20 cm wide and ovate to cordate in shape and are unusually thick for aquatic plants, between 161 and 196 μm

(Han *et al.* 2020) (Figure 5.7). The upper and lower epidermal cells of *O. alismoides* are separated by mesophyll cells and large air spaces (about 22% of the leaf volume). Both the epidermal and mesophyll cells possess chloroplasts; however, the chloroplasts in the latter have particularly large starch grains. This could be compatible with an internal source of inorganic carbon derived from C_4 photosynthesis; the mesophyll cells being equivalent to the bundle-sheath cells of terrestrial C_4 plants. Moreover, the cell size of about 60 μm for the epidermal cells and 75 μm for the mesophyll cells is greater than the critical size of 50 μm that allows separation of carboxylation and decarboxylation within a cell (Juric *et al.* 2017). It is currently unclear if C_4 photosynthesis in *O. alismoides* occurs within a single cell, and which cell(s), or across two cell types. The air spaces may also act as reservoirs of CO_2 derived from respired, photorespired or decarboxylated CO_2 and be an additional source of CO_2 to the mesophyll cells. Refixing photorespired CO_2 via glycine metabolism has been suggested as an intermediate step in the evolution of C_4 photosynthesis (Sage *et al.* 2018). While C_4 in the mature leaves is constitutive, in the juvenile leaves, that are linear to lanceolate in shape and 2 to 4 cm wide (Figure 5.7), this C_4 syndrome is facultative and only present when grown under low CO_2 (Huang *et al.* 2018). In addition to these relatively well-studied species, there is some evidence for C_4 photosynthesis in other aquatic macrophytes, so far all within the monocots (Bowes *et al.* 2002). Evidence for constitutive C_4 photosynthesis has also been found in *O. acuminata* (Zhang *et al.* 2014) and possibly *O. cordata* (Huang *et al.* 2020b).

Figure 5.7. *Shapes of mature (left) and juvenile (right) leaves of* Ottelia alismoides. *The scale bar represents 5 cm (Huang* et al. *2018)*

Figure 5.8 shows that there is a gradient of PEPC:Rubisco activity ratios in freshwater plants. Species such as *H. verticillata* and *O. alismoides* have PEPC:Rubisco ratios equivalent to terrestrial C_4 plants while the ratio is slightly lower in *E. densa*, which is also believed to operate C_4 photosynthesis. There is a gradient of PEPC:Rubisco activity ratios in the remaining species, which may indicate other roles for PEPC, including anaplerotic metabolism (Cousins *et al.* 2007) and, in some species, CAM (see below).

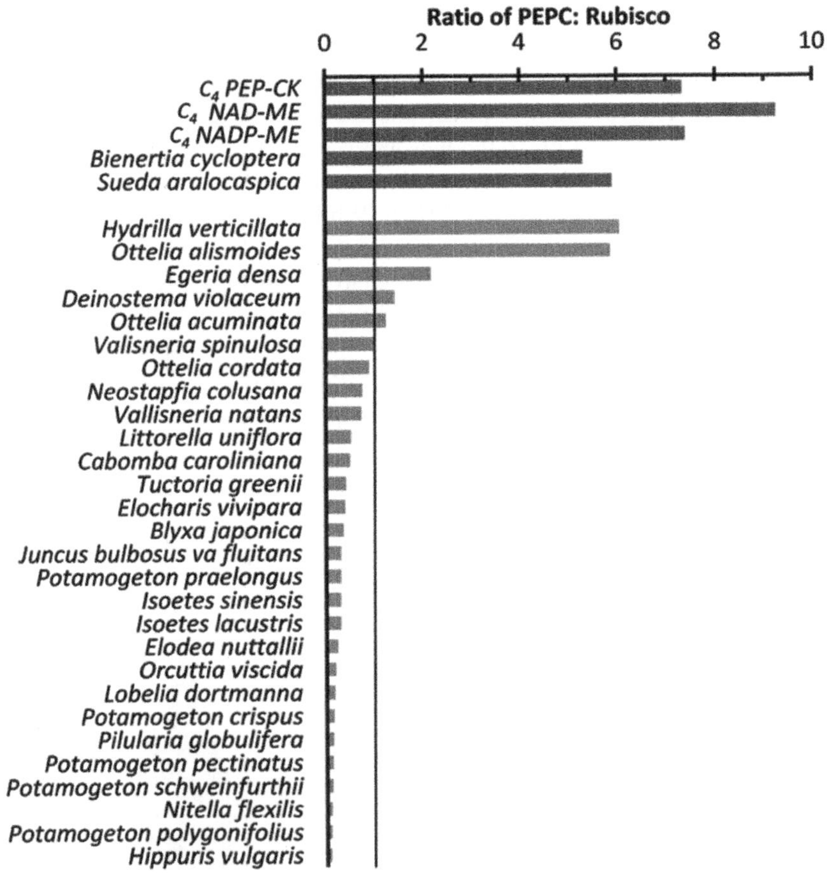

Figure 5.8. *Gradation in the ratio of PEPC to Rubisco activity in freshwater plants (light green) (Holaday and Bowes 1980; Farmer* et al. *1986; Magnin et al. 1997; Keeley 1998a; Ueno 1998; Zhang* et al. *2014; Huang* et al. *2020b; Lara* et al. *2002; Yin* et al. *2017) and terrestrial plants with C_4 photosynthesis (dark green) (Kanai and Edwards 1999; Voznesenskaya* et al. *2001). The vertical line shows a ratio of one. For a color version of this figure, see www.iste.co.uk/maberly/photosynthesis.zip*

In marine angiosperms, the seagrasses, there is less evidence for C_4 photosynthesis, although it has been suggested to be present in *Cymodocea nodosa, Halophila stipulacea* and *Thalassia testudinum* (Beer *et al.* 1980a; 1980b; Koch *et al.* 2013). Earlier conclusions that some seagrass species may possess C_4 photosynthesis, based on low stable isotope discrimination, are now known to result from carbon limitation (Raven *et al.* 2002). However, Beer *et al.* (1980a) and Beer *et al.* (1980b) showed that *Cymodocea nodosa* might have a C_4 metabolism based on pulse-chase experiments and measurement of enzyme activities. The difficulty of extracting active enzymes from these species hampers studies on their biochemistry, but clearly more research is needed.

5.3.2.2. *C_4 in macroalgae*

Udotea flabellum is a calcifying siphonous, green marine macroalga (Reiskind *et al.* 1988; Reiskind and Bowes 1991) that possesses a C_4 metabolism based on PEP carboxykinase (PEPCK) that interconverts PEP and CO_2 into oxaloacetate using one molecule of ATP. This enzyme is located in the cytosol where it allows formation of a C_4 compound that is then decarboxylated in the chloroplast thereby elevating CO_2 concentration at the site of Rubisco. This macroalga lacks Kranz anatomy but the PEPCK activity results in reduced O_2 inhibition and low photorespiration. The genus *Udotea* evolved about 81 Ma ago (Lagourgue and Payri 2021), so if C_4 photosynthesis was ancestral, it possibly evolved in *Udotea* before terrestrial plants. Interestingly, the genus *Udotea,* like some other coenocytic green algae, has plastids of two types; some act as chloroplasts and others are starch filled (Hillis-Colinvaux 1980). This is reminiscent of the dimorphic chloroplasts seen in many C_4 plants (Edwards *et al.* 2001) but any functional significance of the different plastids in *Udotea* remains to be determined.

Evidence for C_4-like photosynthesis in brown marine macroalgae is largely equivocal (Raven 1997; Raven and Hurd 2012; Koch *et al.* 2013; Raven and Giordano 2017). Recently, Shao *et al.* (2019) measured transcripts and enzyme activities in the brown macroalga *Saccharina japonica* (Phaeophyceae) grown under different conditions. They present tentative evidence for a form of C_4 photosynthesis in this species that assumes carbon fixation by PEPC and decarboxylation by PEPCK but since both enzymes were located in the mitochondria it is unclear how a futile cycle will be prevented. The red macroalga *Pyropia* (Bangiophyceae) has been suggested to have a similar form of C_4 photosynthesis based on mitochondrial carbon fixation with pyruvate carboxylase and decarboxylation in the same organelle (Zhang *et al.* 2020) and also *Porphyra haitanensis* based on analysis of expressed sequence tags (Fan *et al.* 2007). A number of papers have recently suggested that the green macroalga, *Ulva,* may possess C_4 photosynthesis. Valiella *et al.* (2018) suggested this based on ^{13}C values being in the range diagnostic for terrestrial C_4 plants (Valiella *et al.* 2018). However, in water, this is more likely to result from a biophysical CCM based on bicarbonate use and possibly strong carbon limitation in some of the dense blooms they studied (Maberly *et al.* 1992; Stepien 2015). The possibility of C_4 photosynthesis in *Ulva* has also been claimed based on increased activity of enzymes involved in C_4 photosynthesis including NAD-ME (Zhou *et al.* 2020), PEPCK (Liu *et al.* 2020) and PPDK (Xu *et al.* 2012) in response to low inorganic carbon or stress conditions. However, pulse-chase experiments by Beer and Israel (1986) on *Ulva* sp. found no evidence for C_4 photosynthesis (Beer and Israel 1986) and C_4 enzymes play numerous additional roles (Aubry *et al.* 2011). Nevertheless, it is clear that many marine macroalgae have different and interesting enzyme activities and biochemical characteristics that involve C_4 acids, but the precise role that they play in carbon fixation, and whether or not this represents C_4 photosynthesis, remains to be resolved (Raven and Giordano 2017). Given their other roles, the possession of genes or activity of enzymes required in C_4 photosynthesis is not by itself adequate to conclude that C_4 photosynthesis is occurring (Beardall and Raven 2020).

5.3.2.3. *C_4 in microalgae*

Early research (e.g. Beardall 1989; Descolas-Gros and Oriol 1992) showed that carboxylating enzymes producing C_4 compounds including PEPC, PEPCK and pyruvate carboxylase (PC, interconverts pyruvate, bicarbonate and ATP into oxaloacetate and ADP) are present in some marine microalgae. These enzymes have important anaplerotic roles, as mentioned above, so their presence is not definitive evidence for C_4 photosynthesis.

Diatoms are ubiquitous photosynthetic microalgae found in aquatic environments that contribute about 40% of the phytoplankton primary production in the ocean and about 20% to global primary production (Falkowski *et al.* 2000; Armbrust 2009). Beyond this essential role in the environment, some species are used in a biotechnological context for their metabolic capacities as lipid producers, for instance (Mekhalfi *et al.* 2012; Levitan *et al.* 2014; Prioretti *et al.* 2017). There is no doubt that the presence of CCMs is one of the characteristics that allows diatoms to thrive in many environments since their Rubisco has an affinity constant for CO_2 between 23 and 68 μM (Young *et al.* 2016) that is too high to be saturated under air-equilibrium concentrations of CO_2 in both marine and freshwater environments. There is good evidence that many diatoms possess a biophysical CCM that could help to overcome carbon limitation (Hopkinson *et al.* 2011; Matsuda *et al.* 2017). However, in 2000, a landmark, but controversial, paper was published by Reinfelder *et al.* (2000) claiming that the marine centric diatom *Thalassiosira weissflogii* (now known as *Conticriba weissflogii*) performed C_4 photosynthesis. The authors observed an increase in PEPC activity when this diatom was grown at a low concentration of CO_2 versus high concentration of CO_2. Moreover, an increase in PEPC activity was also observed in Zn-limited cells (with 6 pM Zn), conditions that impair carbonic anhydrase, an essential component of biophysical and biochemical CCMs (see Chapters 3 and 6), and thus inorganic carbon accumulation in this diatom. Similarly, in the presence of a CA inhibitor, PEPC activity increased (Reinfelder *et al.* 2000). Pulse-chase experiments also showed that part of the carbon from CO_2 was incorporated into malate rather than into 3-phosphoglycerate, also suggesting the presence of a C_4-type (Reinfelder *et al.* 2000) or C_3-C_4 intermediate type (Roberts *et al.* 2007) mechanism in this diatom. Measurements of the activity of C_4 enzymes at high and low CO_2 are consistent with C_4 photosynthesis in this species (Clement *et al.* 2017). In addition, in *T. weissflogii*, the compound 3,3-dichloro-2-dihydroxyphosphinoylmethyl-2-propenoate (DCIP) known to inhibit PEPC, inhibits the ability to assimilate mineral carbon, unlike in *C. reinhardtii* (Reinfelder *et al.* 2004).

In other diatoms, however, the presence of this C_4 syndrome is less likely (Raven 2010). In *T. pseudonana*, pulse-chase experiments have been carried out but do not show the incorporation of CO_2 in 4-carbon compounds, rather suggesting a C_3-type mechanism (Roberts *et al.* 2007). However. transcriptomic study and physiological data from other groups did not allow the absence of this mechanism to be confirmed or rejected. Indeed, at the transcriptomic level, some data show an increase by a factor of about three in the PEPC transcript (McGinn and Morel 2008; Kustka *et al.* 2014) when cells were grown at a low versus high CO_2 concentration. Other data in the literature show the opposite pattern (Tanaka *et al.* 2014) since the localization of the putative C_4 enzymes as well as the affinity for dissolved organic carbon under high and low CO_2 did not support an occurrence of the functional C_4-type biochemical CCM in *T. pseudonana*. Moreover, although PEPC inhibition by DCIP or of PEPC kinase (PEPCK) by 3-mercaptopicolinic acid (3-MPA) reduced photosynthesis in *T. pseudonana* (Reinfelder *et al.* 2004; McGinn and Morel 2008), this might be caused by toxic effects on other metabolic pathways (Tanaka *et al.* 2014). Recent data also suggest that the kinetic properties of this species results from a biophysical CCM and not from a C_4-type metabolism, with the activities of PEPC and PPDK lower in cells grown at low than at high CO_2. The ratios of PEPC and PPDK to Rubisco were also not in favor of a C_4 metabolism as they were substantially lower than one, even at low CO_2 (Clement *et al.* 2016).

In *P. tricornutum*, there is no regulation of PEPC and PEPCK regardless of CO_2 concentration, but 3-MPA inhibits photosynthetic activity (McGinn and Morel 2008). A study using interfering RNA (RNAi) shows that inhibition of the transcription of the PPDK gene in *P. tricornutum* does not affect photosynthetic activity, thus negating the presence of a C_4 metabolism (Haimovich-Dayan *et al.* 2013). The absence of a C_4 syndrome was also confirmed using activity measurement (Clement *et al.* 2017) and by the intracellular distribution of C_4 enzymes (Ewe *et al.* 2018). In the former study, the authors also show that in diatoms, such as *Navicula pelliculosa* and the freshwater species *Asterionella formosa* that are less studied at the molecular level, a C_4 metabolism seems to be absent. Presently it seems that C_4 photosynthesis is possible in *T. weissflogii* but not widespread in the limited number of diatom species that have been studied so far. Furthermore, it is currently unclear how C_4 photosynthesis can occur in diatoms that have a complex architecture of secondary chloroplasts with multiple membranes (see Chapter 3).

5.4. The CAM CO_2 concentrating mechanism

5.4.1. *Terrestrial CAM*

Crassulacean acid metabolism (CAM) is similar to C_4 metabolism since the first product of carbon fixation in both pathways leads to the formation of C_4 organic acids. The term CAM was coined to acknowledge Heyne's observations that the succulent Crassulaceae *Bryophyllum calycinum* had a more acid taste in the morning than in the evening (Black and Osmond 2003). In CAM (Figure 5.3), there is a temporal, rather than a spatial, separation of carboxylation by PEPC and refixation by Rubisco following decarboxylation. At night, PEPC fixes CO_2 producing a C_4 organic acid (predominantly malate) that is stored in the vacuole. During the day, the malate is decarboxylated, generating CO_2 that is fixed by Rubisco. Like C_4 photosynthesis, CAM has evolved multiple times and is found in 35 families (Silvera *et al.* 2010). It evolved before terrestrial C_4, maybe as early as the Triassic (252–201 million years ago) (Griffiths 1992). Since it does not require anatomical or sub-cellular modifications to the basic C_3 structure, it is regarded as a simpler syndrome to evolve (Edwards and Ogburn 2012).

5.4.2. *Aquatic CAM*

CAM was regarded solely as a water conservation measure until the unexpected discovery of aquatic CAM in the freshwater lycophyte *Isoetes howellii* (Isoetaceae) by Keeley (1981). He found that in this species, CO_2 fixation occurred at night, and malate was the first product formed and it then accumulated. First hypothesizing that malate came from anaerobic respiration (Keeley 1988), he later disproved this because there was not a greater oxygen deficit in underground corms than in leaves. He finally concluded that *I. howellii* was a CAM plant with the hallmark of night-time acidification and daytime de-acidification (Keeley and Busch 1984). CAM is present in most aquatic lycophytes in the genus *Isoetes* (Keeley 1998b). The genus *Isoetes* evolved in the Triassic (Hetherington *et al.* 2019), at a similar time to the evolution of terrestrial CAM, and so aquatic CAM is likely to be an ancient syndrome.

Aquatic CAM has subsequently been found in numerous freshwater angiosperms (Keeley 1998b) including *Crassula helmsii, Littorella uniflora, Sagittaria subulata* and *Vallisneria americana*; the number of known aquatic CAM plants is increasing (Yin *et al.* 2017) and has been found in about 9% of species that have been tested, comparable to the percentage estimated for terrestrial plants (Silvera *et al.* 2010). Figure 5.9 shows a gradation in the maximum reported statistically significant change in titratable acidity for 33 aquatic species plus 36 species of *Isoetes*. The maximal recorded diel acidity change was 290 µequiv H^+ g^{-1} fresh weight for *Isoetes howellii* and 169 µequiv H^+ g^{-1} fresh weight for *Crassula peduncularis*. However, many species have much smaller diel changes in titratable acidity that are unlikely to be sufficiently large to make a major contribution to net carbon gain. In species with negative values in Figure 5.9, titratable acidity is greater at the end of the day suggesting that some of these acidity changes are linked to other functions than CAM. The primary acid that is accumulated is malic acid, and on average this accounts for about 92% of the acidity, but citric acid can also be accumulated in lower amounts (Pedersen *et al.* 2011).

Species with substantial changes in diel acidity are likely to be performing CAM. CAM-based carbon fixation at night can trap respiratory CO_2, stimulate photosynthesis during the day when the malate is decarboxylated and potentially extend the period of net carbon uptake (Klavsen and Maberly 2010). Pedersen *et al.* (2011) showed for *I. australis* at rate-limiting air-equilibrium concentrations of CO_2 that rates of oxygen evolution at the beginning of the day, when malate is available to be decarboxylated, were about twice that at the end of the day when malate supplies had been exhausted. Sensitivity to elevated concentrations of oxygen was also reduced at the start of the day consistent with malate decarboxylation producing CO_2. The contribution of CAM to the carbon economy of a plant will depend in part on the amount of malate that can be stored. In *Crassula helmsii*, CAM could contribute as much as 42% of net carbon fixation, depending on submersion depth and season, and the uptake capacity at night could be sufficient to trap all the respired CO_2 (Klavsen and Maberly 2009). The efficiency of converting decarboxylated malic acid into organic carbon is as high as 98% in *L. uniflora* and *I. lacustris*, potentially aided by gas storage in the extensive lacunae (Madsen 1987b).

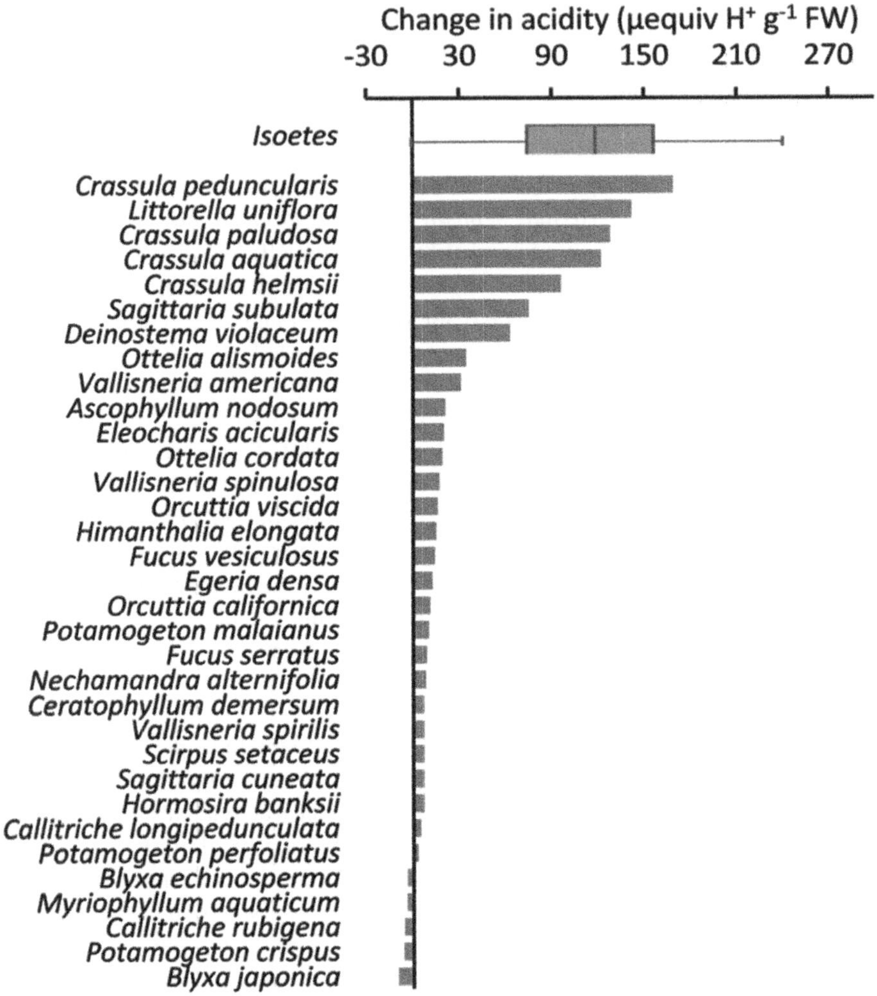

Figure 5.9. *Gradation in statistically significant diel change in titratable acidity (dawn minus dusk) in aquatic photoautotrophs. The box and whisker plot for 36* Isoetes *species shows the median (central line) interquartile range (box) and the error bars are the smallest and largest value up to 1.5 times the interquartile range. The horizontal bars of 33 other species correspond to green-lineage/embryophytes (green) and phaeophytes (orange). References: (Keeley 1998b; Pedersen* et al. *2011; Zhang* et al. *2014 ; Yin* et al. *2017 ; Huang* et al. *2020b). For a color version of this figure, see www.iste.co.uk/maberly/photosynthesis.zip*

Like some forms of aquatic C_4 plants, aquatic CAM is phenotypically variable in some species. Thus, CAM is present in leaves of the amphibious plants *Isoetes howellii* and *Littorella uniflora* when grown in water but not when grown in air (Keeley and Busch 1984; Robe and Griffiths 2000), the opposite to a water-conserving measure. However, CAM was not abolished in *I. sinensis* grown in air (Yang and Liu 2015; Yin *et al.* 2017). Low light and high concentrations of CO_2 also down-regulate CAM in *L. uniflora* (Madsen 1987a). In *O. alismoides,* CAM is also facultative and, in contrast to C_4 photosynthesis in this species, is absent when grown at high concentrations of CO_2 (Zhang *et al.* 2014). In contrast, there was no evidence for diel acidity variation in *O. acuminata* or in juvenile leaves of *O. alismoides*, whatever the conditions. These results suggested both a form of C_4 metabolism and CAM in *O. alismoides*.

5.5. Conclusions and perspectives

Most terrestrial plants lack a CCM while some possess either CAM (about 6% of angiosperms) or C_4 photosynthesis (about 2.5% of angiosperms). Sage discussed possible reasons why the operation of CAM and C_4 in the same leaf might be incompatible (Sage 2002). However, there are a small and growing number of terrestrial plants with both C_4 and CAM in the Portulaceae such as *Portulaca grandiflora* (Guralnick *et al.* 2002) that do possess both pathways, although not in the same cells, and more recently possibly in the stems of *Trianthema portulacastrum* in the Aizoaceae (Winter *et al.* 2020). In freshwater plants, species that possess a well-developed CAM (Iversen *et al.* 2019), such as *Isoetes* spp., *Littorella uniflora, Crassula* spp., *Sagittaria subulata* and *Deinostema violaceum* (Figure 5.9), do not have the ability to use HCO_3^-. The absence of HCO_3^- use could, at least partly, relate to the ecology of these plants since many species with CAM grow at sites with low concentrations of HCO_3^-. In contrast, all known species with an ability to operate C_4 photosynthesis such as *H. verticillata, E. densa, O. acuminata, O. alismoides* and *O. cordata*, also have the ability to use HCO_3^- (Iversen *et al.* 2019; Huang *et al.* 2020a). There is a possible physiological link between HCO_3^- uptake and C_4 photosynthesis since PEPC fixes HCO_3^-, although CO_2 entering by diffusion would be rapidly converted to HCO_3^- by any internal carbonic anhydrase. There is also a possible ecological explanation, since C_4 photosynthesis and HCO_3^- use are both most beneficial at productive sites where HCO_3^- concentrations tend to be relatively high and where CO_2 concentrations can be low and oxygen concentrations can be high. Plants growing at very high concentrations of CO_2 tend to lack the ability to use HCO_3^- (Maberly *et al.* 2015). *O. alismoides*, and possibly *O. cordata*, are unusual in possessing three CCMs: uptake of HCO_3^-, C_4 and CAM (Zhang *et al.* 2014; Huang *et al.* 2020a, 2020b). This raises important questions about how these different carbon concentrating processes are co-ordinated. With regard to CAM, aquatic plants are not restricted by water loss and, unlike land plants, lack the controlling effect of stomata on CO_2 uptake. Continuous PEPC activity, along with regulation of the fate of the C_4 compound produced, stored as malate in the vacuole during the night and exported and decarboxylated around Rubisco during the day, could permit both CAM and C_4 to operate, allowing 24-hour gross carbon fixation. CAM is only active at low CO_2 in *O. alismoides* and in *O. cordata* though at a lesser extent, which suggests that this is the most expensive, least beneficial of the CCMs, perhaps because of a limited capacity to store malic acid in these species.

All known freshwater plants with C_4 photosynthesis belong to the monocots, in contrast to terrestrial C_4 plants where both monocots and dicots have C_4 representatives with more dicot than monocot lineages (Sage *et al.* 2011). In fresh waters, most C_4 plants belong to the family Hydrocharitaceae within the largely aquatic order the Alismatales (Ross *et al.* 2016). The Hydrocharitaceae appeared between 55 and 73 Ma (Chen *et al.* 2012) and currently comprises about 120 species. *Stratiotes* is believed to be the first diverging lineage and it would be interesting to study the physiology and genetics of this species. Future research could increase the number of aquatic species studied since only a fraction has been investigated for possessing a biochemical CCM. Detailed mechanistic studies, which are beginning to be carried out on marine diatoms, should be applied to a wider range of aquatic photoautotrophs.

The need to produce more food for a rapidly increasing human population has attracted interest in the possibility of introducing C_4 biochemistry into important C_3 crops, such as rice, in order to increase productivity and efficiency of nitrogen and water use (von Caemmerer *et al.* 2012). However, a single-cell C_4 system might be easier to engineer as it would not require anatomical changes (Burnell 2011). Bowes *et al.* (2007) argued that knowledge of the single-cell C_4 system in *H. verticillata* will help to develop successful strategies to design an effective C_4 rice, and especially the operation of metabolite transporters in the chloroplast membranes and their permeability to CO_2. Von Caemmerer *et al.* (2014) points out that the inducible system in *H. verticillata* would also be beneficial as the additional energy costs of C_4 would be absent when photorespiration was low.

5.6. Acknowledgments

SCM was supported by a visiting scholarship funded by Aix-Marseille Université. BG thanks CNRS and Aix-Marseille Université for supporting her research.

5.7. References

Akhani, H., Barroca, J., Koteeva, N., Voznesenskaya, E., Franceschi, V., Edwards, G., Ghaffari, S.M., Ziegler, H. (2005). *Bienertia sinuspersici* (Chenopodiaceae): A new species from southwest Asia and discovery of a third terrestrial C4 plant without Kranz anatomy. *Systematic Botany*, 30(2), 290–301.

Akhani, H., Chatrenoor, T., Dehghani, M., Khoshravesh, R., Mahdavi, P., Matinzadeh, Z. (2012). A new species of *Bienertia* (Chenopodiaceae) from Iranian salt deserts: A third species of the genus and discovery of a fourth terrestrial C-4 plant without Kranz anatomy. *Plant Biosystems*, 146(3), 550–559.

Armbrust, E.V. (2009). The life of diatoms in the world's oceans. *Nature*, 459(7244), 185–192.

Aubry, S., Brown, N.J., Hibberd, J.M. (2011). The role of proteins in C3 plants prior to their recruitment into the C4 pathway. *Journal of Experimental Botany*, 62(9), 3049–3059.

Badger, M.R. and Bek, E.J. (2008). Multiple Rubisco forms in proteobacteria: Their functional significance in relation to CO(2) acquisition by the CBB cycle. *Journal of Experimental Botany*, 59(7), 1525–1541.

Bauwe, H. and Chollet, R. (1986). Kinetic-properties of phosphoenolpyruvate carboxylase from C3, C4, and C3-C4 intermediate species of *Flaveria* (Asteraceae). *Plant Physiology*, 82(3), 695–699.

Beardall, J. (1989). Photosynthesis and photorespiration in marine-phytoplankton. *Aquatic Botany*, 34(1–3), 105–130.

Beardall, J. and Raven, J.A. (2020). Structural and biochemical features of carbon acquisition in algae. In *Photosynthesis in Algae: Biochemical and Physiological Mechanisms*, Larkum, A.W.D., Grossman, A.R., Raven, J.A. (eds). Springer International Publishing, Cham.

Beer, S. and Israel, A. (1986). Photosynthesis of Ulva sp. III. O2 Effects, carboxylase activities, and the CO_2 incorporation pattern. *Plant Physiology*, 81(3), 937–938.

Beer, S., Eshel, A., Waisel, Y. (1980a). Carbon metabolism in seagrasses 3. Activities of carbon-fixing enzymes in relation to internal salt concentrations. *Journal of Experimental Botany*, 31(123), 1027–1033.

Beer, S., Shomerilan, A., Waisel, Y. (1980b). Carbon metabolism in seagrasses 2. Patterns of photosynthetic CO_2 incorporation. *Journal of Experimental Botany*, 31(123), 1019–1026.

Berry, J.A. and Farquhar, G.D. (1978). The CO_2 concentrating function of C4 photosynthesis: A biochemical model. In *The Proceedings of the Fourth International Congress on Photosynthesis*, Hall, D., Coombs, J., Goodwin, T. (eds). Biochemical Society of London, London.

Black, C.C. and Osmond, C.B. (2003). Crassulacean acid metabolism photosynthesis: "Working the night shift". *Photosynthesis Research*, 76(1–3), 329–341.

Blätke, M.-A. and Bräutigam, A. (2019). Evolution of C4 photosynthesis predicted by constraint-based modelling. *Elife*, 8, e49305. DOI: 10.7554/eLife.49305.

Borum, J., Pedersen, O., Kotula, L., Fraser, M.W., Statton, J., Colmer, T.D., Kendrick, G.A. (2016). Photosynthetic response to globally increasing CO_2 of co-occurring temperate seagrass species. *Plant Cell and Environment*, 39(6), 1240–1250.

Bowes, G. (2011). Single-cell C_4 photosynthesis in aquatic plants. In *C4 Photosynthesis and Related CO_2 Concentrating Mechanisms*, Raghavendra, A.S. and Sage, R.F. (eds). Springer, Dordrecht.

Bowes, G., Rao, S.K., Estavillo, G.M., Reiskind, J.B. (2002). C-4 mechanisms in aquatic angiosperms: Comparisons with terrestrial C-4 systems. *Functional Plant Biology*, 29(2–3), 379–392.

Bowes, G., Rao, S.K., Reiskind, J.B., Estavillo, G.M., Rao, V.S. (2007). Hydrilla: Retrofitting a C3 leaf with a single cell C4 NADP-ME system. In *Charting New Pathways to C4 Rice*, Sheehy, J.E., Mitchell, D.C., Hardy, B. (eds). IRRI, The Philippinnes.

Bräutigam, A., Schliesky, S., Kuelahoglu, C., Osborne, C.P., Weber, A.P.M. (2014). Towards an integrative model of C-4 photosynthetic subtypes: Insights from comparative transcriptome analysis of NAD-ME, NADP-ME, and PEP-CK C-4 species. *Journal of Experimental Botany*, 65(13), 3579–3593.

Burnell, J.N. (2011). Hurdles to engineering greater photosynthetic rates in crop plants: C4 rice. In *C4 Photosynthesis and Related CO_2 Concentrating Mechanisms*, Raghavendra, A.S. and Sage, R.F. (eds). Springer, Dordrecht.

Casati, P., Lara, M.V., Andreo, C.S. (2000). Induction of a C-4-like mechanism of CO_2 fixation in *Egeria densa*, a submersed aquatic species. *Plant Physiology*, 123(4), 1611–1621.

Chen, L.Y., Chen, J.M., Gituru, R.W., Wang, Q.F. (2012). Generic phylogeny, historical biogeography and character evolution of the cosmopolitan aquatic plant family Hydrocharitaceae. *BMC Evolutionary Biology*, 12(30) [Online]. Available at: https://doi.org/10.1186/1471-2148-12-30.

Christin, P.-A., Sage, T.L., Edwards, E.J., Ogburn, R.M., Khoshravesh, R., Sage, R.F. (2011). Complex evolutionary transitions and the significance of C3-C4 intermediate forms of photosynthesis in Molluginaceae. *Evolution*, 65(3), 643–660.

Clement, R., Dimnet, L., Maberly, S.C., Gontero, B. (2016). The nature of the CO_2 – Concentrating mechanisms in a marine diatom, *Thalassiosira pseudonana*. *New Phytologist*, 209(4), 1417–1427.

Clement, R., Jensen, E., Prioretti, L., Maberly, S.C., Gontero, B. (2017). Diversity of CO_2^- concentrating mechanisms and responses to CO_2 concentration in marine and freshwater diatoms. *Journal of Experimental Botany*, 68(14), 3925–3935.

Colman, B., Huertas, I.E., Bhatti, S., Dason, J.S. (2002). The diversity of inorganic carbon acquisition mechanisms in eukaryotic microalgae. *Functional Plant Biology*, 29(2–3), 261–270.

Cousins, A.B., Baroli, I., Badger, M.R., Ivakov, A., Lea, P.J., Leegood, R.C., Von Caemmerer, S. (2007). The role of phosphoenolpyruvate carboxylase during C4 photosynthetic isotope exchange and stomatal conductance. *Plant Physiology*, 145(3), 1006–1017.

Descolas-Gros, C. and Oriol, L. (1992). Variations in carboxylase activity in marine phytoplankton cultures. β-carboxylation in carbon flux studies. *Marine Ecology Progress Series*, 85(1/2), 163–169.

Doncaster, H.D. and Leegood, R.C. (1987). Regulation of phosphoenolpyruvate carboxylase activity in maize leaves. *Plant Physiology*, 84(1), 82–87.

Edwards, E.J. (2019). Evolutionary trajectories, accessibility and other metaphors: The case of C-4 and CAM photosynthesis. *New Phytologist*, 223(4), 1742–1755.

Edwards, E.J. and Ogburn, R.M. (2012). Angiosperm responses to a low-CO_2 world: CAM and C4 photosynthesis as parallel evolutionary trajectories. *International Journal of Plant Sciences*, 173(6), 724–733.

Edwards, G.E., Franceschi, V.R., Ku, M.S.B., Voznesenskaya, E.V., Pyankov, V.I., Andreo, C.S. (2001). Compartmentation of photosynthesis in cells and tissues of C-4 plants. *Journal of Experimental Botany*, 52(356), 577–590.

Edwards, G.E., Franceschi, V.R., Voznesenskaya, E.V. (2004). Single-cell C-4 photosynthesis versus the dual-cell (Kranz) paradigm. *Annual Review of Plant Biology*, 55, 173–196.

Edwards, E.J., Osborne, C.P., Stroemberg, C.A.E., Smith, S.A., Bond, W.J., Christin, P.-A., Cousins, A.B., Duvall, M.R., Fox, D.L., Freckleton, R.P., Ghannoum, O., Hartwell, J., Huang, Y., Janis, C.M., Keeley, J.E., Kellogg, E.A., Knapp, A.K., Leakey, A.D.B., Nelson, D.M., Saarela, J.M., Sage, R.F., Sala, O.E., Salamin, N., Still, C.J., Tipple, B., Consortium, C.G. (2010). The origins of C-4 grasslands: Integrating evolutionary and ecosystem science. *Science*, 328(5978), 587–591.

Edwards, G.E. and Voznesenskya, E.V. (2011). C4 Photosynthesis: Kranz forms and single-cell C4 in terrestrial Plants. In *C4 Photosynthesis and Related CO_2 Concentrating Mechanisms*, Raghavendra, A.S. and Sage, R.F. (eds). Springer, Dordrecht.

Erb, T.J. and Zarzycki, J. (2016). Biochemical and synthetic biology approaches to improve photosynthetic CO_2-fixation. *Current Opinion in Chemical Biology*, 34, 72–79.

Estavillo, G.M., Rao, S.K., Reiskind, J.B., Bowes, G. (2007). Characterization of the NADP malic enzyme gene family in the facultative, single-cell C-4 monocot *Hydrilla verticillata*. *Photosynthesis Research*, 94(1), 43–57.

Ewe, D., Tachibana, M., Kikutani, S., Gruber, A., Bartulos, C.R., Konert, G., Kaplan, A., Matsuda, Y., Kroth, P.G. (2018). The intracellular distribution of inorganic carbon fixing enzymes does not support the presence of a C4 pathway in the diatom *Phaeodactylum tricornutum*. *Photosynthesis Research*, 137(2), 263–280.

Falkowski, P., Scholes, R.J., Boyle, E., Canadell, J., Canfield, D., Elser, J., Gruber, N., Hibbard, K., Hogberg, P., Linder, S., Mackenzie, F.T., Moore, B., Pedersen, T., Rosenthal, Y., Seitzinger, S., Smetacek, V., Steffen, W. (2000). The global carbon cycle: A test of our knowledge of earth as a system. *Science*, 290(5490), 291–296.

Fan, X., Fang, Y., Hu, S., Wang, G. (2007). Generation and analysis of 5318 expressed sequence tags from the filamentous sporophyte of *Porphyra haitanensis* (Rhodophyta). *Journal of Phycology*, 43(6), 1287–1294.

Farmer, A.M., Maberly, S.C., Bowes, G. (1986). Activities of carboxylation enzymes in freshwater macrophytes. *Journal of Experimental Botany*, 37(183), 1568–1573.

Farquhar, G.D., Caemmerer, S.V., Berry, J.A. (1980). A biochemical-model of photosynthetic CO_2 assimilation in leaves of C3 species. *Planta*, 149(1), 78–90.

Flamholz, A.I., Prywes, N., Moran, U., David, D., Bar-On, Y.M., Oltrogge, L.M., Alves, R., Savage, D., Milo, R. (2019). Revisiting trade-offs between Rubisco kinetic parameters. *Biochemistry*, 58(31), 3365–3376.

Furbank, R.T. (2011). Evolution of the C-4 photosynthetic mechanism: Are there really three C-4 acid decarboxylation types? *Journal of Experimental Botany*, 62(9), 3103–3108.

Genkov, T., Meyer, M., Griffiths, H., Spreitzer, R.J. (2010). Functional hybrid Rubisco enzymes with plant small subunits and algal large subunits: Engineered rbcS cDNA for expression in chlamydomonas. *Journal of Biological Chemistry*, 285(26), 19833–19841.

Giordano, M., Beardall, J., Raven, J.A. (2005). CO_2 concentrating mechanisms in algae: Mechanisms, environmental modulation, and evolution. *Annual Review of Plant Biology*, 56, 99–131. DOI: 10.1146/annurev.arplant.56.032604.144052.

Griffiths, H. (1992). Carbon isotope discrimination and the integration of carbon assimilation pathways in terrestrial CAM plants. *Plant Cell and Environment*, 15(9), 1051–1062.

Gu, S., Yin, L.-Y., Wang, Q.-F. (2015). Phosphoenolpyruvate carboxylase in the stem of the submersed species *Egeria densa* may be involved in an inducible C-4-like mechanism. *Aquatic Botany*, 125, 1–8.

Guralnick, L.J., Edwards, G., Ku, M.S.B., Hockema, B., Franceschi, V.R. (2002). Photosynthetic and anatomical characteristics in the C-4-crassulacean acid metabolism-cycling plant, *Portulaca Grandiflora. Functional Plant Biology*, 29(6), 763–773.

Haimovich-Dayan, M., Garfinkel, N., Ewe, D., Marcus, Y., Gruber, A., Wagner, H., Kroth, P.G., Kaplan, A. (2013). The role of C4 metabolism in the marine diatom Phaeodactylum tricornutum. *New Phytologist*, 197(1), 177–185.

Han, S., Maberly, S.C., Gontero, B., Xing, Z., Li, W., Jiang, H., Huang, W. (2020). Structural basis for C-4 photosynthesis without Kranz anatomy in leaves of the submerged freshwater plant *Ottelia alismoides*. *Annals of Botany*, 125(6), 869–879.

Hatch, M.D. and Slack, C.R. (1966). Photosynthesis by sugar-cane leaves – A new carboxylation reaction and pathway of sugar formation. *Biochemical Journal*, 101(1), 103–111.

Hetherington, A.J., Dimichele, W.A., Lucas, S.G., Voigt, S. (2019). Tiny rhizomorphic rooting systems from the early Permian Abo Formation of New Mexico, USA. *International Journal of Plant Sciences*, 180(6), 504–512.

Hibberd, J.M. and Quick, W.P. (2002). Characteristics of C-4 photosynthesis in stems and petioles of C-3 flowering plants. *Nature*, 415(6870), 451–454.

Hillis-Colinvaux, L. (1980). Ecology and taxonomy of halimeda: Primary producer of coral reefs. In *Advances in Marine Biology*, Blaxter, J.H.S., Russell, F.S., Yonge, M. (eds). Academic Press.

Holaday, A.S. and Bowes, G. (1980). Acid metabolism and dark CO_2 fixation in a submersed aquatic macrophyte (*Hydrilla verticillata*). *Plant Physiology*, 65(2), 331–335.

Hopkinson, B.M., Dupont, C.L., Allen, A.E., Morel, F.M.M. (2011). Efficiency of the CO_2-concentrating mechanism of diatoms. *Proceedings of the National Academy of Sciences of the United States of America*, 108(10), 3830–3837.

Hsu, J.S., Powell, J., Adler, P.B. (2012). Sensitivity of mean annual primary production to precipitation. *Global Change Biology*, 18(7), 2246–2255.

Huang, W.M., Shao, H., Zhou, S.N., Zhou, Q., Fu, W.L., Zhang, T., Jiang, H.S., Li, W., Gontero, B., Maberly, S.C. (2018). Different CO_2 acclimation strategies in juvenile and mature leaves of Ottelia alismoides. *Photosynthesis Research*, 138(2), 219–232.

Huang, W.M., Han, S., Jiang, H., Gu, S., Li, W., Gontero, B., Maberly, S.C. (2020a). External alpha-carbonic anhydrase and solute carrier 4 are required for bicarbonate uptake in a freshwater angiosperm. *Journal of Experimental Botany*, 71(19), 6004–6014.

Huang, W.M., Han, S.J., Xing, Z.F., Li, W. (2020b). Responses of leaf anatomy and CO_2 concentrating mechanisms of the aquatic plant *Ottelia cordata* to variable CO_2. *Frontiers in Plant Science*, 11, 1261 [Online]. Available at: https://doi.org/10.3389/fpls.2020.01261.

Hwang, S.R. and Tabita, F.R. (1989). Cloning and expression of the chloroplast-encoded rbcL and rbcS genes from the marine diatom *Cylindrotheca* sp. strain N1. *Plant Molecular Biology*, 13(1), 69–79.

Hwang, S.R. and Tabita, F.R. (1991). Cotranscription, deduced primary structure, and expression of the chloroplast-encoded rbcL and rbcS genes of the marine diatom *Cylindrotheca* sp. strain-N1. *Journal of Biological Chemistry*, 266(10), 6271–6279.

Iniguez, C., Capo-Bauca, S., Niinemets, U., Stoll, H., Aguilo-Nicolau, P., Galmes, J. (2020). Evolutionary trends in RuBisCO kinetics and their co-evolution with CO_2 concentrating mechanisms. *Plant Journal*, 101(4), 897–918.

Iversen, L.L., Winkel, A., Baastrup-Spohr, L., Hinke, A.B., Alahuhta, J., Baattrup-Pedersen, A., Birk, S., Brodersen, P., Chambers, P.A., Ecke, F., Feldmann, T., Gebler, D., Heino, J., Jespersen, T.S., Moe, S.J., Riis, T., Sass, L., Vestergaard, O., Maberly, S.C., Sand-Jensen, K., Pedersen, O. (2019). Catchment properties and the photosynthetic trait composition of freshwater plant communities. *Science*, 366(6467), 878–881.

Juric, I., Gonzalez-Perez, V., Hibberd, J.M., Edwards, G., Burroughs, N.J. (2017). Size matters for single-cell C-4 photosynthesis in Bienertia. *Journal of Experimental Botany*, 68(2), 255–267.

Kanai, R. and Edwards, G. (1999). The biochemistry of C4 photosynthesis. In *C4 Plant Biology*, Sage, R. and Monson, R. (eds). Academic Press, San Diego.

Keeley, J.E. (1981). *Isoetes Howellii* a submerged aquatic CAM plant. *American Journal of Botany*, 68(3), 420–424.

Keeley, J.E. (1988). Photosynthesis in quillworts, or why are some submerged plants similar to cacti? *Plants Today*, 1, 127–132.

Keeley, J.E. (1998a). C-4 photosynthetic modifications in the evolutionary transition from land to water in aquatic grasses. *Oecologia*, 116(1–2), 85–97.

Keeley, J.E. (1998b). CAM photosynthesis in submerged aquatic plants. *Botanical Review*, 64(2), 121–175.

Keeley, J.E. and Busch, G. (1984). Carbon assimilation characteristics of the aquatic cam plant, *Isoetes howellii. Plant Physiology*, 76(2), 525–530.

Klavsen, S.K. and Maberly, S.C. (2009). Crassulacean acid metabolism contributes significantly to the *in situ* carbon budget in a population of the invasive aquatic macrophyte *Crassula helmsii. Freshwater Biology*, 54(1), 105–118.

Klavsen, S.K. and Maberly, S.C. (2010). Effect of light and CO_2 on inorganic carbon uptake in the invasive aquatic CAM-plant *Crassula helmsii. Functional Plant Biology*, 37(8), 737–747.

Knoll, A.H. and Nowak, M.A. (2017). The timetable of evolution. *Science Advances*, 3, e1603076. DOI: 10.1126/SCIADV.1603076.

Koch, M., Bowes, G., Ross, C., Zhang, X.-H. (2013). Climate change and ocean acidification effects on seagrasses and marine macroalgae. *Global Change Biology*, 19(1), 103–132.

Kortschak, H.P., Hartt, C.E., Burr, G.O. (1965). Carbon dioxide fixation in sugarcane leaves. *Plant Physiology*, 40(2), 209–213.

Koteyeva, N.K., Voznesenskaya, E.V., Berry, J.O., Cousins, A.B., Edwards, G.E. (2016). The unique structural and biochemical development of single cell C-4 photosynthesis along longitudinal leaf gradients in *Bienertia sinuspersici* and *Suaeda aralocaspica* (Chenopodiaceae). *Journal of Experimental Botany*, 67(9), 2587–2601.

Kustka, A.B., Milligan, A.J., Zheng, H.Y., New, A.M., Gates, C., Bidle, K.D., Reinfelder, J.R. (2014). Low CO_2 results in a rearrangement of carbon metabolism to support C-4 photosynthetic carbon assimilation in *Thalassiosira pseudonana. New Phytologist*, 204(3), 507–520.

Lagourgue, L. and Payri, C.E. (2021). Large-scale diversity reassessment, evolutionary history, and taxonomic revision of the green macroalgae family Udoteaceae (Bryopsidales, Chlorophyta). *Journal of Systematics and Evolution*. DOI: 10.1111/jse.12716.

Lara, M.V., Casati, P., Andreo, C.S. (2002). CO_2 concentrating mechanisms in *Egeria densa*, a submersed aquatic plant. *Physiologia Plantarum*, 115(4), 487–495.

Larkum, A.W.D., Davey, P.A., Kuo, J., Ralph, P.J., Raven, J.A. (2017). Carbon-concentrating mechanisms in seagrasses. *Journal of Experimental Botany*, 68(14), 3773–3784.

Levitan, O., Dinamarca, J., Hochman, G., Falkowski, P.G. (2014). Diatoms: A fossil fuel of the future. *Trends in Biotechnology*, 32(3), 117–124.

Liu, D.Y., Ma, Q., Valiela, I., Anderson, D.M., Keesing, J.K., Gao, K.S., Zhen, Y., Sun, X.Y., Wang, Y.J. (2020). Role of C_4 carbon fixation in *Ulva prolifera*, the macroalga responsible for the world's largest green tides. *Communications Biology*, 3, 494 [Online]. Available at: https://doi.org/10.1038/s42003-020-01225-4.

Maberly, S.C. (1990). Exogenous sources of inorganic carbon for photosynthesis by marine macroalgae. *Journal of Phycology*, 26(3), 439–449.

Maberly, S.C. and Gontero, B. (2017). Ecological imperatives for aquatic CO_2-concentrating mechanisms. *Journal of Experimental Botany*, 68(14), 3797–3814.

Maberly, S.C. and Gontero, B. (2018). Trade-offs and synergies in the structural and functional characteristics of leaves photosynthesizing in aquatic environments. In *The Leaf: A Platform for Performing Photosynthesis, Advances in Photosynthesis and Respiration* 44, Adams III, W.W. and Terashima, I. (eds). Springer International Publishing AG [Online]. Available at: https://doi.org/2010.1007/2978-2013-2319-93594-93592_93511.

Maberly, S.C., Raven, J.A., Johnston, A.M. (1992). Discirmination between C_{12} AND C_{13} by marine plants. *Oecologia*, 91(4), 481–492.

Maberly, S.C., Ball, L.A., Raven, J.A., Sültemeyer, D. (2009). Inorganic carbon acquisition by chrysophytes. *Journal of Phycology*, 45(5), 1052–1061.

Maberly, S.C., Berthelot, S.A., Stott, A.W., Gontero, B. (2015). Adaptation by macrophytes to inorganic carbon down a river with naturally variable concentrations of CO_2. *Journal of Plant Physiology*, 172, 120–127.

Madsen, T.V. (1987a). The effect of different growth-conditions on dark and light carbon assimilation in *Littorella uniflora*. *Physiologia Plantarum*, 70(2), 183–188.

Madsen, T.V. (1987b). Interactions between internal and external CO_2 pools in the photosynthesis of the aquatic CAM plants *Littorella uniflora* (l) Aschers and *Isoetes lacustris* L. *New Phytologist*, 106(1), 35–50.

Magnin, N.C., Cooley, B.A., Reiskind, J.B., Bowes, G. (1997). Regulation and localization of key enzymes during the induction of Kranz-less, C_4-type photosynthesis in *Hydrilla verticillata*. *Plant Physiology*, 115(4), 1681–1689.

Matsuda, Y., Hopkinson, B.M., Nakajima, K., Dupont, C.L., Tsuji, Y. (2017). Mechanisms of carbon dioxide acquisition and CO_2 sensing in marine diatoms: A gateway to carbon metabolism. *Philosophical Transactions of the Royal Society B – Biological Sciences*, 372(1728).

Maurino, V.G., Drincovich, M.F., Andreo, C.S. (1996). NADP-malic enzyme isoforms in maize leaves. *Biochemistry and Molecular Biology International*, 38(2), 239–250.

McGinn, P.J. and Morel, F.M.M. (2008). Expression and inhibition of the carboxylating and decarboxylating enzymes in the photosynthetic C-4 pathway of marine diatoms. *Plant Physiology*, 146(1), 300–309.

Mekhalfi, M., Avilan, L., Lebrun, R., Botebol, H., Gontero, B. (2012). Consequences of the presence of 24-epibrassinolide, on cultures of a diatom, Asterionella formosa. *Biochimie*, 94(5), 1213–1220.

Meyer, M.T. and Griffiths, H. (2013). Origins and diversity of eukaryotic CO_2 concentrating mechanisms: Lessons for the future. *Journal of Experimental Botany*, 64(3), 769–786.

Meyer, M.T., Genkov, T., Skepper, J.N., Jouhet, J., Mitchell, M.C., Spreitzer, R.J., Griffiths, H. (2012). Rubisco small-subunit alpha-helices control pyrenoid formation in Chlamydomonas. *Proceedings of the National Academy of Sciences of the United States of America*, 109(47), 19474–19479.

Minge, M.A., Shalchian-Tabrizi, K., Torresen, O.K., Takishita, K., Probert, I., Inagaki, Y., Klaveness, D., Jakobsen, K.S. (2010). A phylogenetic mosaic plastid proteome and unusual plastid-targeting signals in the green-colored dinoflagellate *Lepidodinium chlorophorum*. *BMC Evolutionary Biology*, 10.

Moody, N.R., Christin, P.A., Reid, J.D. (2020). Kinetic modifications of C4 PEPC are qualitatively convergent, but larger in *Panicum* than in *Flaveria*. *Frontiers in Plant Science*, 11, 1014 [Online]. Available at: https://doi.org/10.3389/fpls.2020.01014.

Muñoz-Clares, R.A., González-Segura, L., Juárez-Díaz, J.A., Mújica-Jiménez, C. (2020). Structural and biochemical evidence of the glucose 6-phosphate-allosteric site of maize C4-phosphoenolpyruvate carboxylase: Its importance in the overall enzyme kinetics. *Biochemical Journal*, 477(11), 2095–2114.

Offermann, S., Okita, T.W., Edwards, G.E. (2011). Resolving the compartmentation and function of C_4 photosynthesis in the single-cell C_4 species *Bienertia sinuspersici*. *Plant Physiology*, 155(4), 1612–1628.

Pedersen, O., Rich, S.M., Pulido, C., Cawthray, G.R., Colmer, T.D. (2011). Crassulacean acid metabolism enhances underwater photosynthesis and diminishes photorespiration in the aquatic plant *Isoetes australis*. *New Phytologist*, 190(2), 332–339.

Prioretti, L., Lebrun, R., Gontero, B., Giordano, M. (2016). Redox regulation of ATP sulfurylase in microalgae. *Biochemical and Biophysical Research Communications*, 478(4), 1555–1562.

Prioretti, L., Avilan, L., Carrière, F., Montané, M., Field, B., Gregori, G., Menand, B., Gontero, B. (2017). The inhibition of TOR in the model diatom *Phaeodactylum Tricornutum* promotes a get-fat growth regime. *Algal Research*, 26, 265–274.

Rao, S.K., Magnin, N.C., Reiskind, J.B., Bowes, G. (2002). Photosynthetic and other phosphoenolpyruvate carboxylase isoforms in the single-cell, facultative C_4 system of *Hydrilla verticillata*. *Plant Physiology*, 130(2), 876–886.

Rao, S.K., Reiskind, J.B., Bowes, G. (2008). Kinetic analyses of recombinant isoforms of phosphoenolpyruvate carboxylase from *Hydrilla verticillata* leaves and the impact of substituting a C_4-signature serine. *Plant Science*, 174(4), 475–483.

Raven, J.A. (1997). Inorganic carbon acquisition by marine autotrophs. In *Advances in Botanical Research, Vol 27*, Callow, J.A. (eds). Academic Press, London.

Raven, J.A. (2009). Contributions of anoxygenic and oxygenic phototrophy and chemolithotrophy to carbon and oxygen fluxes in aquatic environments. *Aquatic Microbial Ecology*, 56(2–3), 177–192.

Raven, J.A. (2010). Inorganic carbon acquisition by eukaryotic algae: Four current questions. *Photosynthesis Research*, 106(1–2), 123–134.

Raven, J.A. and Giordano, M. (2017). Acquisition and metabolism of carbon in the Ochrophyta other than diatoms. *Philosophical Transactions of the Royal Society B – Biological Sciences*, 372, 20160400 [Online]. Available at: https://doi.org/10.1098/rstb.2016.0400.

Raven, J.A. and Hurd, C.L. (2012). Ecophysiology of photosynthesis in macroalgae. *Photosynthesis Research*, 113(1–3), 105–125.

Raven, J.A., Johnston, A.M., Kübler, J.E., Korb, R., McInroy, S.G., Handley, L.L., Scrimgeour, C.M., Walker, D.I., Beardall, J., Vanderklift, M., Fredriksen, S., Dunton, K.H. (2002). Mechanistic interpretation of carbon isotope discrimination by marine macroalgae and seagrasses. *Functional Plant Biology*, 29(3), 355–378.

Raven, J.A., Giordano, M., Beardall, J., Maberly, S.C. (2011). Algal and aquatic plant carbon concentrating mechanisms in relation to environmental change. *Photosynthesis Research*, 109(1–3), 281–296.

Raven, J.A., Giordano, M., Beardall, J., Maberly, S.C. (2012). Algal evolution in relation to atmospheric CO_2: Carboxylases, carbon-concentrating mechanisms and carbon oxidation cycles. *Philosophical Transactions of the Royal Society B – Biological Sciences*, 367(1588), 493–507.

Reinfelder, J.R., Kraepiel, A.M.L., Morel, F.M.M. (2000). Unicellular C-4 photosynthesis in a marine diatom. *Nature*, 407(6807), 996–999.

Reinfelder, J.R., Milligan, A.J., Morel, F.M.M. (2004). The role of the C-4 pathway in carbon accumulation and fixation in a marine diatom. *Plant Physiology*, 135(4), 2106–2111.

Reiskind, J.B. and Bowes, G. (1991). The role of phosphoenolpyruvate carboxykinase in a marine macroalga with C4-like photosynthetic characteristics. *Proceedings of the National Academy of Sciences of the United States of America*, 88(7), 2883–2887.

Reiskind, J.B., Seamon, P.T., Bowes, G. (1988). Alternative methods of photosynthetic carbon assimilation in marine macroalgae. *Plant Physiology*, 87(3), 686–692.

Reiskind, J.B., Berg, R.H., Salvucci, M.E., Bowes, G. (1989). Immunogold localization of primary carboxylases in leaves of aquatic and a C3-C4 intermediate species. *Plant Science*, 61(1), 43–52.

Reiskind, J.B., Madsen, T.V., Vanginkel, L.C., Bowes, G. (1997). Evidence that inducible C_4 type photosynthesis is a chloroplastic CO_2 concentrating mechanism in *Hydrilla*, a submersed monocot. *Plant Cell and Environment*, 20(2), 211–220.

Robe, W.E. and Griffiths, H. (2000). Physiological and photosynthetic plasticity in the amphibious, freshwater plant, *Littorella uniflora*, during the transition from aquatic to dry terrestrial environments. *Plant Cell and Environment*, 23(10), 1041–1054.

Roberts, K., Granum, E., Leegood, R.C., Raven, J.A. (2007). C-3 and C-4 pathways of photosynthetic carbon assimilation in marine diatoms are under genetic, not environmental, control. *Plant Physiology*, 145(1), 230–235.

Ross, T.G., Barrett, C.F., Soto Gomez, M., Lam, V.K.Y., Henriquez, C.L., Les, D.H., Davis, J.I., Cuenca, A., Petersen, G., Seberg, O., Thadeo, M., Givnish, T.J., Conran, J., Stevenson, D.W., Graham, S.W. (2016). Plastid phylogenomics and molecular evolution of Alismatales. *Cladistics*, 32(2), 160–178.

Sage, R.F. (2002). Are crassulacean acid metabolism and C-4 photosynthesis incompatible? *Functional Plant Biology*, 29(6), 775–785.

Sage, R.F., Christin, P.-A., Edwards, E.J. (2011). The C-4 plant lineages of planet Earth. *Journal of Experimental Botany*, 62(9), 3155–3169.

Sage, R.F., Sage, T.L., Kocacinar, F. (2012). Photorespiration and the evolution of C_4 photosynthesis. *Annual Review of Plant Biology*, 63, 19–47.

Sage, R.F., Monson, R.K., Ehleringer, J.R., Adachi, S., Pearcy, R.W. (2018). Some like it hot: The physiological ecology of C-4 plant evolution. *Oecologia*, 187(4), 941–966.

Salvucci, M.E. and Bowes, G. (1981). Induction of reduced photorespiratory activity in submersed and amphibious aquatic macrophytes. *Plant Physiology*, 67(2), 335–340.

Shao, Z.R., Wang, W.L., Zhang, P.Y., Yao, J.T., Wang, F.H., Duan, D.L. (2019). Genome-wide identification of genes involved in carbon fixation in *Saccharina japonica* and responses of putative C-4-related genes to bicarbonate concentration and light intensity. *Plant Physiology and Biochemistry*, 137, 75–83.

Shih, P.M., Occhialini, A., Cameron, J.C., Andralojc, P.J., Parry, M.A.J., Kerfeld, C.A. (2016). Biochemical characterization of predicted Precambrian RuBisCO. *Nature Communications*, 7, 10382 [Online]. Available at: https://doi.org/10.1038/ncomms10382.

Silvera, K., Neubig, K.M., Whitten, W.M., Williams, N.H., Winter, K., Cushman, J.C. (2010). Evolution along the crassulacean acid metabolism continuum. *Functional Plant Biology*, 37(11), 995–1010.

Spreitzer, R.J. (2003). Role of the small subunit in ribulose-1,5-bisphosphate carboxylase/oxygenase. *Archives of Biochemistry and Biophysics*, 414(2), 141–149.

Stepien, C.C. (2015). Impacts of geography, taxonomy and functional group on inorganic carbon use patterns in marine macrophytes. *Journal of Ecology*, 103(6), 1372–1383.

Still, C.J., Berry, J.A., Collatz, G.J., Defries, R.S. (2003). Global distribution of C-3 and C-4 vegetation: Carbon cycle implications. *Global Biogeochemical Cycles*, 17(1).

Tabita, F.R., Hanson, T.E., Li, H., Satagopan, S., Singh, J., Chan, S. (2007). Function, structure, and evolution of the RubisCO-like proteins and their RubisCO homologs. *Microbiology and Molecular Biology Reviews*, 71(4), 576–599.

Tabita, F.R., Satagopan, S., Hanson, T.E., Kreel, N.E., Scott, S.S. (2008). Distinct form I, II, III, and IV Rubisco proteins from the three kingdoms of life provide clues about Rubisco evolution and structure/function relationships. *Journal of Experimental Botany*, 59(7), 1515–1524.

Tanaka, R., Kikutani, S., Mahardika, A., Matsuda, Y. (2014). Localization of enzymes relating to C-4 organic acid metabolisms in the marine diatom, *Thalassiosira pseudonana*. *Photosynthesis Research*, 121(2–3), 251–263.

Tortell, P.D. (2000). Evolutionary and ecological perspectives on carbon acquisition in phytoplankton. *Limnology and Oceanography*, 45(3), 744–750.

Ueno, O. (1998). Induction of Kranz anatomy and C(4)-like biochemical characteristics in a submerged amphibious plant by abscisic acid. *Plant Cell*, 10(4), 571–583.

Valiela, I., Liu, D., Lloret, J., Chenoweth, K., Hanacek, D. (2018). Stable isotopic evidence of nitrogen sources and C4 metabolism driving the world's largest macroalgal green tides in the Yellow Sea. *Scientific Reports*, 8(1), 17437.

Von Caemmerer, S., Quick, W.P., Furbank, R.T. (2012). The development of C4 rice: Current progress and future challenges. *Science*, 336(6089), 1671–1672.

Von Caemmerer, S., Edwards, G.E., Koteyeva, N., Cousins, A.B. (2014). Single cell C-4 photosynthesis in aquatic and terrestrial plants: A gas exchange perspective. *Aquatic Botany*, 118, 71–80.

Voznesenskaya, E.V., Franceschi, V.R., Kiirats, O., Freitag, H., Edwards, G.E. (2001). Kranz anatomy is not essential for terrestrial C-4 plant photosynthesis. *Nature*, 414(6863), 543–546.

Winter, K., Garcia, M., Virgo, A., Ceballos, J., Holtum, J.A.M. (2020). Does the C-4 plant *Trianthema portulacastrum* (Aizoaceae) exhibit weakly expressed crassulacean acid metabolism (CAM)? *Functional Plant Biology*, 48(7), 655–665. DOI: 10.1071/FP20247.

Xu, J., Fan, X., Zhang, X., Xu, D., Mou, S., Cao, S., Zheng, Z., Miao, J., Ye, N. (2012). Evidence of coexistence of C_3 and C_4 photosynthetic pathways in a green-tide-forming alga, *Ulva prolifera*. *PLOS ONE*, 7(5), e37438 [Online]. Available at: https://doi.org/10.1371/journal.pone.0037438.

Yang, T. and Liu, X. (2015). Comparing photosynthetic characteristics of *Isoetes sinensis* Palmer under submerged and terrestrial conditions. *Scientific Reports*, 5, 17783 [Online]. Available at: https://doi.org/10.1038/srep17783.

Yin, L., Li, W., Madsen, T.V., Maberly, S.C., Bowes, G. (2017). Photosynthetic inorganic carbon acquisition in 30 freshwater macrophytes. *Aquatic Botany*, 140, 48–54.

Young, J.N., Heureux, A.M.C., Sharwood, R.E., Rickaby, R.E.M., Morel, F.M.M., Whitney, S.M. (2016). Large variation in the Rubisco kinetics of diatoms reveals diversity among their carbon-concentrating mechanisms. *Journal of Experimental Botany*, 67(11), 3445–3456.

Zhang, Y., Yin, L., Jiang, H.-S., Li, W., Gontero, B., Maberly, S.C. (2014). Biochemical and biophysical CO_2 concentrating mechanisms in two species of freshwater macrophyte within the genus *Ottelia* (Hydrocharitaceae). *Photosynthesis Research*, 121, 285–297.

Zhang, B., Xie, X., Liu, X., He, L., Sun, Y., Wang, G. (2020). The carbonate concentration mechanism of *Pyropia yezoensis* (Rhodophyta): Evidence from transcriptomics and biochemical data. *BMC Plant Biology*, 20, 424 [Online]. Available at: https://doi.org/10.1186/s12870-020-02629-4.

Zhou, L., Gao, S., Huan, L., Wu, S., Wang, G., Gu, W. (2020). Enzyme activities suggest that the NAD-ME C4 type CCM exist in *Ulva* sp. *Algal Research*, 47, 101809.

6

Carbonic Anhydrase

Yusuke MATSUDA, Hermanus NAWALY, Kohei YONEDA
Department of Bioscience, Kwansei Gakuin University, Sanda, Hyogo, Japan

6.1. Overview

Carbonic anhydrase (CA) is a metalloenzyme, often involving zinc, which catalyzes the reversible reaction of CO_2 hydration. Nine different subclasses of CA are currently known, from α to ι, based on their primary sequences. Mammalian α-CA, the first described CA, maintains pH homeostasis. α-CAs are widespread in photoautotrophs, and some are vital factors for photosynthesis and CO_2-concentrating mechanisms (CCMs). For example, an α-CA, CAH3, in the green alga, *Chlamydomonas reinhardtii*, resides in the thylakoid lumen within the pyrenoid structure (see Chapter 7) in the chloroplast and facilitates an ample CO_2 flux towards the CO_2 fixing enzyme, ribulose-1,5-carboxylase/oxygenase, Rubisco. Photoautotrophs, in general, also possess β-subclass CAs within the chloroplast, sometimes in the cytosol and the mitochondrion. Some plant β-CAs are vital for photosynthesis and the operation of the CCM. Most γ-CAs in plants and diatoms are present in mitochondria, although their CA activity is not yet confirmed. On the other hand, γ-CAs in cyanobacteria and some chemoautotrophs are present in carboxysomes (see Chapter 7) where they dehydrate accumulated HCO_3^- producing CO_2 close to Rubisco. Besides these classical α-, β-, and γ-CAs, numerous new subclasses of CA have been discovered in the secondary endosymbiotic photoautotrophs, stramenopiles and haptophytes: that is, δ-, ζ-, θ- and ι-subclasses. δ-CA has an active site that is very similar to that of α-CA, while ζ- and θ-CAs have a very similar active site to that of β-CA. Of these, ζ- and ι-subclasses use cadmium and manganese, respectively, for their active center, indicating a wide competency of CA active sites for metals. A θ-CA in a marine diatom was recently localized within the thylakoid lumen, and this discovery strongly suggests that CAs locate in the interior and exterior of the thylakoid tube within the pyrenoid structure that commonly plays a vital role in the operation of the algal CCM and photosynthesis. Recent studies also showed the localization of numerous CAs, including the newest subclass, ι-CA, within the four-layered chloroplast envelope system of the secondary endosymbiotic phototrophs. The prevalence of these envelope-type CAs in the CCM indicates their importance as a part of a leakage barrier mechanism of CO_2 from the secondary chloroplast.

6.2. Introduction

Carbonic anhydrase (CA) is a ubiquitous metalloenzyme that catalyzes the slowest key step within the carbonate equilibrium system with extremely high catalytic efficiency, k_{cat}, the reversible reaction of the hydration of carbon dioxide (CO_2) to carbonic acid (H_2CO_3) (bold reactions in equation [6.1]). As H_2CO_3 can quickly dissociate to form a proton and bicarbonate (HCO_3^-), CA essentially equilibrates CO_2 and HCO_3^- equation [6.1]). Equilibrium of dissolved inorganic carbon (DIC) species is highly dependent on pH as it releases a proton and at higher pH values, HCO_3^- further dissociates into a proton and carbonate (CO_3^{2-}):

$$\mathbf{CO_2 + H_2O \rightleftarrows H_2CO_3} \rightleftarrows HCO_3^- + H^+ \rightleftarrows CO_3^{2-} + 2H^+ \qquad [6.1]$$

CA was first discovered during physiological investigations into pH control in mammalian erythrocytes (Brinkman *et al.* 1932; Meldrum, 1932). Later, this CA was designated as an α-subclass CA because of the discovery of different, phylogenetically independent CAs. α-CA also occurs in microalgae, some of which are now known to play vital roles in photosynthesis (Funke *et al.* 1997). Three subclasses of the CA enzymes are well known, namely α-, β- and γ-subclasses. Beta and γ-CAs do not occur in vertebrates, but β-CA occurs in plants, algae, cyanobacteria and non-photosynthetic bacteria, while γ-CA occurs in archaea and photosynthetic bacteria (DiMario *et al.* 2018). These CAs do not share any sequence homology and hence result from convergent evolution (Hewett-Emmett and Tashian, 1996; Roberts *et al.* 1997; So *et al.* 2004; Lane *et al.* 2005). Since their active sites had a zinc molecule without exception, it was believed that Zn^{2+} chelation is essential for CA activities. However, investigations of marine photoautotrophic algae from the late 1990s have demonstrated that there are many other convergent evolution subclasses of CAs and that Zn is not necessarily an exclusive element for their active site structure (Xu *et al.* 2008; Jensen *et al.* 2019). To date, besides the above-mentioned three subclasses of CA, six unique subclasses, δ-, ε-, ζ-, η-, θ- and ι-, have been reported to occur in cyanobacteria (ε-CA), stramenopiles and Haptophyta (δ-, ζ-, θ- and ι), and Apicomplexa (η-CA) (Roberts *et al.* 1997; So *et al.* 2004; Xu *et al.* 2008; Del Prete *et al.* 2014; Kikutani *et al.* 2016; Jensen *et al.* 2019). Of particular interest in the occurrence of very diverse CA sequences is that most of the new subclasses of CA have been discovered from secondary symbiotic photoautotrophs, strongly suggesting the importance of this enzyme activity in the process of aquatic photosynthesis in the organisms that recently evolved during the Phanerozoic. In this review, we introduce the wide variety of CAs and summarize their key features in Table 6.1 and describe their roles in aquatic photosynthetic organisms.

6.3. Types of CA

6.3.1. *Alpha CA*

Fifteen phylogenetic isoforms of CA occur in mammals localized in almost all tissues ubiquitously, but there are no other CA gene subclasses in mammals (Alterio *et al.* 2012). The first identified CA was isolated from mammalian erythrocytes in the 1930s and was later classified as α-CA (Brinkman *et al.* 1932; Meldrum, 1932). Of the 15 isoforms, 12 have catalytic activity, while the three isoforms (hCA VIII, X and XI) lack the CA activity due to histidine ligand substitution. They lack one (hCA VIII; H94R), two (hCA X; H94R, H119Q) or all three (hCAXI; H94R, H96L, H119Q) zinc ligands from the active site (hCA I numbering) (Alterio *et al.* 2012). Active CAs, in principle, work to maintain DIC equilibrium and pH homeostasis, but their functions are diverse depending upon their location within cells and tissues. Deficiency of CA thus often causes disease (Baranauskienė and Matulis, 2019). α-CA genes also widely occur in animals, terrestrial plants, eubacteria, mammal-hosted viruses and green algae (Hewett-Emmett, 2000). A plant α-CA gene was first identified and cloned in the green microalga, *Chlamydomonas reinhardtii* (Fukuzawa *et al.* 1990). *C. reinhardtii* is now known to possess three α-CAs: CAH1 and CAH2 are located in the periplasmic space, and CAH3 is located in the thylakoid lumen (Fukuzawa *et al.* 1990; Karlsson *et al.* 1998). Of particular interest, the luminal CAH3 plays a vital role in photosynthesis under a CO_2-limiting environment as the lack of this enzyme results in a high-CO_2 requiring phenotype (Funke *et al.* 1997). Pronina and Semenenko (1984) and Raven (1997) pointed out that thylakoid luminal CA should be essential to support CO_2 formation within the chloroplast during the operation of the algal biophysical CCM (see Chapter 4). It enables algal cells to operate high DIC affinity photosynthesis, and CAH3 is one of the key components for the CCM in *C. reinhardtii* (Funke *et al.* 1997; Raven, 1997). Diatoms also contain numerous α-CAs mostly grouped close to those in Apicomplexa and Euglenozoa, a distant clade from mammals and green plants (Tachibana *et al.* 2011). In the marine diatom *Phaeodactylum tricornutum*, all α-CAs are localized within the four-layered chloroplastic membrane system (Tachibana *et al.* 2011). The function of α-CAs in terrestrial plants is largely unknown, but their particular location patterns, such as in leaf trichomes and root nodules, strongly suggest that their functions are highly organ-specific (DiMario *et al.* 2017). The active center of α-CA is composed of zinc with three chelating histidine residues and one water molecule associated with Zn (Figure 6.1; Liljas *et al.* 1972). They work as a monomer, but some mammalian and algal α-CAs are known to be dimers (Whittington *et al.* 2001; Suzuki *et al.* 2011; DiMario *et al.* 2018).

6.3.2. *Beta CA*

The first cDNA sequence data for β-CAs were obtained from terrestrial plants, spinach (Burnell *et al.* 1990) and pea (Roeske and Ogren, 1990), which encode stromal CAs. This CA subclass ubiquitously occurs in prokaryotes and eukaryotes except for mammals. It can be separated into six phylogenetic clades A to E, and newly found F (Pander *et al.* 2019). These phylogenetic groups are all composed of sequences from bacteria, archaea, algae, plants and/or fungi and thus do not represent evolutionary taxa (Pander *et al.* 2019), strongly suggesting the involvement of extensive horizontal gene transfer (HGT) events in the evolution of β-CAs. Clade A comprises β-CAs from the red microalga *Porphyridium purpureum*, fungi and mainly bacteria (Pander *et al.* 2019). Two sequences from the marine diatom *P. tricornutum*, PtCA1 and PtCA2, also belong to this clade (Harada and Matsuda, 2005). β-CAs from terrestrial plants and cyanobacteria exclusively belong to clade B (Pander *et al.* 2019). Other bacterial sequences, for instance, those from *Escherichia coli*, *Helicobacter pylori* and *Rhodobacter capsulatus,* are also grouped in this clade (Harada and Matsuda, 2005; Pander *et al.* 2019). Clade C includes β-CAs from *C. reinhardtii* (Harada and Matsuda, 2005; Pander *et al.* 2019). There are no photosynthetic organisms included in the remaining clades D to F except for that in alpha-cyanobacteria for clade E (Pander *et al.* 2019). In terrestrial plants, β-CAs are located in the cytosol, mitochondrion and chloroplast stroma (Fabre *et al.* 2007; Jiang *et al.* 2014; DiMario *et al.* 2016). In the cytosol of guard cells of *Arabidopsis thaliana*, two β-CAs (CA1 and CA4) are involved in sensing CO_2 concentration (DiMario *et al.* 2016). In *C. reinhardtii*, β-CAs are located in the mitochondrion, partly contributing to the anaplerotic pathway and in part to the CCM (Giordano *et al.* 2003). PtCA1 and PtCA2 in the diatom *P. tricornutum* are located in or near the pyrenoid (Harada and Matsuda, 2005; Tachibana *et al.* 2011), a large protein body in which most of the CO_2 fixing enzyme, Rubisco, occurs (Griffiths, 1970) (see Chapter 7). PtCA1 and PtCA2 are inducible under CO_2 limitation and play a critical role in the diatom CCM (Satoh *et al.* 2001; Harada and Matsuda, 2005). These pyrenoidal CAs are also shown to be activated by reducing power in the light by the action of chloroplastic thioredoxin (Kikutani *et al.* 2012). However, the β-CA sequence is not widespread among diatoms but is limited to particular radial centric and araphid pennate species (Nonoyama *et al.* 2019).

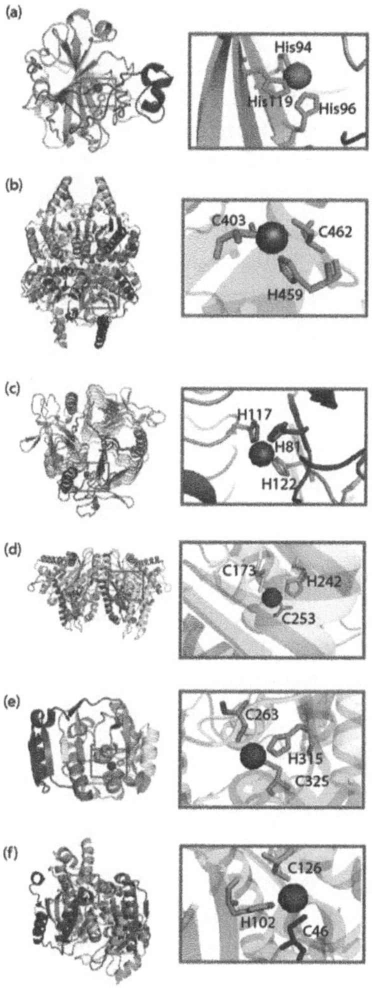

Figure 6.1. *Typical structures of α-, β-, γ-, ε-, ζ- and θ-carbonic anhydrases with their active site architecture. (a) α-CA from human, hCA1 (PDB ID: 1CA2); (b) β-CA from* P. purpureum *(PDB ID:1DDZ); (c) γ-CA from* M. thermophila *(PDB ID: 1THJ); (d) ε-CA from* H. neapolitanus *(PDB ID: 2FGY); (e) ζ-CA from* T. weissflogii *(PDB ID: 3BOC); and (f) θ-CA from* P. tricornutum *(PDB ID: 5B5Y). Red box indicates the enlarged locations of each protein structure to display their active-site architecture in the right. Sphere indicates divalent metals in the active site; for specific types of metal, see in text and Table 6.1. CA protein structures and active-site images were generated using Pymol (www.pymol.org). For a color version of this figure, see www.iste.co.uk/maberly/photosynthesis.zip*

β-CA, in principle, forms dimers, producing a unit structure of two active sites with a zinc atom chelated by two Cys residues, one His residue and a water molecule (Hewett-Emmett, 2000). This unit structure can further dimerize and tetramerize to form tetramers and octamers, but the way to assemble these higher 2^n tertiary structures seems diverse (Kimber and Pai, 2000; Rowlett, 2010). Relating to this potential diversity of the tertiary structure of CAs, the above-mentioned PtCA1 and PtCA2 from diatoms are known to form further large clumping structures, which are ensured by the function of the C-terminal amphipathic α-helix of each subunit (Kitao and Matsuda, 2009). Another case of diverse structure, β-CA in the red microalga, *P. purpureum,* has two duplicated β-CA domain sequences in one polypeptide, forming a pseudo-dimer structure with one polypeptide. Its active center uses an additional amino acid, Asp, instead of water; thus, the Zn-coordination amino acids are four in total, two His, one Cys and one Asp, and Asp is provided from the other domain amino acid when the protein is folded (Figure 6.1) (Mitsuhashi *et al.* 2000).

6.3.3. *Gamma CA*

γ-subclass CA sequences have been found in archaebacteria, eubacteria, cyanobacteria, green algae, Euglenozoa, Alveolata, stramenopiles and terrestrial plants (Smith *et al.* 1999; Tachibana *et al.* 2011). The first protein, which was purified and had its activity confirmed, was from the methanogen archaeon *Methanosarcina thermophila* (Alber and Ferry, 1994). Since the activity of γ-CAs has also been confirmed by some species of archaea, eubacteria, and cyanobacteria (Alber and Ferry, 1994; Peña *et al.* 2010; Vullo *et al.* 2017), γ-CA was initially thought to be the most ancient CA (Tripp *et al.* 2001), but recent studies with terrestrial plants do not necessarily support this hypothesis. In *A. thaliana,* multiple γ-CA ortholog proteins were located in the mitochondria (Parisi *et al.* 2004), and they were shown to be associated with complex I in the respiratory chain (Klodmann *et al.* 2010). However, there is currently no direct evidence for the CA activity of these proteins, although they were required for the assembly of complex I (Soto *et al.* 2015) and could bind CO_2/HCO_3^- (Martin *et al.* 2009). This evidence suggests an alternative possibility that γ-CA activity has evolved recently, only in particular organisms, but that the protein itself may have some other original function, which may be deeply related to mitochondrial respiration.

γ-CA from *M. thermophila* comprises a homotrimer structure with three β-sheet barrel subunits (Kisker *et al.* 1996). The active center of γ-CA resides in the cleft between each subunit and three His residues provided from two adjacent subunits chelate Zn; thus, one molecule of γ-CA has three CA active sites (Figure 6.1)

(Alber *et al.* 1999). It has been suggested that in the γ-CA from *M. thermophila*, Fe and Co can also substitute for Zn (Alber *et al.* 1999; Tripp *et al.* 2004). In contrast to the archaeal enzyme, γ-CAs that locate in the mitochondria of plants and algae constitute a heterotrimer (Maldonado *et al.* 2020; Soufari *et al.* 2020; Klusch *et al.* 2021). Interestingly, only one active site binds to Zn in a heterotrimer structure (Maldonado *et al.* 2020; Soufari *et al.* 2020; Klusch *et al.* 2021). Multiple γ-CA sequences are also found in the marine diatoms, *P. tricornutum* and *T. pseudonana*. Their subcellular locations are mostly the mitochondria except for one cytosolic factor in *T. pseudonana* (Tachibana *et al.* 2011; Samukawa *et al.* 2014). γ-CA in *P. tricornutum* is part of complex I of mitochondria, and it is predicted to form a heterotrimer similarly to that of plants (Cainzos *et al.* 2021). A *bona fide* γ-CA, CcmM, in the cyanobacterium *Thermosynechococcus elongatus* is a component of the carboxysome (see Chapter 7), the protein structure constituting the focal point of the cyanobacterial biophysical CCM. CcmM needs to be oxidized to form a disulfide bond for activation, indicating that the mis-assembled CcmM outside the carboxysome will be inactivated by a reduction in the cytosol (Peña *et al.* 2010).

6.3.4. *Delta CA*

After discovering α-, β- and γ-subclasses of CA, no new subclasses of CA were discovered among early model organisms belonging to primary endosymbionts, Opisthokonta and Plantae. In the late 1990s, Roberts *et al.* (1997) reported in secondary endosymbionts the occurrence of a new convergent evolution class of CA from the marine centric diatom, *Thalassiosira weissflogii* (Roberts *et al.* 1997). A new CA class, δ-subclass, was proposed in *Thalassiosira*, and this CA was denoted as TWCA1 (Roberts *et al.* 1997). The occurrence of δ-CAs has also been confirmed by the Haptophyta, *Emiliania huxleyi*, the dinoflagellate, *Lingulodinium polyedrum*, the diatom *Thalassiosira pseudonana* and the Pelagophyta, *Aureococcus anophagefferens* (Soto *et al.* 2006; Lapointe *et al.* 2008; McGinn and Morel, 2008). Thus, the occurrence of δ-CA so far has only been reported in the Alveolata, stramenopiles and Haptophyta. However, the sequence also exists in some Prasinophyta (Samukawa *et al.* 2014), which are believed to be involved in the evolution of the chloroplast of stramenopiles around 12–14 x 10^8 years ago (Moustafa *et al.* 2009; Dorrell *et al.* 2017). δ-CA would then be one of the green-originated genes, which are known to be abundant in the diatom nuclear genome (Armbrust *et al.* 2004) and is a widespread CCM factor among diatoms (Shen *et al.* 2017). Genome information on diatoms (Armbrust *et al.* 2004; Bowler *et al.* 2008) revealed no genes for δ-CA in *P. tricornutum* but four δ-CA sequences in

T. pseudonana (Armbrust *et al.* 2004). These four putative δ-CAs share the same cluster with other δ-CAs of Alveolata, stramenopiles and Haptophyta in the phylogenetic tree (Samukawa *et al.* 2014). Tagging of these genes with the green fluorescent protein (*gfp*) gene and following the transformation of *T. pseudonana* showed the localization of these gene products in the periplasmic space, mitochondria and the periplastidal compartment, which is the matrix between the outer two membranes (chloroplast ER: CER) and the inner two membranes (chloroplast envelope: CE) of the secondary chloroplast (Samukawa *et al.* 2014). Similarly, in the dinoflagellate, *L. polyedrum*, δ-CA was localized in the periplasmic space. These extracellular δ-CAs are supposed to assist the CO_2 formation from HCO_3^- in the periplasmic space (Lapointe *et al.* 2008; Tsuji *et al.* 2017a).

The structure of the active site of TWCA1 was partially solved by X-ray absorption analysis, and it was demonstrated that the active site of TWCA1 is strikingly similar to that of bovine α-CAII and three His residues most probably constitute the Zn ligand (Cox *et al.* 2000). The similarity between these two subclasses of CA was also supported by the finding of esterase activity and Co substitution competency for Zn in TWCA1, as these characteristics are also observed in α-CA (Lane and Morel 2000a, b; Lee *et al.* 2013).

6.3.5. *Epsilon CA*

The epsilon subclass of CA is only found in a few types of organisms. Epsilon CA, CsoS3 (later renamed as CsoSCA for carboxysome shell carbonic anhydrase), was first identified in the chemolithotrophic bacterium *Halothiobacillus neapolitanus* as a novel carboxysomal CA (So *et al.* 2004). The homologs of CsoSCA are only conserved among non-photosynthetic chemoautotrophic bacteria and alpha cyanobacteria, both of which possess α-carboxysomes (for details on α- and β- carboxysome, see the review by Kerfeld and Melnicki (2016)). The primary amino acid sequence of ε-CA is not homologous to any other CA, but the crystal structure of CsoSCA revealed a Zn coordination pattern similar to that of β-CA; that is, two cysteine and one histidine residues serve as a Zn ligand (Figure 6.1), and an aspartic acid and arginine dyad are presumably well conserved to fulfill several catalytic steps (Sawaya *et al.* 2006).

6.3.6. *Eta CA*

The eta subclass of CAs is also only found in a few types of organisms. The η-CA, PfCA1, was first described as a novel subclass of CA in the protozoan malaria pathogen *Plasmodium falciparum* (Del Prete *et al.* 2014), which belongs to the Alveolata. Since η-CA was specifically found in the genus *Plasmodium,* it is regarded as a promising drug target for malaria disease (Reungprapavut *et al.* 2004; Supuran and Capasso, 2015). The PfCA1 was initially reported as an α-CA due to the occurrence of three His residues (His299, His301, His323) for the putative Zn ligand, which is a particular feature of α-CA (Reungprapavut *et al.* 2004). However, amino acid sequences of PfCA1 and conventional α-CAs do not align with each other because of the numerous indels (insertions and deletions) of the PfCA1 sequence relative to those of the human α-subclasses CA-I and CA-II (Del Prete *et al.* 2014; De Simone *et al.* 2015). Thus, the Zn ligands of PfCA1 were then reconsidered. Later, the three-dimensional model of partial PfCA1 was computationally solved using the crystal structure of the α-CA from *Thermovibrio ammonificans* (TaCA), displaying that Zn-coordinating residues were reliably predicted as His299, His301 and Gln320 instead of His323 with an overall structure similar to that of α-CA. Due to the unique pattern of Zn coordination (two His and one Gln), PfCA1 is categorized as η-class CA (De Simone *et al.* 2015).

6.3.7. *Zeta CA*

The occurrence of cadmium-containing CA (CDCA1) was suggested in the marine diatom *T. weissflogii* (Lane and Morel 2000; Lane *et al.* 2005). Later, the recombinant CDCA1 was produced, and it was shown that the active site could bind either Cd or Zn (Xu *et al.* 2008). X-ray crystallography further demonstrated that CDCA1 is a structural mimic of the β-CA dimer even though there is little similarity between their primary structure (Figure 6.1) (Xu *et al.* 2008). CDCA is now categorized as the ζ-subclass, and the gene sequence encoding this CA subclass is widespread among diatoms and occurs in green algae and Haptophyta (Park *et al.* 2007). CDCA1 has three repeat domains of 178 amino acid residues in one polypeptide, while the ζ-CA gene in *T. pseudonana* encodes a single domain type sequence of 238 amino acid residues, strongly suggesting that each domain encodes functional CA, but that the repeat number is diverse (Park *et al.* 2007). The ζ-CA gene sequence is present in the genome of the pennate diatom, *P. tricornutum,* but the gene is not transcribed, and thus is very likely a pseudogene, which may have lost its promoter function in the process of diversification of raphid pennate species (Samukawa *et al.* 2014). In *T. pseudonana,* in contrast, ζ-CA (TpζCA1) mRNA is highly accumulated under atmospheric CO_2 concentrations but repressed at high CO_2 concentrations (0.4% in air) (Samukawa *et al.* 2014). TpζCA1 occurs in the periplasm (Tachibana *et al.* 2011; Samukawa *et al.* 2014). The CCM of *T. pseudonana* is known to be sensitive to acetazolamide, an extracellular CA inhibitor (Tsuji *et al.* 2017a), strongly suggesting that this enzyme is a part of the CO_2 acquisition system in this species.

6.3.8. *Theta CA*

θ-subclass CA is a new subclass of CAs, which was first identified in 2016 in the marine diatom *P. tricornutum* (Kikutani *et al.* 2016). Before that, these proteins were known as low-CO_2-inducible (LCI) B-E factors, vital for survival under air-levels of CO_2 in the green alga, *C. reinhardtii* (Yamano *et al.* 2010; Wang and Spalding, 2014). LCIB and LCIC form hetero-hexamers, which are located at the peripheral pyrenoid area (Yamano *et al.* 2010). The LCIB/C complex was initially hypothesized to work with the stromal β-CA, CAH6, to trap the unfixed stromal CO_2, which leaks out of the plastid (Yamano *et al.* 2010).

The primary sequence of *C. reinhardtii* LCIB, LCIC and LCIE is relatively well conserved among green algae and cyanobacteria but is absent in terrestrial plants (Yamano *et al.* 2010; Kikutani *et al.* 2016). There are orthologous sequences that also occur in stramenopiles, Haptophyta, Alveolata and Cryptophyta (Nonoyama *et al.* 2019), but their primary sequences are dissimilar except for one domain, which comprises about 110 amino acid residues. This domain is rich in Cys, Gly and His residues and thus was denoted as the Cys–Gly–His-rich (CGHR) domain (Kikutani *et al.* 2016). There are four orthologous genes tandemly aligned on chromosome 1 of the diatom *P. tricornutum* (Kikutani *et al.* 2016). Two of these gene products have CA activity (Kikutani *et al.*, 2016; Jin *et al.*, 2016), and one occurs in the mitochondria (Pt43232, hereafter PtθCA2) and the other occurs in the lumen of the thylakoid (Pt43233, hereafter PtθCA1), which penetrates the pyrenoid of the *P. tricornutum* chloroplast (Kikutani *et al.* 2016). RNAi knockdown of PtθCA1 gene expression resulted in retarded growth and photosynthesis (Kikutani *et al.* 2016), strongly suggesting its vital role in the primary metabolisms in the pyrenoid.

X-ray crystallography of the above-described mitochondrial PtθCA2, and LCIB and LCIBC from *C. reinhardtii* (CrLCIB and CrLCIC), had a striking similarity of the active center within the CGHR domain in these proteins to that of β-CA, in which two Cys and one His residues coordinate Zn (Figure 6.1) (Jin *et al.* 2016). Even though there is a striking structural similarity, the little primary sequence between β-subclass and θ-subclass CAs is conserved. Apart from this analogous structure of the active site to that of β-CA, the recombinant diatom θ-CA (Pt43233) revealed esterase activity (Kikutani *et al.* 2016), which was previously known to be characteristic of α- and δ-subclass CAs. Although θ-CA is widespread among red-chloroplast-type secondary endosymbionts, the predominant occurrence of chloroplast-targeted θ-CA is likely a specific evolutional innovation of diatoms (Nonoyama *et al.* 2019).

6.3.9. *Iota CA*

Another independent evolutionary line of CA was recently discovered in the marine diatom *T. pseudonana* (Jensen *et al.* 2019). This new ι-subclass of CA was initially denoted as low CO_2-inducible protein 63 (LCIP63) and was localized in the peripheral chloroplast area, most probably within the four-layered chloroplast membranes (Jensen *et al.* 2019; Jensen *et al.* 2021). Overexpression of this gene conferred high-CO_2-grown *T. pseudonana* cells with significantly higher photosynthetic DIC affinity relative to that of cells carrying an empty vector, strongly suggesting that LCIP63 functions to maintain a high concentration of chloroplastic DIC by preventing CO_2 leakage at the chloroplastic membrane (Jensen *et al.* 2019). Homologous LCIP63 sequences widely occur in stramenopiles, green algae, Haptophyta, cyanobacteria and other bacteria (Jensen *et al.* 2019; Del Prete *et al.* 2020). Again, similarly to θ-CA, many ι-CA sequences possess chloroplast transit peptide in diatoms. The putative chloroplastic form is evenly widespread in other organisms, unlike θ-CA, strongly suggesting that chloroplast targeting of ι-CA was an innovation that preceded the establishment of diatom chloroplast (Nonoyama *et al.* 2019).

The primary sequences of ι-CA contain a calcium-calmodulin protein kinase II-association domain (CaMKII-AD) followed by a consensus (H)HHSS motif at the C-terminus of each CaMKII-AD (Jensen *et al.* 2019). This CaMKII-AD:(H)HHSS is a repeated sequence and a number of times that varies depending on the species: four times in the genomes of *T. pseudonana*, *Thalassiosira oceanica* and *Fistulifera solaris*; three times in *Fragilariopsis cylindrus* and *Pseudo-nitzschia multiseries*; and two times in *P. tricornutum* (Jensen *et al.* 2019). Interestingly, LCIP63 gained its highest CA activity with Mn^{2+} and a moderate level of activity was observed with Ca^{2+}, Co^{2+} and Mg^{2+}, while in contrast, activity was little with Zn^{2+}, displaying an unusual metal dependency (Jensen *et al.* 2019). Interestingly, a novel variant of ι-CAs isolated from the cyanobacterium, *Anabaena* sp. PCC7120, and the chlorarachniophyte, *Bigelowiella natans,* lacks a metal cofactor but still retains irreversible CO_2 hydration activity (Hirakawa *et al.* 2021). The crystal structures of this ι-CA form homodimer (Hirakawa *et al.* 2021) and homotetramer in solution (Hirakawa *et al.* 2021; Jensen *et al.* 2021). It is hypothesized that LCIP63 is a homotetramer taking a "drone-like" shape with a core formed by the first two repeat sequences (domains 1 and 2), four protruding arms formed by repeat sequences 3 and 4 (domains 3 and 4). In addition, repeated sequence 1 (domain 1) and repeated sequence 2 (domain 2) might be related to Mn^{2+} binding for CA activity (Jensen *et al.* 2021).

6.3.10. *Subclasses and primary sequences*

CAs are divided into nine subclasses, α-, β-, γ-, δ-, ε-, ζ-, η-, θ- and ι-, according to their primary sequences. The structure of active sites of these subclasses could be categorized into several types; that is, β-, ε-, ζ- and θ-subclasses showed a strikingly similar active center with two His and one Cys (Figure 6.1) and with slightly different selectivity for the divalent metal; α- and δ-subclasses show the similarity in structure for the Zn coordination site even though they have a mirror shape orientation of three His (Figure 6.1). These examples of similarity of active site structure with a totally independent primary sequence indicate the occurrence of an extensive convergent evolution among these CAs. The active site structure of γ- and ι-subclass is so far unique, but considering the very extensive examples of convergent evolution of α- and β-type active-site structures among many subclasses of CA, the γ-type active site is also a future candidate of a reference structure to constitute CA activity. Further, given the very unique metal selectivity of ι-subclass CA, there is a possibility of the occurrence of another new type structure of CA active site. Eta-subclass CA is an exceptional case, which occurred by divergent evolution from α-subclass CA but using unique Zn ligand Gln instead of one of three His. Divergence of the utilization of the Zn ligand is also observed in β-CA in the red microalga, *P. purpureum,* which coordinates Zn with Asp in addition to the typical β-type ligand of two His and one Cys. These atypical examples strongly suggest not only convergent evolution but also the potential of divergent evolution within each subclass of CA. Given these potential complexities of CA evolution, it would be so far the clearest to classify CAs based on their primary sequence rather than the active site structure.

6.4. The functions of CAs in aquatic photoautotrophs

The functions of CAs in photosynthetic organisms are well studied in cyanobacteria and the green alga *C. reinhardtii*, but are much less clarified in terrestrial plants. As for major marine photoautotrophs, research on diatoms has made significant progress in the last two decades. Diatom evolution involved red-type chloroplasts via secondary endosymbiosis, but genome information strongly indicates the occurrence of precedential green-type chloroplast in the common ancestor of stramenopiles (Moustafa *et al.* 2009; Dorrell *et al.* 2017). Thus, molecular information of diatom photosynthesis must include many clues to understand the evolution of red-type plastid modified by transferred green- and bacterial-type genes (Moustafa *et al.* 2009; Dorrell *et al.* 2017).

The diatom pyrenoid is a lens-shaped body surrounded by three-layered thylakoid membranes and two-layered thylakoid membranes that penetrate the central part of the pyrenoid in *P. tricornutum* (Borowitzka and Volcani, 1978; see Chapter 7). Pyrenoids are known to sequester much of the Rubisco enzyme in the chloroplast. It is the location of CO_2 fixation and is thus thought to be the focal point of the algal biophysical CCM (Jenks and Gibbs, 2000). As CCMs accumulate DIC intracellularly as a membrane-impermeable form, HCO_3^-, CAs are essential to convert HCO_3^- rapidly to CO_2, the substrate of Rubisco, and also to prevent CO_2 leakage by converting CO_2 to HCO_3^- either/both in the stroma or the chloroplast membranes (Figure 6.2). Recent investigations have clarified the flux control mechanisms of DIC within the chloroplast involving the pyrenoid in the green alga *C. reinhardtii,* and the marine diatom *P. tricornutum*. Interestingly, it has been shown that there are luminal CAs in the thylakoid system in both species, which reside within the pyrenoid body (Funke *et al.* 1997; Kikutani *et al.* 2016). In *C. reinhardtii*, the luminal CA is α-class, CAH3, while θ-CA, PtθCA1, occurs in *P. tricornutum*. This strongly suggests a common mechanism relating to the DIC equilibrium reaction in the thylakoid membrane within the pyrenoid body (Raven, 1997; Tsuji *et al.* 2017b) for independently acquired CAs, implying that they developed as a result of convergent evolution (Tsuji *et al.* 2017b). At the exterior of the pyrenoid-associating (penetrating) thylakoid membrane, there are also other CAs; putative θ-subclass CA, LCIB/C hetero hexamer in *C. reinhardtii* (Yamano *et al.* 2010); β-subclass PtCA1 and PtCA2 in *P. tricornutum* (Harada and Matsuda, 2005; Tachibana *et al.* 2011), once again strongly suggesting that the acquisition of CAs from wide origins resulted in an invention of the common structural feature and mechanism to control DIC flow at the areas of the pyrenoid (Tsuji *et al.* 2017b). Interestingly, the LCIB/C complex, PtCA1 and PtCA2 do not uniformly localize throughout the pyrenoid. They occur as clumped particles or as a sheath-like structure within or around the pyrenoid (Karlsson *et al.* 1998; Harada and Matsuda, 2005; Tachibana *et al.* 2011). The current evidence suggests that DIC accumulates as HCO_3^- within the stroma partly catalyzed by CAs under alkaline pH during active photosynthesis. The HCO_3^- is transported into the lumen of the pyrenoid-associated thylakoid, where light-driven proton accumulation at the lumen provides the dehydration of HCO_3^- with a function of luminal CAs that create ample flux of CO_2 to Rubisco in the pyrenoid (Raven, 1997; Kikutani *et al.* 2016) (Figure 6.2).

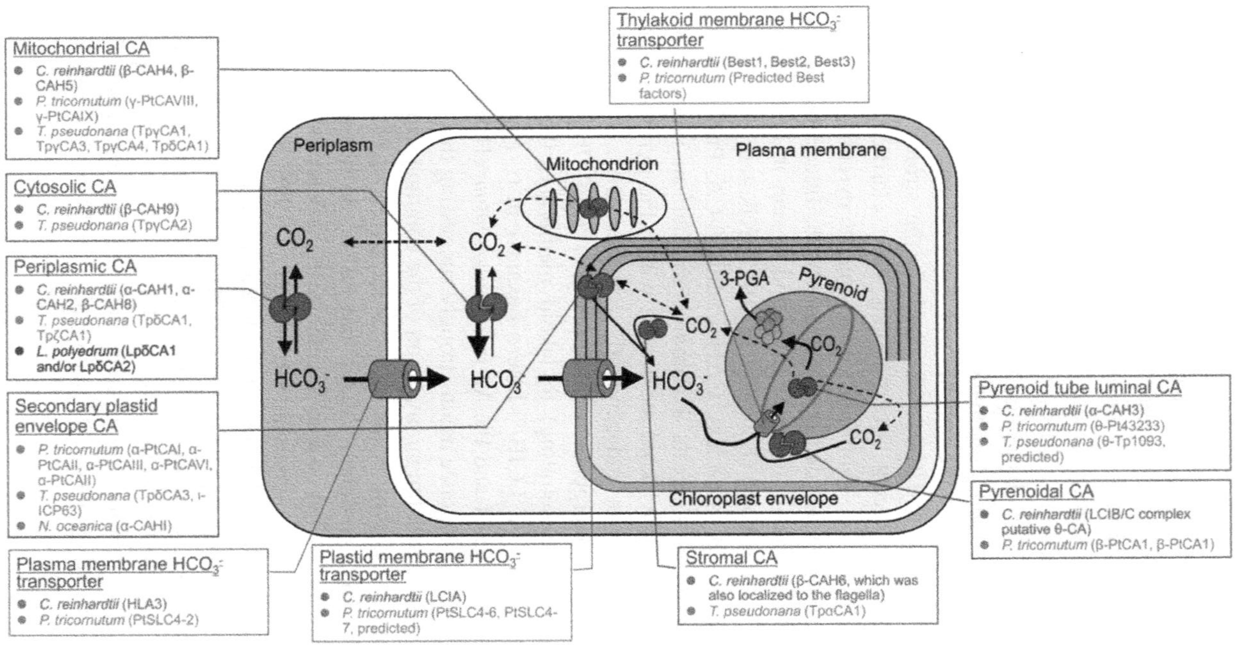

Figure 6.2. *A model of an algal CCM showing locations of CA and DIC transport systems which are so far defined or predicted at molecular levels. Green lineage algae are shown in green and red lineage algae are shown in orange (diatoms) or red (dinoflagellates). Best stands for Bestrophin. For a color version of this figure, see www.iste.co.uk/maberly/photosynthesis.zip*

A system transporting HCO_3^- across the thylakoid membrane was unknown until a recent study revealed that the products of three anion-transporter bestrophin family genes in *C. reinhardtii*, *BST1*, *BST2* and *BST3*, were localized at the thylakoid membrane close to the pyrenoid. It played a role in high-DIC-affinity photosynthesis when CO_2 concentration was low (Mukherjee *et al.* 2019). They were little involved in the photosystems (Mukherjee *et al.* 2019), suggesting the substrate for these proteins to be an accumulated HCO_3^-. Bestrophin family protein is one of thylakoidal HCO_3^- transporters, which works for the CCM together with the above-described CAs (Figure 6.2). Nonetheless, the subclass of CA is not a crucial issue for these systems, and different organisms likely have employed different subclasses of pyrenoidal and luminal CAs in their evolution. Interestingly, transcripts of both thylakoid luminal CAs, PtθCA1 and CAH3, are constitutively accumulated independently of the CO_2 conditions (Yamano *et al.* 2010; Kikutani *et al.* 2016), while those of pyrenoidal CAs, PtCA1/2 and LCIB/C, are far more abundant under atmospheric level CO_2 relative to that in high CO_2 (Yamano *et al.* 2010; Ohno *et al.* 2012). These results strongly suggest that luminal CAs are important for photoautotrophic growth in general in addition to the growth under CO_2 limitation. While pyrenoidal CAs are the relatively specific factor for the CCM in low CO_2. The fact that a similar mechanism has evolved in distant algal taxa suggests the occurrence of a strong convergent co-evolution between the functions of pyrenoid and thylakoid membranes in aquatic photosynthesis.

6.5. Regulation of CO_2 efflux by CA at the chloroplast envelope

As described above, pyrenoidal CAs act to rehydrate unfixed CO_2 into HCO_3^- provided to the thylakoid-membrane HCO_3^- transporter. It prevents DIC from leaking from the pyrenoid and maintaining the DIC flow cycle between the pyrenoid and the thylakoid lumen (Figure 6.2). A stable stromal DIC concentration is essential to replenish CO_2, which has been lost by fixation, and thus uncontrolled leakage of CO_2 from the stroma would immediately dissipate the biophysical CCM. Unlike the double-layered chloroplast envelope in primary endosymbionts, secondary chloroplast in stramenopiles and others possess a four-layered chloroplast membrane with a periplastidal compartment in the middle, a space inherited from the ancestral red alga (Gibbs, 1979). This structure can provide a much longer biophysical distance between the cytosol and the stroma than the primary chloroplast. An efficient transport system for HCO_3^- is required for the DIC intake into the stroma (Hopkinson *et al.* 2011), while multiple layers of the envelope could confer efficient resistance to CO_2 leakage on the chloroplast. Of particular importance regarding such a system are the CAs localized to the matrices between the chloroplast envelopes. In *P. tricornutum* and *T. pseudonana*, there are numerous CAs between the four-layered chloroplast envelope (Tachibana *et al.* 2011; Samukawa *et al.* 2014) (Figure 6.2). In *P. tricornutum,* all of the αCAs were localized in the CER or PPC (Tachibana *et al.* 2011), while in *T. pseudonana*, δCA was in the PPC (Samukawa *et al.* 2014). Recently, the newest ι-subclass CA, LCIP63, was localized at the periphery of the chloroplast of *T. pseudonana.* It was demonstrated that over-expression of LCIP63 conferred *T. pseudonana* cells a higher photosynthetic DIC affinity relative to cells carrying an empty vector (Jensen *et al.* 2019). In another Stramenopile, the eustigmatophyte, *Nannochloropsis oceanica*, an α-subclass CA, CAH1 was localized in the CER, and knock-out CAH1 resulted in dissipation of the CCM in *N. oceanica* cells (Gee and Niyogi, 2017). These investigations clearly indicate that the four-layered chloroplast envelope system in stramenopiles contains CAs that function to reduce CO_2 leakage from the chloroplast, working as a part of the CCM (Figure 6.2). CAs in the matrices between the chloroplast envelopes presumably work to hydrate CO_2 into HCO_3^-, providing HCO_3^- to the inward transporters. It is currently unclear what the pH levels are in each matrix and where the HCO_3^- forming area within the chloroplast envelope system is located.

CO_2 leakage from the chloroplast causes dissipation of the CCM, while controlled leakage of unfixed CO_2 could be part of a protection system against excess light energy in algae by constituting a futile cycle of simultaneous DIC uptake and leakage (Kaplan and Reinhold, 1999; Nakajima *et al.* 2013). Thus, it is probable that DIC switching of inward and outward fluxes is rapidly and tightly regulated in the biophysical CCM (Tsuji *et al.* 2017a). Such a switching mechanism within the pyrenoid and across a four-layered chloroplast membrane is so far unknown, but there is some information about the post-translational regulation of CAs in algae. It is known in *C. reinhardtii* that thylakoid-luminal CAH3 is activated by phosphorylation (Blanco-Rivero *et al.* 2012), and the LCIB/C complex is assembled within the pyrenoid only when CO_2 is limited and light energy is supplied (Mitra *et al.* 2004; Yamano *et al.* 2010). In these conditions, Ca^{2+} binding protein (CAS) is recruited to the pyrenoid tube and facilitates a retrograde signal from the pyrenoid in the chloroplast to the nucleus that controls the transcription factor of genes related to the CCM in *C. reinhardtii* (Wang *et al.* 2016). Regulator proteins that had the same function to maintain CCM transcription factors in diatoms have not yet been identified. However, there is some evidence that both PtCA1 and PtCA2 are regulated by chloroplastic thioredoxins (Trxs), suggesting an activation of PtCA1 and PtCA2 under the reducing environment of the pyrenoid in the light (Kikutani *et al.* 2012).

6.6. Summary: CAs in red and green photosynthesis

CA reaction produces two inorganic carbon species, CO_2 and HCO_3^-, which reveal contrasting natures in regard to membrane permeability. The very weak membrane permeability of HCO_3^- allows photoautotrophic cells to concentrate DIC inside the plasma membrane and chloroplast envelope if the conversion from CO_2 to HCO_3^- is rapid enough, catalyzed by CA. Such a leakage barrier system for intracellular DIC promotes photosynthesis under CO_2-limited environments as a part of the CCM and thus is believed to be an evolutionary adaptation to a low CO_2 environment after the Carboniferous period. Microalgae dramatically diversified in this period and perhaps have evolved their CCMs, acquiring unique genes from diverse origins. In fact, they use extremely diverse CA sequences in inconsistent cellular locations. Among nine subclasses of CA sequences, five subclasses (δ, ε, ζ, θ and ι) are likely specific inventions in aquatic photoautotrophs, strongly suggesting the importance of CA reaction in water under low atmospheric pCO_2 for photosynthesis. The Apicomplexa, which is also supposed to have originated from photosynthetic secondary endosymbiont, possesses a η-subclass CA, suggesting that CAs from a wide variety of origins occurred in stramenopiles and Alveolata. However, the relation to secondary endosymbiosis is not clear. As depicted in Figure 6.2, the subclasses of CA do not define their function at all, but the location of each CA is very important to understand their functions. The recent discovery of thylakoid luminal θ-CA in diatoms indicated a strong convergent aspect of the chloroplastic CA function in algae. Both *C. reinhardtii* and *P. tricornutum* produce CO_2 for Rubisco by thylakoid-luminal CA and prevent CO_2 leakage from the pyrenoid by pyrenoidal CA (Figure 6.2), strongly suggesting the common use of proton motive force with CAs for the final step of the CCM in distant taxa. Recent pieces of evidence are quite supportive of the role of layered chloroplast envelope as a leakage barrier of CO_2 with the function of CAs in the envelope system of the red-type secondary chloroplast. Apart from these, the importance of extracellular and cytosolic CAs is pointed out for DIC acquisition in green algae and stramenopiles (Hopkinson *et al.* 2011; Moroney *et al.* 2011; Tsuji *et al.* 2017b). However, occurrences of CAs of these locations are inconsistent and molecular evidence yet exists; thus, the role of these CAs awaits to be studied.

Class	Organisms (subcellular location)	Metal (metal ligands at the active site)	Quaternary structure	Reference
α	Terrestrial plants (stroma, thylakoid)	Unsolved	Unsolved	Friso *et al.* (2004) and Villarejo *et al.* (2005)
	Chlorophytes (periplasm, cytosol, thylakoid lumen)	Zn (3 His)	Monomeric (with some exception of dimer)	Fukuzawa *et al.* (1990), Fisher *et al.* (1996), Karlsson *et al.* (1998), Satoh *et al.* (1998) and Suzuki *et al.* (2011)
	Rhodophytes (endoplasmic reticulum, vacuole)	Unsolved	Unsolved	Razzak *et al.* (2019)
	Stramenopiles (chloroplast membrane matrices)	Unsolved	Unsolved	Tachibana *et al.* (2011), Samukawa *et al.* (2014) and Gee and Niyogi (2017)
β	Terrestrial plants (cytosol, mitochondria, stroma)	Zn (1 His, 2 Cys)	Tetrameric, Octameric	Roeske and Ogren (1990), Fabre *et al.* (2007), Jiang *et al.* (2014) and DiMario *et al.* (2016)
	Chlorophytes (periplasm, mitochondria)	Zn (1 His, 2 Cys)	Dimeric	Giordano *et al.* (2003)
	Rhodophytes (unsolved)	Zn (1 His, 2 Cys, 1 Asp)	Dimeric (pseudo-tetrameric structure with 2 monomers with 2 tandem β-CA domains)	Mitsuhashi *et al.* (2000) and Razzak *et al.* (2019)
	Stramenopiles (pyrenoid, mitochondria)	Zn (1 His, 2 Cys)	Polymeric of dimeric form forming clump	Satoh *et al.* (2001), Harada and Matsuda (2005), Tachibana *et al.* (2011) and Nonoyama *et al.* (2019)
γ	Terrestrial plants (mitochondria)	Zn (3 His)	Heterotrimeric (a part of complex I of respiratory chain, activity not confirmed)	Martin *et al.* (2009), Klodmann *et al.* (2010), Maldonado *et al.* (2020), Soufari *et al.* (2020) and Klusch *et al.* (2021)
	Rhodophytes (cytosol, mitochondria in prediction)	Unsolved	Unsolved	Razzak *et al.* (2019)
	Archaea (unsolved)	Zn, Co, Fe (3 His)	Homotrimeric	Alber and Ferry (1994)
	Cyanobacteria (carboxysome)	Zn (3 His)	Homotrimeric	Peña *et al.* (2010)
	Stramenopiles (mostly mitochondria, one cytosolic reported)	Unsolved	Unsolved	Tachibana *et al.* (2011) and Samukawa *et al.* (2014)
	Euglenozoa (mitochondria)	Unsolved	Unsolved	Perez *et al.* (2014)

δ	Stramenopiles (periplasm, mitochondria, chloroplast membrane matrices)	Zn, Co (3 His, similar to α-CA)	Unsolved	Cox *et al.* (2000) and Samukawa *et al.* (2014)
	Dinoflagellates (periplasm)	Zn (3 His)	Unsolved	Lapointe *et al.* (2008)
	Haptophytes (unsolved)	Unsolved	Unsolved	Soto *et al.* (2006)
ε	Proteobacteria (carboxysome)	Zn (1 His, 2 Cys)	Dimeric or polymeric of the dimeric form	So *et al.* (2004)
	Cyanobacteria (carboxysome)	Unsolved	Unsolved	Sawaya *et al.* (2006)
ζ	Stramenopiles (periplasm)	Cd, Zn (1 His, 2 Cys)	Can be monomeric but diverse domain repeat number (active site structure is similar to β-CA)	Park *et al.* (2007), Xu *et al.* (2008), Tachibana *et al.* (2011) and Samukawa *et al.* (2014)
	Chlorophytes (unsolved)	Unsolved	Unsolved	Park *et al.* (2007)
	Haptophytes (unsolved)	Unsolved	Unsolved	Park *et al.* (2007)
η	Apicomplexa (unsolved)	Zn (2 His, 1 Gln)	Unsolved	Reungprapavut *et al.* (2004) and De Simone *et al.* (2015)
θ	Chlorophytes (peripheral pyrenoid)	Zn (1 His, 2 Cys)	Heterohexameric (active site structure is similar to β-CA)	Yamano *et al.* (2010), Wang and Spalding (2014) and Jin *et al.* (2016)
	Stramenopiles (thylakoid lumen, mitochondria, unsolved peripheral layers of chloroplast)	Zn (1 His, 2 Cys)	Dimeric (active site structure is similar to β-CA)	Jin *et al.* (2016), Kikutani *et al.* (2016) and Nonoyama *et al.* (2019)
	Haptophyte (unsolved)	Unsolved	Unsolved	Nonoyama *et al.* (2019)
	Dinoflagellates (unsolved)	Unsolved	Unsolved	Nonoyama *et al.* (2019)
	Cryptophytes (unsolved)	Unsolved	Unsolved	Nonoyama *et al.* (2019)

ι	Stramenopiles (peripheral chloroplast)	Mn (Unsolved could be substituted by Co, Mg)	homotetrameric with "drone shape"	Jensen *et al.* (2019, 2021)
	Chlorophytes (unsolved)	Unsolved	Unsolved	Jensen *et al.* (2019)
	Haptophytes (unsolved)	Unsolved	Unsolved	Jensen *et al.* (2019)
	Chlorarachniophytes (stroma, mitochondria)	Not required	Dimeric	Hirakawa *et al.* (2021)
	Cyanobacteria (unsolved)	Not required	Dimeric	Jensen *et al.* (2019) and Hirakawa *et al.* (2021)

Table 6.1. *Classes of carbonic anhydrase with determined/predicted subcellular localizations, and their characteristics so far studied mainly in photoautotrophs*

6.7. References

Alber, B.E. and Ferry, J.G. (1994). A carbonic anhydrase from the archaeon *Methanosarcina thermophila*. *Proceedings of the National Academy of Sciences of the United States of America*, 91(15), 6909–6913.

Alber, B.E., Colangelo, C.M., Dong, J., Stalhandske, C.M., Baird, T.T., Tu, C., Fierke, C.A., Silverman, D.N., Scott, R.A., Ferry, J.G. (1999). Kinetic and spectroscopic characterization of the gamma-carbonic anhydrase from the methanoarchaeon *Methanosarcina thermophila*. *Biochemistry*, 38(40), 13119–13128.

Alterio, V., Di Fiorev A., D'Ambrosio, K., Supuran, C.T., De Simone, G. (2012). Multiple binding modes of inhibitors to carbonic anhydrases: How to design specific drugs targeting 15 different isoforms? *Chemical Reviews*, 112(8), 4421–4468.

Armbrust, E.V., Berges, J.A., Bowler, C., Green, B.R., Martinez, D., Putnam, N.H., Zhou, S., Allen, A.E., Apt, K.E., Bechner, M. *et al.* (2004). The genome of the diatom *Thalassiosira pseudonana*: Ecology, evolution, and metabolism. *Science*, 306(5693), 79–86.

Baranauskienė, L. and Matulis, D. (2019). Overview of human carbonic anhydrases. In *Carbonic Anhydrase as Drug Target*, Matulis, D. (ed.). Springer, Cham.

Blanco-Rivero, A., Shutova, T., Román, M.J., Villarejo, A., Martinez, F. (2012). Phosphorylation controls the localization and activation of the lumenal carbonic anhydrase in *Chlamydomonas reinhardtii*. *PLOS ONE*, 7(11), e49063.

Borowitzka, M.A. and Volcani, B.E. (1978). The polymorphic diatom *Phaeodactylum tricornutum*: Ultrastructure of ITS morphotypes. *Journal of Phycology*, 14(1), 10–21.

Bowler, C., Allen, A.E., Badger, J.H., Grimwood, J., Jabbari, K., Kuo, A., Maheswari, U., Martens, C., Maumus, F., Otillar, R.P. (2008). The *Phaeodactylum* genome reveals the evolutionary history of diatom genomes. *Nature*, 456(7219), 239–244.

Brinkman, R., Margaria, R., Meldrum, N., Roughton, F. (1932). The CO_2 catalyst present in blood. *Journal of Physiology*, 75(3).

Burnell, J.N., Gibbs, M.J., Mason, J.G. (1990). Spinach chloroplastic carbonic anhydrase: Nucleotide sequence analysis of cDNA. *Plant Physiology*, 92(1), 37–40.

Cainzos, M., Marchetti, F., Popovich, C., Leonardi, P., Pagnussat, G., Zabaleta, E. (2021). Gamma carbonic anhydrases are subunits of the mitochondrial complex I of diatoms. *Molecular Microbiology*, 116(1), 109–125.

Cox, E.H., McLendon, G.L., Morel, F.M., Lane, T.W., Prince, R.C., Pickering, I.J., George, G.N. (2000). The active site structure of *Thalassiosira weissflogii* carbonic anhydrase 1. *Biochemistry*, 39(40), 12128–12130.

De Simone, G., Di Fiore, A., Capasso, C., Supuran, C.T. (2015). The zinc coordination pattern in the η-carbonic anhydrase from *Plasmodium falciparum* is different from all other carbonic anhydrase genetic families. *Bioorganic & Medicinal Chemistry Letters*, 25(7), 1385–1389.

Del Prete, S., Vullo, D., Fisher, G.M., Andrews, K.T., Poulsen, S.A., Capasso, C., Supuran, C.T. (2014). Discovery of a new family of carbonic anhydrases in the malaria pathogen *Plasmodium falciparum* – The η-carbonic anhydrases. *Bioorganic & Medicinal Chemistry Letters*, 24(18), 4389–4396.

Del Prete, S., Nocentini, A., Supuran, C.T., Capasso, C. (2020). Bacterial ι-carbonic anhydrase: A new active class of carbonic anhydrase identified in the genome of the Gram-negative bacterium *Burkholderia territorii*. *Journal of Enzyme Inhibition and Medicinal Chemistry*, 35(1), 1060–1068.

DiMario, R.J., Quebedeaux, J.C., Longstreth, D.J., Dassanayake, M., Hartman, M.M., Moroney, J.V. (2016). The cytoplasmic carbonic anhydrases βCA2 and βCA4 are required for optimal plant growth at low CO_2. *Plant Physiology*, 171(1), 280–293.

DiMario, R.J., Clayton, H., Mukherjee, A., Ludwig, M., Moroney, J.V. (2017). Plant carbonic anhydrases: Structures, locations, evolution, and physiological roles. *Molecular Plant*, 10(1), 30–46.

DiMario, R.J., Machingura, M.C., Waldrop, G.L., Moroney, J.V. (2018). The many types of carbonic anhydrases in photosynthetic organisms. *Plant Science*, 268, 11–17.

Dorrell, R.G., Gile, G., Mccallum, G., Méheust, R., Bapteste, E.P., Klinger, C.M., Brillet-Guéguen, L., Freeman, K.D., Richter, D.J., Bowler, C. (2017). Chimeric origins of ochrophytes and haptophytes revealed through an ancient plastid proteome. *Elife*, 6, e23717.

Fabre, N., Reiter, I.M., Becuwe-Linka, N., Genty, B., Rumeau, D. (2007). Characterization and expression analysis of genes encoding α and β carbonic anhydrases in *Arabidopsis*. *Plant, Cell & Environment*, 30(5), 617–629.

Fisher, M., Gokhman, I., Pick, U., Zamir, A. (1996). A salt-resistant plasma membrane carbonic anhydrase is induced by salt in *Dunaliella salina*. *Journal of Biological Chemistry*, 271(30), 17718–17723.

Friso, G., Giacomelli, L., Ytterberg, A.J., Peltier, J.-B., Rudella, A., Sun, Q., Van Wijk, K.J. (2004). In-depth analysis of the thylakoid membrane proteome of *Arabidopsis thaliana* chloroplasts: New proteins, new functions, and a plastid proteome database. *The Plant Cell*, 16(2), 478–499.

Fukuzawa, H., Fujiwara, S., Yamamoto, Y., Dionisio-Sese, M.L., Miyachi, S. (1990). cDNA cloning, sequence, and expression of carbonic anhydrase in *Chlamydomonas reinhardtii*: Regulation by environmental CO_2 concentration. *Proceedings of the National Academy of Sciences of the United States of America*, 87(11), 4383–4387.

Funke, R.P., Kovar, J.L., Weeks, D.P. (1997). Intracellular carbonic anhydrase is essential to photosynthesis in *Chlamydomonas reinhardtii* at atmospheric levels of CO_2 – Demonstration via genomic complementation of the high-CO_2-requiring mutant ca-1. *Plant Physiology*, 114(1), 237–244.

Gee, C.W. and Niyogi, K.K. (2017). The carbonic anhydrase CAH1 is an essential component of the carbon-concentrating mechanism in *Nannochloropsis oceanica. Proceedings of the National Academy of Sciences of the United States of America*, 114(17), 4537–4542.

Gibbs, S.P. (1979). The route of entry of cytoplasmically synthesized proteins into chloroplasts of algae possessing chloroplast ER. *Journal of Cell Science*, 35, 253–266.

Giordano, M., Norici, A., Forssen, M., Eriksson, M., Raven, J.A. (2003). An anaplerotic role for mitochondrial carbonic anhydrase in *Chlamydomonas reinhardtii. Plant Physiology*, 132(4), 2126–2134.

Griffiths, D.J. (1970). The pyrenoid. *The Botanical Review*, 36(1), 29–58.

Harada, H. and Matsuda, Y. (2005). Identification and characterization of a new carbonic anhydrase in the marine diatom *Phaeodactylum tricornutum. Canadian Journal of Botany*, 83(7), 909–916.

Hewett-Emmett, D. (2000). Evolution and distribution of the carbonic anhydrase gene families. In *The Carbonic Anhydrases*, Chegwidden, W.R., Carter, N.D., Edwards, Y.H. (eds). Birkhäuser, Basel.

Hewett-Emmett, D. and Tashian, R.E. (1996). Functional diversity, conservation, and convergence in the evolution of the α-, β-, and γ-carbonic anhydrase gene families. *Molecular Phylogenetics and Evolution*, 5(1), 50–77.

Hirakawa, Y., Senda, M., Fukuda, K., Yu, H.Y., Ishida, M., Taira, M., Kinbara, K., Senda, T. (2021). Characterization of a novel type of carbonic anhydrase that acts without metal cofactors. *BMC Biology*, 19(1), 1–15.

Hopkinson, B.M., Dupont, C.L., Allen, A.E., Morel, F.M. (2011). Efficiency of the CO_2-concentrating mechanism of diatoms. *Proceedings of the National Academy of Sciences of the United States of America*, 108(10), 3830–3837.

Jenks, A. and Gibbs, S.P. (2000). Immunolocalization and distribution of form II Rubisco in the pyrenoid and chloroplast stroma of *Amphidinium carterae* and form I Rubisco in the symbiont-derived plastids of *Peridinium foliaceum* (dinophyceae). *Journal of Phycology*, 36(1), 127–138.

Jensen, E.L., Clement, R., Kosta, A., Maberly, S.C., Gontero, B. (2019). A new widespread subclass of carbonic anhydrase in marine phytoplankton. *The ISME Journal*, 13(8), 2094–2106.

Jensen, E.L., Receveur-Brechot, V., Hachemane, M., Wils, L., Barbier, P., Parsiegla, G., Gontero, B., Launay, H. (2021). Structural contour map of the iota carbonic anhydrase from the diatom *Thalassiosira pseudonana* using a multiprong approach. *International Journal of Molecular Sciences*, 22(16), 8723.

Jiang, C., Tholen, D., Xu, J.M., Xin, C., Zhang, H., Zhu, X., Zhao, Y. (2014). Increased expression of mitochondria-localized carbonic anhydrase activity resulted in an increased biomass accumulation in *Arabidopsis thaliana. Journal of Plant Biology*, 57(6), 366–374.

Jin, S., Sun, J., Wunder, T., Tang, D., Cousins, A.B., Sze, S.K., Mueller-Cajar, O., Gao, Y.-G. (2016). Structural insights into the LCIB protein family reveals a new group of β-carbonic anhydrases. *Proceedings of the National Academy of Sciences of the United States of America*, 113(51), 14716–14721.

Kaplan, A. and Reinhold, L. (1999). CO_2 concentrating mechanisms in photosynthetic microorganisms. *Annual Review of Plant Biology*, 50(1), 539–570.

Karlsson, J., Clarke, A.K., Chen, Z.Y., Hugghins, S.Y., Park, Y.I., Husic, H.D., Moroney, J.V., Samuelsson, G. (1998). A novel α-type carbonic anhydrase associated with the thylakoid membrane in *Chlamydomonas reinhardtii* is required for growth at ambient CO_2. *The EMBO Journal*, 17(5), 1208–1216.

Kerfeld, C.A. and Melnicki, M.R. (2016). Assembly, function and evolution of cyanobacterial carboxysomes. *Current Opinion in Plant Biology*, 31, 66–75.

Kikutani, S., Tanaka, R., Yamazaki, Y., Hara, S., Hisabori, T., Kroth, P.G., Matsuda, Y. (2012). Redox regulation of carbonic anhydrases via thioredoxin in chloroplast of the marine diatom *Phaeodactylum tricornutum*. *Journal of Biological Chemistry*, 287(24), 20689–20700.

Kikutani, S., Nakajima, K., Nagasato, C., Tsuji, Y., Miyatake, A., Matsuda, Y. (2016). Thylakoid luminal θ-carbonic anhydrase critical for growth and photosynthesis in the marine diatom *Phaeodactylum tricornutum*. *Proceedings of the National Academy of Sciences of the United States of America*, 113(35), 9828–9833.

Kimber, M.S. and Pai, E.F. (2000). The active site architecture of *Pisum sativum* β-carbonic anhydrase is a mirror image of that of α-carbonic anhydrases. *The EMBO Journal*, 19(7), 1407–1418.

Kisker, C., Schindelin, H., Alber, B.E., Ferry, J.G., Rees, D.C. (1996). A left-hand beta-helix revealed by the crystal structure of a carbonic anhydrase from the archaeon *Methanosarcina thermophila*. *The EMBO Journal*, 15(10), 2323–2330.

Kitao, Y. and Matsuda, Y. (2009). Formation of macromolecular complexes of carbonic anhydrases in the chloroplast of a marine diatom by the action of the C-terminal helix. *Biochemical Journal*, 419, 681–688.

Klodmann, J., Sunderhaus, S., Nimtz, M., Jänsch, L., Braun, H.-P. (2010). Internal architecture of mitochondrial complex I from *Arabidopsis thaliana*. *The Plant Cell*, 22(3), 797–810.

Klusch, N., Senkler, J., Yildiz, Ö., Kühlbrandt, W., Braun, H.-P. (2021). A ferredoxin bridge connects the two arms of plant mitochondrial complex I. *The Plant Cell*, 33(6), 2072–2091.

Lane, T.W. and Morel, F.M.M. (2000a). A biological function for cadmium in marine diatoms. *Proceedings of the National Academy of Sciences of the United States of America*, 97(9), 4627–4631.

Lane, T.W. and Morel, F.M.M. (2000b). Regulation of carbonic anhydrase expression by zinc, cobalt, and carbon dioxide in the marine diatom *Thalassiosira weissflogii*. *Plant Physiology*, 123(1), 345–352.

Lane, T.W., Saito, M.A., George, G.N., Pickering, I.J., Prince, R.C., Morel, F.M. (2005). Biochemistry: A cadmium enzyme from a marine diatom. *Nature*, 435(7038), 455–456.

Lapointe, M., MacKenzie, T.D.B., Morse, D. (2008). An external δ-carbonic anhydrase in a free-living marine dinoflagellate may circumvent diffusion-limited carbon acquisition. *Plant Physiology*, 147(3), 1427–1436.

Lee, R.B.Y., Smith, J.A.C., Rickaby, R.E. (2013). Cloning, expression and characterization of the δ-carbonic anhydrase of *Thalassiosira weissflogii* (Bacillariophyceae). *Journal of Phycology*, 49(1), 170–177.

Liljas, A., Kannan, K.K., Bergsten, P.C., Waara, I., Fridborg, K., Strandberg, B., Carlbom, U., Jarup, L., Lovgren, S., Petef, M. (1972). Crystal structure of human carbonic anhydrase C. *Nature New Biology*, 235(57), 131–137.

Maldonado, M., Padavannil, A., Zhou, L., Guo, F., Letts, J.A. (2020). Atomic structure of a mitochondrial complex I intermediate from vascular plants. *Elife*, 9, e56664.

Martin, V., Villarreal, F., Miras, I., Navaza, A., Haouz, A., González-Lebrero, R.M., Kaufman, S.B., Zabaleta, E. (2009). Recombinant plant gamma carbonic anhydrase homotrimers bind inorganic carbon. *FEBS Letters*, 583(21), 3425–3430.

McGinn, P.J. and Morel, F.M. (2008). Expression and regulation of carbonic anhydrases in the marine diatom *Thalassiosira pseudonana* and in natural phytoplankton assemblages from Great Bay, New Jersey. *Physiologia Plantarum*, 133(1), 78–91.

Meldrum, N. (1932). Some properties of carbonic anhydrase, the CO_2 enzyme present in blood. *Journal of Physiology*, 75, 15–16.

Mitra, M., Lato, S.M., Ynalvez, R.A., Xiao, Y., Moroney, J.V. (2004). Identification of a new chloroplast carbonic anhydrase in *Chlamydomonas reinhardtii*. *Plant Physiology*, 135(1), 173–182.

Mitsuhashi, S., Mizushima, T., Yamashita, E., Yamamoto, M., Kumasaka, T., Moriyama, H., Ueki, T., Miyachi, S., Tsukihara, T. (2000). X-ray structure of β-carbonic anhydrase from the red alga, *Porphyridium purpureum*, reveals a novel catalytic site for CO_2 hydration. *Journal of Biological Chemistry*, 275(8), 5521–5526.

Moroney, J.V., Ma, Y., Frey, W.D., Fusilier, K.A., Pham, T.T., Simms, T.A., DiMario, R.J., Yang, J., Mukherjee, B. (2011). The carbonic anhydrase isoforms of *Chlamydomonas reinhardtii*: Intracellular location, expression, and physiological roles. *Photosynthesis Research*, 109(1–3), 133–149.

Moustafa, A., Beszteri, B., Maier, U.G., Bowler, C., Valentin, K., Bhattacharya, D. (2009). Genomic footprints of a cryptic plastid endosymbiosis in diatoms. *Science*, 324(5935), 1724–1726.

Mukherjee, A., Lau, C.S., Walker, C.E., Rai, A.K., Prejean, C.I., Yates, G., Emrich-Mills, T., Lemoine, S.G., Vinyard, D.J., Mackinder, L.C. (2019). Thylakoid localized bestrophin-like proteins are essential for the CO_2 concentrating mechanism of *Chlamydomonas reinhardtii. Proceedings of the National Academy of Sciences of the United States of America*, 116(34), 16915–16920.

Nakajima, K., Tanaka, A., Matsuda, Y. (2013). SLC4 family transporters in a marine diatom directly pump bicarbonate from seawater. *Proceedings of the National Academy of Sciences of the United States of America*, 110(5), 1767–1772.

Nonoyama, T., Kazamia, E., Nawaly, H., Gao, X., Tsuji, Y., Matsuda, Y., Bowler, C., Tanaka, T., Dorrell, R.G. (2019). Metabolic innovations underpinning the origin and diversification of the diatom chloroplast. *Biomolecules*, 9(8), 322.

Ohno, N., Inoue, T., Yamashiki, R., Nakajima, K., Kitahara, Y., Ishibashi, M., Matsuda, Y. (2012). CO_2-cAMP-responsive cis-elements targeted by a transcription factor with CREB/ATF-like basic zipper domain in the marine diatom *Phaeodactylum tricornutum*. *Plant Physiology*, 158(1), 499–513.

Pander, B., Harris, G., Scott, D.J., Winzer, K., Kopke, M., Simpson, S.D., Minton, N.P., Henstra, A.M. (2019). The carbonic anhydrase of *Clostridium autoethanogenum* represents a new subclass of beta-carbonic anhydrases. *Applied Microbiology and Biotechnology*, 103(17), 7275–7286.

Parisi, G., Perales, M., Fornasari, M.S., Colaneri, A., Gonzalez-Schain, N., Gomez-Casati, D., Zimmermann, S., Brennicke, A., Araya, A., Ferry, J.G. *et al.* (2004). Gamma carbonic anhydrases in plant mitochondria. *Plant Molecular Biology*, 55(2), 193–207.

Park, H., Song, B., Morel, F.M. (2007). Diversity of the cadmium-containing carbonic anhydrase in marine diatoms and natural waters. *Environmental Microbiology*, 9(2), 403–413.

Peña, K.L., Castel, S.E., de Araujo, C., Espie, G.S., Kimber, M.S. (2010). Structural basis of the oxidative activation of the carboxysomal gamma-carbonic anhydrase, CcmM. *Proceedings of the National Academy of Sciences of the United States of America*, 107(6), 2455–2460.

Perez, E., Lapaille, M., Degand, H., Cilibrasi, L., Villavicencio-Queijeiro, A., Morsomme, P., González-Halphen, D., Field, M.C., Remacle, C., Baurain, D. (2014). The mitochondrial respiratory chain of the secondary green alga *Euglena gracilis* shares many additional subunits with parasitic Trypanosomatidae. *Mitochondrion*, 19, 338–349.

Pronina, N.A. and Semenenko, V.E. (1984). Localization of membrane-bound and soluble forms of carbonic anhydrase in the Chlorella cell. *Soviet Plant Physiology*, 31(2), 187–196.

Raven, J. (1997). CO_2-concentrating mechanisms: A direct role for thylakoid lumen acidification? *Plant, Cell & Environment*, 20(2), 147–154.

Razzak, M.A., Lee, J., Lee, D.W., Kim, J.H., Yoon, H.S., Hwang, I. (2019). Expression of seven carbonic anhydrases in red alga *Gracilariopsis chorda* and their subcellular localization in a heterologous system, *Arabidopsis thaliana. Plant Cell Reports*, 38(2), 147–159.

Reungprapavut, S., Krungkrai, S.R., Krungkrai, J. (2004). *Plasmodium falciparum* carbonic anhydrase is a possible target for malaria chemotherapy. *Journal of Enzyme Inhibition and Medicinal Chemistry*, 19(3), 249–256.

Roberts, S.B., Lane, T.W., Morel, F.M.M. (1997). Carbonic anhydrase in the marine diatom *Thalassiosira weissflogii* (Bacillariophyceae). *Journal of Phycology*, 33(5), 845–850.

Roeske, C.A. and Ogren, W.L. (1990). Nucleotide sequence of pea cDNA encoding chloroplast carbonic anhydrase. *Nucleic Acids Research*, 18(11), 3413.

Rowlett, R.S. (2010). Structure and catalytic mechanism of the β-carbonic anhydrases. *Biochimica et Biophysica Acta*, 1804(2), 362–373.

Samukawa, M., Shen, C., Hopkinson, B.M., Matsuda, Y. (2014). Localization of putative carbonic anhydrases in the marine diatom, *Thalassiosira pseudonana. Photosynthesis Research*, 121(2–3), 235–249.

Satoh, A., Iwasaki, T., Odani, S., Shiraiwa, Y. (1998). Purification, characterization and cDNA cloning of soluble carbonic anhydrase from *Chlorella sorokiniana* grown under ordinary air. *Planta*, 206(4), 657–665.

Satoh, D., Hiraoka, Y., Colman, B., Matsuda, Y. (2001). Physiological and molecular biological characterization of intracellular carbonic anhydrase from the marine diatom *Phaeodactylum tricornutum. Plant Physiology*, 126(4), 1459–1470.

Sawaya, M.R., Cannon, G.C., Heinhorst, S., Tanaka, S., Williams, E.B., Yeates, T.O., Kerfeld, C.A. (2006). The structure of beta-carbonic anhydrase from the carboxysomal shell reveals a distinct subclass with one active site for the price of two. *Journal of Biological Chemistry*, 281(11), 7546–7555.

Shen, C., Dupont, C.L., Hopkinson, B.M. (2017). The diversity of CO_2-concentrating mechanisms in marine diatoms as inferred from their genetic content. *Journal of Experimental Botany*, 68(14), 3937–3948.

Smith, K.S., Jakubzick, C., Whittam, T.S., Ferry, J.G. (1999). Carbonic anhydrase is an ancient enzyme widespread in prokaryotes. *Proceedings of the National Academy of Sciences of the United States of America*, 96(26), 15184–15189.

So, A.K., Espie, G.S., Williams, E.B., Shively, J.M., Heinhorst, S., Cannon, G.C. (2004). A novel evolutionary lineage of carbonic anhydrase (ε class) is a component of the carboxysome shell. *Journal of Bacteriology*, 186(3), 623–630.

Soto, A.R., Zheng, H., Shoemaker, D., Rodriguez, J., Read, B.A., Wahlund, T.M. (2006). Identification and preliminary characterization of two cDNAs encoding unique carbonic anhydrases from the marine alga *Emiliania huxleyi. Applied and Environmental Microbiology*, 72(8), 5500–5511.

Soto, D., Córdoba, J.P., Villarreal, F., Bartoli, C., Schmitz, J., Maurino, V.G., Braun, H.P., Pagnussat, G.C., Zabaleta, E. (2015). Functional characterization of mutants affected in the carbonic anhydrase domain of the respiratory complex I in *Arabidopsis thaliana. The Plant Journal*, 83(5), 831–844.

Soufari, H., Parrot, C., Kuhn, L., Waltz, F., Hashem, Y. (2020). Specific features and assembly of the plant mitochondrial complex I revealed by cryo-EM. *Nature Communications*, 11(1), 1–7.

Supuran, C.T. and Capasso, C. (2015). The η-class carbonic anhydrases as drug targets for antimalarial agents. *Expert Opinion on Therapeutic Targets*, 19(4), 551–563.

Suzuki, K., Yang, S.-Y., Shimizu, S., Morishita, E.C., Jiang, J., Zhang, F., Hoque, M.M., Sato, Y., Tsunoda, M., Sekiguchi, T. (2011). The unique structure of carbonic anhydrase αCA1 from *Chlamydomonas reinhardtii. Acta Crystallographica Section D: Structural Biology*, 67(10), 894–901.

Tachibana, M., Allen, A.E., Kikutani, S., Endo, Y., Bowler, C., Matsuda, Y. (2011). Localization of putative carbonic anhydrases in two marine diatoms, *Phaeodactylum tricornutum* and *Thalassiosira pseudonana. Photosynthesis Research*, 109(1–3), 205–221.

Tripp, B.C., Smith, K., Ferry, J.G. (2001). Carbonic anhydrase: New insights for an ancient enzyme. *Journal of Biological Chemistry*, 276(52), 48615–48618.

Tripp, B.C., Bell III, C.B., Cruz, F., Krebs, C., Ferry, J.G. (2004). A role for iron in an ancient carbonic anhydrase. *Journal of Biological Chemistry*, 279(8), 6683–6687.

Tsuji, Y., Mahardika, A., Matsuda, Y. (2017a). Evolutionarily distinct strategies for the acquisition of inorganic carbon from seawater in marine diatoms. *Journal of Experimental Botany*, 3949–3958.

Tsuji, Y., Nakajima, K., Matsuda, Y. (2017b). Molecular aspects of the biophysical CO_2-concentrating mechanism and its regulation in marine diatoms. *Journal of Experimental Botany*, 68(14), 3763–3772.

Villarejo, A., Burén, S., Larsson, S., Déjardin, A., Monné, M., Rudhe, C., Karlsson, J., Jansson, S., Lerouge, P., Rolland, N. (2005). Evidence for a protein transported through the secretory pathway en route to the higher plant chloroplast. *Nature Cell Biology*, 7(12), 1224–1231.

Vullo, D., Del Prete, S., Osman, S.M., AlOthman, Z., Capasso, C., Donald, W.A., Supuran, C.T. (2017). *Burkholderia pseudomallei* γ-carbonic anhydrase is strongly activated by amino acids and amines. *Bioorganic & Medicinal Chemistry Letters*, 27(1), 77–80.

Wang, Y. and Spalding, M.H. (2014). Acclimation to very low CO_2: Contribution of limiting CO_2 inducible proteins, LCIB and LCIA, to inorganic carbon uptake in *Chlamydomonas reinhardtii. Plant Physiology*, 166(4), 2040–2050.

Wang, L., Yamano, T., Takane, S., Niikawa, Y., Toyokawa, C., Ozawa, S.I., Tokutsu, R., Takahashi, Y., Minagawa, J., Kanesaki, Y. *et al.* (2016). Chloroplast-mediated regulation of CO_2-concentrating mechanism by Ca^{2+}-binding protein CAS in the green alga *Chlamydomonas reinhardtii. Proceedings of the National Academy of Sciences of the United States of America*, 113(44), 12586–12591.

Whittington, D.A., Waheed, A., Ulmasov, B., Shah, G.N., Grubbv J.H., Sly, W.S., Christianson, D.W. (2001). Crystal structure of the dimeric extracellular domain of human carbonic anhydrase XII, a bitopic membrane protein overexpressed in certain cancer tumor cells. *Proceedings of the National Academy of Sciences of the United States of America*, 98(17), 9545–9550.

Xu, Y., Feng, L., Jeffrey, P.D., Shi, Y.G., Morel, F.M.M. (2008). Structure and metal exchange in the cadmium carbonic anhydrase of marine diatoms. *Nature*, 452(7183), 56–61.

Yamano, T., Tsujikawa, T., Hatano, K., Ozawa, S., Takahashi, Y., Fukuzawa, H. (2010). Light and low-CO_2-dependent LCIB-LCIC complex localization in the chloroplast supports the carbon-concentrating mechanism in *Chlamydomonas reinhardtii. Plant and Cell Physiology*, 51(9), 1453–1468.

7

Rubisco Microcompartments: The Function of Carboxysomes and Pyrenoids in Aquatic CO_2-Concentrating Mechanisms

Moritz T. MEYER
Department of Molecular Biology, Princeton University, USA

7.1. Introduction

The primary source of carbon supporting terrestrial and aquatic life is CO_2 fixed by Rubisco in the Calvin–Benson–Bassham cycle (CBB). There are at least five alternative carbon assimilation pathways (Berg 2011) but their contribution to the carbon budget is, by current estimates, negligible.

On land, the supply of CO_2 to Rubisco is unaided in 93% of all plants (by number of species). Numerous diffusion barriers reduce the CO_2 concentration between the boundary layer and the active site of Rubisco, limiting the efficiency of photosynthesis. In the remaining 7%, molecular add-ons operate upstream of the CBB. In these mechanisms (known as crassulacean acid metabolism, CAM, and the C_4 dicarboxylic acid pathway, C_4), first bicarbonate (HCO_3^-) and then an organic carbon-carrier molecule serve as intermediates to accumulate an intracellular pool of carbon. Catalytic back-conversion to CO_2 occurs at night (CAM) or in specialized cells (C_4). This reduces CO_2 leakage and supplies Rubisco with CO_2 concentrations higher than possible through diffusion only (see Chapter 5 for C_4 in aquatic photoautotrophs). CAM and C_4 plants account for ~25% of the net primary carbon fixation on land (Kellogg 2013).

In water, access to free dissolved CO_2 is hindered by an ~10^4 lower diffusion factor. By the mid-1970s, experiments with radiolabeled tracers on submerged plants and algae had demonstrated that HCO_3^- can act as a reservoir for extracellular CO_2 (Steemann Nielsen 1952). The C_4 pathway had also been described (Hatch 2002). Equipped with this knowledge, and after testing that no carboxylase other than Rubisco was present in the single-cell species they studied, M. Badger, A. Kaplan and J. Berry deduced that a CO_2-concentrating mechanism (CCM) operates in cyanobacteria and algae (Badger *et al.* 1978).

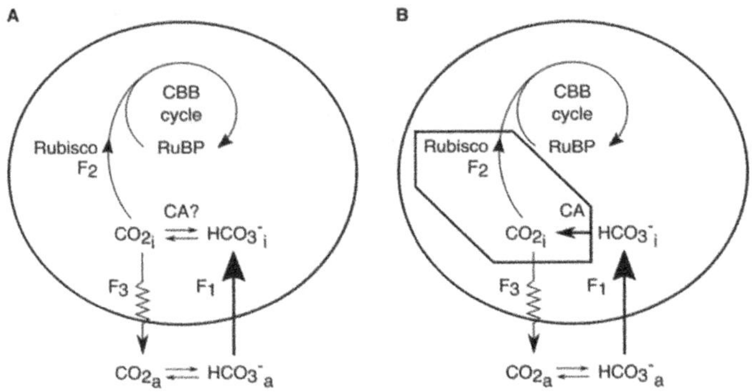

Figure 7.1. *General model of the CO_2-concentrating mechanism in cyanobacteria and microalgae. A. Historical CCM model showing how active HCO_3^- transport (F1) was hypothesized to increase the CO_2-fixation rate (F2). At steady state, when F1 > F2, then intracellular CO_2 > ambient CO_2. The original model explicitly ignored intracellular compartmentation and did not impose a unidirectional conversion of HCO_3^- to CO_2. F3 is the flux rate of the diffuse leak of CO_2 to the medium. Redrawn from* Badger et al. (1978). *B. Augmented CCM model incorporating the catalytic conversion of HCO_3^- to CO_2 by a carbonic anhydrase, and sequestration of Rubisco and the carbonic anhydrase into a leakage-minimizing microcompartment, illustrating the principle of the cyanobacterial carboxysome or algal pyrenoid*

Their model proposed that extracellular HCO_3^- is actively transported into the cell where it is converted to CO_2 for assimilation (Figure 7.1(A)). Unlike the C_4 mechanism, the aquatic CCM makes the economy of not requiring an organic carbon intermediate. It simply taps into the pool of bicarbonate naturally present in aquatic environments (in oceans, bicarbonate ions are ~200 times more abundant than free dissolved CO_2). The model initially showed no barrier between the site of accumulation of bicarbonate and the site of conversion to CO_2. This was supplemented by incorporating a microcompartment for Rubisco, carboxysomes (in cyanobacteria) and pyrenoids (in eukaryotic algae), and by demonstrating that this spatial separation was needed for the aquatic CCM to work efficiently (Figure 7.1(B)). Ectopic expression of a carbonic anhydrase (see Chapter 6) outside the Rubisco microcompartment renders the CCM ineffective (Price and Badger 1989).

Aquatic photoautotrophs assimilate nearly as much CO_2 as land plants with only 10% of the planet's Rubisco (Field *et al.* 1998; Bar-On and Milo 2019). Most of this performance owes to the ability that (seemingly all) cyanobacteria and most algae have to induce a CCM when CO_2 becomes limiting. The acclimation involves in-depth transcriptional changes, and many research efforts are now focusing on understanding the post-transcriptional and metabolic regulation of CCMs (see, for example, Burnap *et al.* (2015) and Launay *et al.* (2020)).

In this chapter, the tremendous progress made in unraveling the molecular details of the aquatic CCM will be described. We will show that cyanobacteria and eukaryotic algae have nearly identical strategies to source and concentrate inorganic carbon. Crystallography and cryogenic electron microscopy have now resolved with atomic-scale precision the fine structure of the Rubisco microcompartments, carboxysomes and pyrenoids. It is also becoming apparent that these organelles assemble in part through liquid–liquid phase separation. The successful introduction of a prokaryotic or eukaryotic CCM into a land plant that normally relies on diffusive CO_2 will be a powerful validation of our understanding of the aquatic CCM. We will show that researchers are getting tantalizingly close to realizing this objective.

7.2. The cyanobacterial CCM

7.2.1. *Cyanobacteria accumulate bicarbonate through high- and low-affinity uptake systems*

In cyanobacteria, extracellular HCO_3^- has to cross three barriers to reach the cytosol, where it accumulates as a pool. The first two barriers, an outer membrane and a layer of peptidoglycan, are permeable to small ions (Kowata *et al.* 2017); the inner (cytoplasmic) membrane is not. Three different bicarbonate transporters have been identified at the inner membrane (Figure 7.2): a five-component ATP-binding cassette transporter BCT1 (Omata *et al.* 1999; Koropatkin *et al.* 2007) and two Na^+-dependent symporters, SbtA (Shibata *et al.* 2002), and the dimeric "elevator-type" transporter BicA (Price *et al.* 2004; Wang *et al.* 2019a). BCT1 and SbtA have a high affinity for HCO_3^- and are transcriptionally induced when CO_2 becomes limiting (McGinn *et al.* 2003). Half-saturating concentrations for HCO_3^- are ~15 μM and

~2 μM; BicA has a lower affinity of ~40 μM. The initial discovery process was hampered by the multiplicity of uptake systems and the small growth effect of single transporter mutants.

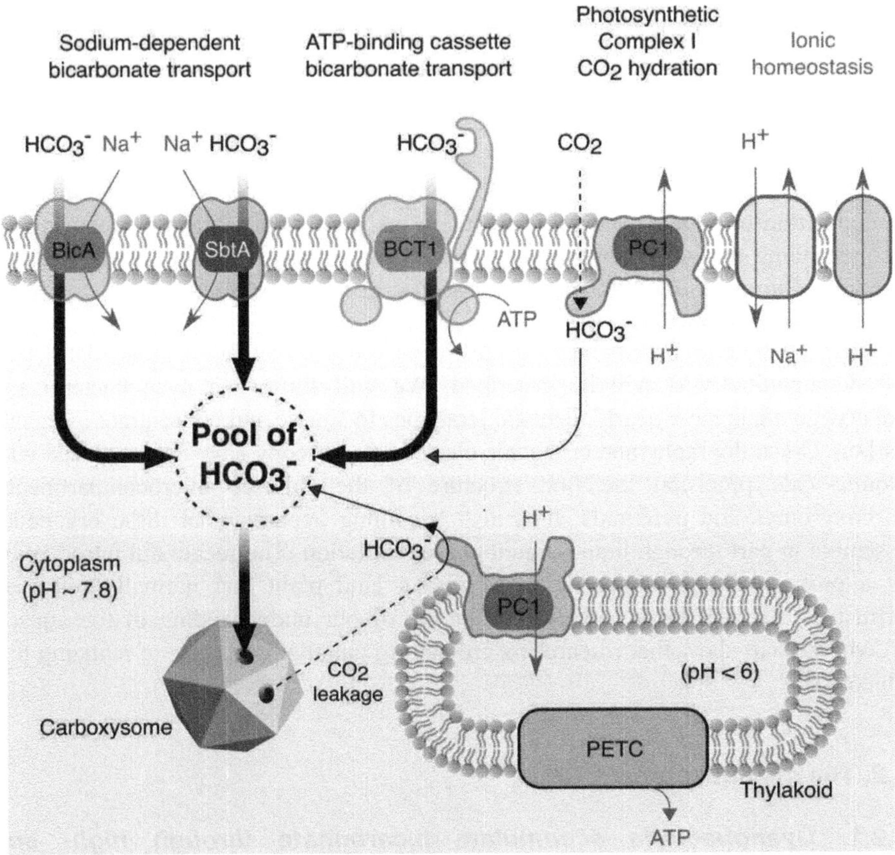

Figure 7.2. *Inorganic carbon uptake and CO_2 recapture systems in cyanobacteria. Schematic model illustrating how inorganic carbon fluxes across the inner cytoplasmic membrane and within the cytosol, around thylakoids and carboxysomes, contribute to creating an intracellular pool of HCO_3^-. Solid black arrows, HCO_3^- flux; dashed black arrows, CO_2 flux; grayed out arrows, other ionic fluxes; and PETC, photosynthetic electron transport chain. Localization of PC1 at the inner cytoplasmic membrane is conjectural. A single carboxysome is shown. Cyanobacteria have ~6 or more carboxysomes. For a color version of this figure, see www.iste.co.uk/maberly/photosynthesis.zip*

Cyanobacteria also have membrane-bound photosynthetic CO_2 hydration complexes resembling NADH dehydrogenases, using ferredoxin instead of NADH as redox donor. These complexes have a cytosol-exposed vectorial carbonic anhydrase (vCA) working against the gradient. An inducible high-affinity complex and a constitutive low-affinity complex have been identified (Shibata *et al.* 2001; Folea *et al.* 2008). Previously annotated as NDH-1_3 and NDH-1_4, they are now designated by PC1. These complexes localize to the thylakoids, although the plasma membrane is sometimes proposed as an alternative localization. The crystal structure of PC1 resolved an internal channel connecting the thylakoid lumen to the vCA moiety (Laughlin *et al.* 2019; Schuller *et al.* 2020). If this was the route followed by CO_2, bicarbonate would need to be supplied to the thylakoid lumen to sustain vCA activity. Such a futile cycle is incompatible with our current understanding of the cyanobacterial CCM and there is no evidence for a bicarbonate transporter in thylakoid membranes (Hagemann and Kaplan 2020). If the complex was at the plasma membrane instead, it could directly pump CO_2 from the inner periplasm and contribute to building the intracellular pool of HCO_3^- (as illustrated in Figure 7.2). This system would be analogous to the plasmalemma DAB complex in non-photosynthetic proteobacteria that have a CBB. DAB is a two-protein complex, with one component pumping periplasmic protons (or sodium) into the cytosol and the other component a cytosol-exposed vCA converting CO_2 to HCO_3^-. This generates an inward concentration gradient for the diffusive entry of periplasmic CO_2 (Desmarais *et al.* 2019). The current model for PC1 favors an all intra-cellular recycling function, where unfixed CO_2 leaking from carboxysomes is recaptured by conversion to HCO_3. This function is compatible with either membrane localization.

When all cyanobacterial bicarbonate pumps and recycling systems are knocked out, mutants cannot grow without the addition of high concentrations of CO_2 to the growth medium (Xu *et al.* 2008). Cyanobacteria therefore require a minimum set of these systems. Whole-genome searches found that the bicarbonate transporters and photosynthetic hydration complexes are widely conserved (Scott *et al.* 2020), with absence/presence associated with environmental preferences. A few, but ecologically extremely important, picoplankton species adapted to nutrient-poor environments (e.g. open ocean *Prochlorococcus marinus* and various Synechococcus species) underwent large genome reductions (Dufresne *et al.* 2003; Badger *et al.* 2006; Sandrini *et al.* 2014). In these species, BCT1 and the low-affinity PC1 isoform are absent.

7.2.2. *Carboxysomes belong to two distinct evolutionary lineages*

Carboxysomes are a particular class of bacterial microcompartments (BMC) specialized in the fixation of CO_2. BMC perform many different metabolic functions across numerous bacterial phyla and are unified by an icosahedral protein shell encoded by an operon (Kerfeld *et al.* 2018). Icosahedral capsids are common in nature (e.g. spherical viral capsids) as the high degree of symmetry facilitates the assembly process and incurs a minimal genetic cost (Twarock and Luque 2019).

Carboxysomes are designated as α or β. The building blocks of the protein shell are nearly identical (Table 7.1), but the chromosomal arrangement of the genes is different and the main cargo enzyme, Rubisco, has a different evolutionary origin (Watson and Tabita 1997; Badger *et al.* 2002).

Function	α-carboxysome	β-carboxysome
Shell proteins		
Hexameric, facet	CsoS1A, CsoS1B, CsoS1C	CcmK1, CcmK2, CcmK3, CcmK4
Pentameric, vertex	CsoS4A, CsoS4B	CcmL
Trimeric, large pore	CsoS1D, CsoS1E	CcmO, CcmP
Enzymatic cargo proteins		
Rubisco large and small	CbbL, CbbS	RbcL, RbcS
Carbonic anhydrase	CsoS3 (= CsoSCA)	CcaA or N-term of CcmM58
Linker proteins		
Rubisco only linker	CsoS2A (short version of CsoS2)	Ccm35 (short version of CcmM)
Rubisco to shell linker	CsoS2B (full-length CsoS2)	-
Rubisco to CA linker	-	Ccm58 (full-length CcmM)
CA to shell linker	Domain on CsoS3	CcmN

Table 7.1. *Functional equivalence of α– and β–carboxysome proteins. α-carboxysomes are encoded by a single polycistronic operon (CsoS). β-carboxysomes are encoded by several carbon-concentrating mechanism (CCM) operons, in close proximity of each other on the cyanobacterial chromosome*

The α-carboxysomes encapsulate a proteobacterial Form IA Rubisco and their capsid is encoded by a single operon. The β-carboxysomes encapsulate the archetypal Form IB Rubisco shared by cyanobacteria, green algae and all land plants, and their capsid is encoded by multiple operons (see Chapter 5). Cyanobacteria with an α-carboxysome include the above-mentioned oligotrophic picoplankton species that underwent a genome contraction. These cyanobacteria inherited an α-carboxysome together with Form IA Rubisco through lateral gene transfer from a proteobacterial donor. The endogenous Rubisco was displaced by the chemoautotrophic variant, alongside a stringent need to retain the proteobacterial BMC operon arrangement. The first carboxysome to be isolated was from a sulfur-oxidizing γ-proteobacterium, *Halothiobacillus neapolitanus* (Shively *et al.* 1973). Several other proteobacteria (α, β, γ), as well as actinobacteria, also have carboxysomes.

7.2.3. *Carboxysomes isolate Rubisco from the rest of the CBB cycle*

Carboxysomes are bound by a protein shell that surrounds an enzymatically active matrix composed of Rubisco and a carbonic anhydrase. All other CBB enzymes are cytosolic. CA dehydrates the cytosolic HCO_3^- and supplies Rubisco with saturating concentrations of CO_2. The shell must therefore allow the trafficking of substrates and products to couple Rubisco to the CBB. It also has to be permeable to HCO_3^-, protons and water to sustain CA activity. The selective permeability is a requirement of the CCM model (Figure 7.3). Nearly all the shell proteins have now been crystallized, revealing the porous nature of carboxysomes. The pores that form during oligomerization have distinct biochemical properties that modulate the fluxes.

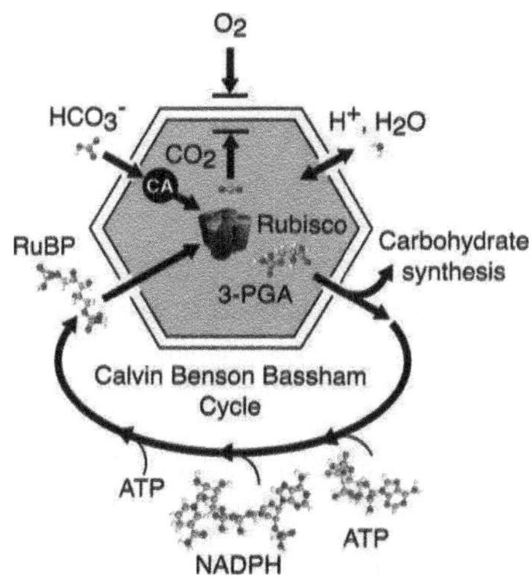

Figure 7.3. *Metabolic pathway of CO_2 assimilation in a carboxysome. The carboxysome shell must facilitate the entry of catalytic substrates (the five-carbon RuBP for Rubisco and HCO_3^- for the carbonic anhydrase) and the exit of products (the three-carbon PGA), while minimizing CO_2 leakage and entry of the competing Rubisco substrate, O_2. Molecules larger than RuBP are probably size excluded. For a color version of this figure, see www.iste.co.uk/maberly/photosynthesis.zip*

7.2.4. *Carboxysome shells are composed of thousands of self-assembling capsid proteins*

The icosahedral geometry of a carboxysome is illustrated in Figure 7.4. Carboxysomes vary with respect to size, regularity and stoichiometry of the different shell and cargo proteins. Vertices are formed by a pentameric protein. Mutants in this gene typically lack carboxysomes (Cai *et al.* 2009). The 20 large (irregular) hexagonal faces are a mostly single-layer lattice of hexameric proteins encoded by a gene family with multiple isoforms (Table 7.1). Thermodynamically less stable heterohexamers can be present and are likely important for protein turnover (Garcia-Alles *et al.* 2019; Sommer *et al.* 2019). Locally, the shell can be double-layered, with the stacking of hexamers encoded by proteins other than those forming the bulk of the shell (Cai *et al.* 2013).

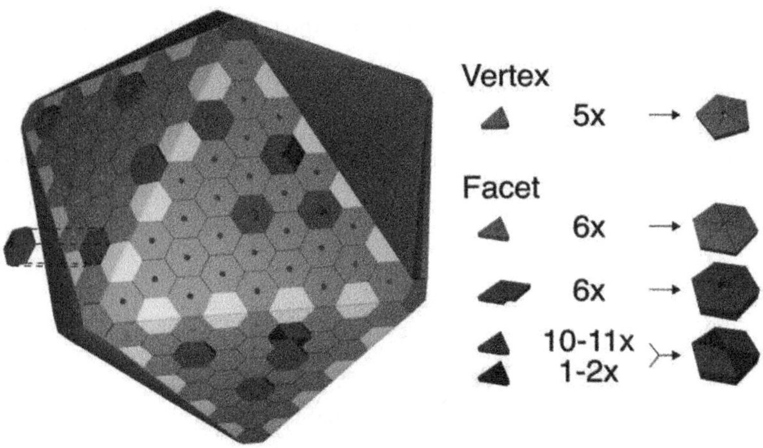

Figure 7.4. *Structure of a β-carboxysome shell. Single-layer facets are composed of homohexamers encoded by a gene family. Heterohexamers capping a homohexamer (dark blue) and stacked trimers forming a wider pore (purple) can also be present. The number and position of stacked capsomeres is not known precisely. Carboxysome facets are probably not as planar as illustrated here. If carboxysomes are slightly convex, the need for folded hexamers at the edges (green) is eliminated. The hollow lumen packs the enzymatic cargo (not shown) via dedicated linkers. The smaller α-carboxysomes differ in detail but not in general principle. For a color version of this figure, see www.iste.co.uk/maberly/photosynthesis.zip*

Hexamers have a concave side turned towards the cytosol and a convex side opening into the lumen. A central pore traverses each hexamer and the electrostatic properties at opposite ends are different. Properties vary across isoforms. Synthetic hexagonal lattices assemble in *E. coli* by the incorporation of preformed hexamers (Sutter *et al.* 2017). Importantly, hexamers are incorporated in a uniform orientation. The outer and the inner shell surfaces are therefore isotropic, with properties determined by the pore biochemistry and stoichiometry of different shell proteins. For example, the β-carboxysomal CcmK2 homohexamer has a mostly negatively charged concave side, a positively charged central pore, and a positively charged convex side. Molecular dynamics simulation predicts that it is energetically favorable for negatively charged HCO_3^- to move through the ~7 Å pore (Faulkner *et al.* 2020). These simulations also predict which residues are important for the electrostatic charge-dependent selective permeability. For example, the mutation of a serine on the convex side of the CcmK2 pore led to an ~40% decrease in CO_2 fixation in air-grown mutants. The larger 3-phosphoglyceric acid (3-PGA) and ribulose-1,5-bisphosphate (RuBP) are probably beyond the carrying capacity of this CcmK isoform but the double-stacked CcmP, with a pore twice as wide, could accommodate these molecules (Cai *et al.* 2013).

The self-assembly into a highly organized and globally stable structure requires cargo proteins and maybe chaperones, pending further experimentation. Individual capsid proteins bind through weak but numerous electrostatic interactions. Small symmetrical and stable carboxysomes, a tenth the size of wild-type carboxysomes, can form in *E. coli* by expressing only two genes coding for hexamers and the gene coding for the pentameric vertex (Sutter *et al.* 2019). The self-assembly into a larger structure requires a fine control of capsid protein stoichiometry, scaffolding proteins and a nucleation site in the form of a cargo protein aggregate. This requirement is common to many other spherical capsids (Li *et al.* 2018).

7.2.5. *Alpha and beta carboxysomes package the enzymatic cargo with different protein linkers*

α-carboxysomes are 70–160 nm in diameter and contain ~250 individual loosely arranged Rubisco molecules (Schmid *et al.* 2006; Iancu *et al.* 2007; Hantke *et al.* 2014). The larger β-carboxysomes (up to 400 nm in diameter) contain about seven times more Rubisco and the packing is often paracrystalline (Gantt and Conti 1969; Faulkner *et al.* 2017). α- and β-carboxysomes use different linker proteins to condense Rubisco into a matrix and to anchor the enzymatic cargo to the shell, but there are striking parallels between the mechanisms in the two types of carboxysomes (Table 7.1).

The α-carboxysomal Rubisco linker, CsoS2, is a large protein (~90 kDa) with three roughly equally sized regions. The N-terminal domain has four copies of a short basic motif that binds to the electronegative ridge between adjacent Rubisco large subunit dimers (Oltrogge *et al.* 2020) (see also comparative Figure 7.7 below). Only Rubisco holoenzymes can bind as the interaction involves residues on a proximal small subunit. The C-terminal domain binds to capsid proteins. C-truncation mutants are incapable of assembling carboxysomes (Chaijarasphong *et al.* 2016). A translational frameshifting site produces a variant of the linker that lacks the capsid-binding domain; it only binds Rubisco.

α-carboxysome assembly starts with a partial shell formation, with full-length variants of the linker tethering nascent carboxysome shell facets. The short variant of the linker condenses Rubisco into a matrix. The docking of cargo with a partially formed shell allows shell closure. This model is consistent with observations that the shell of α-carboxysomes starts to form independently of Rubisco aggregates (Menon *et al.* 2008; Iancu *et al.* 2010). Incorporation of the α-carboxysomal CA proceeds independently. The CA has its own shell binding domain (So *et al.* 2004), preventing it from being trapped inside the Rubisco matrix. The exact anchoring point and timing of incorporation during carboxysome assembly have yet to be determined experimentally.

The gene coding for the β-carboxysomal Rubisco linker, *CcmM*, also has a translational frameshifting site, producing a short and a full-length form of the protein. Both forms have multiple electropositive C-terminal repeats that bind, similarly to the α-carboxysome linker, to the electronegative pocket between adjacent Rubisco large subunit dimers, albeit in a more equatorial position (Wang *et al.* 2019b) (Figure 7.7). An earlier model speculated that a cross-linking of β-carboxysomal Rubisco could also occur if the C-terminal repeats of the linker displaced Rubisco small subunits (Long *et al.* 2007; Niederhuber *et al.* 2017). This was not unreasonable given sequence similarly and predicted protein fold, but crystallization revealed that the linker lacked critical residues needed for Rubisco holoenzyme stability, and could therefore not replace the small subunit (Ryan *et al.* 2019). The full-length form of the protein has an N-terminal domain that either binds to a separately encoded CA (McGurn *et al.* 2016) or is itself a CA (Peña *et al.* 2010). A second linker (or assembly) protein tethers the CA to the carboxysome shell (Kinney *et al.* 2012). Another difference with α-carboxysomes is that β-carboxysomes assemble the Rubisco matrix before encapsulation by shell proteins (Cameron *et al.* 2013; Chen *et al.* 2013).

7.3. The algal CCM

The term "algae" masks a hugely diverse and polyphyletic grouping of eukaryotic photoautotrophs (see Chapter 3). There is physiological and molecular evidence of CCM activity in most algal phyla, with a few exceptions such as the chrysophytes (Maberly *et al.* 2009). For a discussion of the methods to diagnose the presence/absence of a CCM, see Beardall and Raven (2020). One key difference with cyanobacteria is that extracellular inorganic carbon has to cross at least one more barrier (the eukaryotic plasma membrane) and quite a few more barriers when the chloroplast was inherited from a serial endosymbiosis. How the different uptake and accumulation systems are coordinated across multiple sub-cellular compartments is poorly understood. Another difference is that the eukaryotic Rubisco microcompartment, the pyrenoid, usually lacks a peripheral boundary, exposing Rubisco to the chloroplast stroma. This seemingly conflicts with the CCM paradigm (see Figure 7.1(B)). Green algae co-opted the thylakoid lumen to achieve a compartmentation functionally analogous to cyanobacterial carboxysomes. Whether all algae with a pyrenoid operate on a similar basis remains speculative. Not all algae with a CCM have a pyrenoid but the most abundant species in phytoplankton populations, like diatoms and common coccolithophores like *Emiliania huxleyi*, typically have a pyrenoid (Meyer *et al.* 2020a).

7.3.1. *The model alga* Chlamydomonas *has multiple acclimation states*

The molecular characterization of an algal CCM is most advanced in the unicellular green alga *Chlamydomonas reinhardtii* (Figure 7.5(A)), one of the best established model systems for the study of aquatic photosynthesis (Sasso *et al.* 2018). The Chlamydomonas CCM, similarly to the cyanobacterial CCM, is induced when CO_2 is limiting and switched off when concentrations are high (>0.5% CO_2 in air). Chlamydomonas has two distinct low-CO_2 acclimation states, very-low CO_2 (<0.02% CO_2 in air) and air-level CO_2, eliciting nearly identical transcriptomes (Fang *et al.* 2012). This may be a Chlamydomonas idiosyncrasy, an adaptation to a terrestrial habitat. Chlamydomonas is a soil-dwelling alga, with a higher affinity for CO_2 than HCO_3^-. In air-acclimated cells, $K_{1/2}$ for CO_2 is ~1 μM vs ~80 μM for HCO_3^- (Amoroso *et al.* 1998). The exact physiological importance is not fully understood, and it has not been systematically investigated if other algae have multiple low-CO_2 acclimation states.

7.3.2. Chlamydomonas *has a cooperative* CO_2*–*HCO_3^- *uptake system*

HLA3 (for high-light-activated) and LCI1 (for low-CO_2-inducible) code for transmembrane proteins that form a supramolecular assemblage at the plasmalemma (Figure 7.5(B1,2)). The localization was validated for each transporter individually by expressing *in vivo* fluorescent fusion proteins (Atkinson *et al.* 2016). The supramolecular assembly was deduced from immunoprecipitation mass-spectrometry data; FLAG-tagged LCI1 pulled down HLA3 and reciprocally (Mackinder *et al.* 2017). HLA3 belongs to the ubiquitous superfamily of ATP-binding cassette transporters and is primarily active when CO_2 is very low (Gao *et al.* 2015). The capacity to transport HCO_3^- was validated in *Xenopus* oocytes (Atkinson *et al.* 2016), an expression system commonly used to characterize plant transporters. LCI1 is a transmembrane protein with no homolog in other algae (Ohnishi *et al.* 2010). While mutants affected in LCI1 display no discernible growth phenotype, LCI1 is believed to facilitate CO_2 uptake. The crystal structure resolved a homotrimer with each subunit delimiting a channel lined with residues allowing a directional passage of CO_2, but not charged ions (Kono *et al.* 2020).

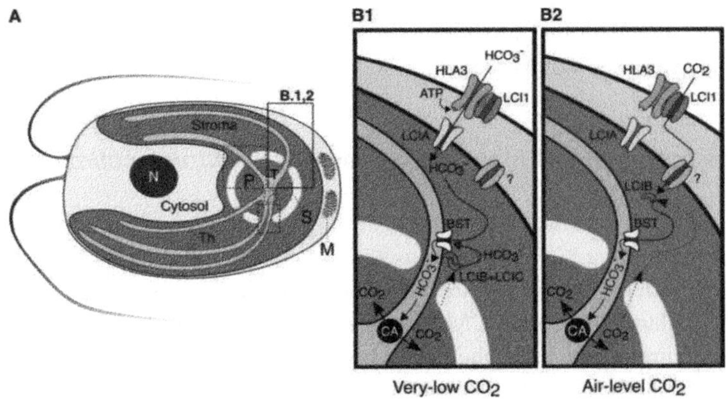

Figure 7.5. *Inorganic carbon uptake in* Chlamydomonas reinhardtii. *A, Illustration of the flagellated unicellular model green alga Chlamydomonas. Extracellular inorganic carbon must reach the pyrenoid (P). Densely packed Rubisco (gray) assembles around tubules (T), continuous with thylakoids (Th). Starch plates (S) enclose the Rubisco matrix. Nucleus (N). Mitochondria (M) could be a source of* CO_2 *through cellular respiration, but this is not discussed in the present chapter. Dashed boxes refer to Figure 7.6. B1, Preferred* HCO_3^- *uptake route under very-low* CO_2*. Converted* CO_2 *leaking from the pyrenoid is recycled by LICB+LCIC (after* Mackinder 2018*). B2, Preferred* CO_2 *uptake route under air-level* CO_2*. Stromal* CO_2 *(actively delivered through membrane channels or leaking from the pyrenoid) is converted to* HCO_3^- *by a soluble LCIB, for uptake by a bestrophin (BST). For a color version of this figure, see www.iste.co.uk/maberly/photosynthesis.zip*

LCIA and CIA8 code transmembrane proteins at, or predicted to be at, the inner chloroplast envelope (Figure 7.5(B1,2)). The localization of LCIA was validated by fluorescent tagging (Yamano *et al.* 2015). LCIA belongs to the ubiquitous family of nitrate transporters and is, like HLA3, primarily active when CO_2 concentrations are very low (Wang and Spalding 2014). The capacity to transport HCO_3^- was validated using the same protocol as for HLA3 (Mariscal *et al.* 2006). The exact function (and localization) of CIA8 is unknown, but deletion mutants show that it plays an important role in acclimation to air-level CO_2 (Machingura *et al.* 2017).

The current model of the Chlamydomonas CCM proposes a coordinated uptake of HCO_3^- and CO_2 through serially arranged transporters and channels in the plasma and chloroplast membranes. When CO_2 is severely limiting (Figure 7.5(B1)), HLA3 and LCIA pump extracellular HCO_3^- into the chloroplast stoma. Under air-level CO_2 (Figure 7.5(B2)), a unidirectional influx of CO_2 via LCI1 and a CO_2 channel in the chloroplast membrane is active (Kono and Spalding 2020), and the more costly HCO_3^- uptake system is likely tuned down. Further experiments are needed to determine whether CIA8 is a suitable candidate for the chloroplast CO_2 channel and whether the model identified all (and the correct) uptake channels. Compellingly, *hla3/lcia* double mutants cannot grow under very-low CO_2 or alkaline conditions, when HCO_3^- is the predominant form of dissolved inorganic carbon (Yamano *et al.* 2015).

7.3.3. Chlamydomonas *has a stromal vCA to capture CO_2*

The function of the stromal protein LCIB is still debated. The overall fold and active site architecture are those of a carbonic anhydrase but recombinant proteins have no CA activity when assayed *in vitro* (Jin *et al.* 2016), which could also mean that LCIB is under some unknown allosteric control. LCIB is included in most Chlamydomonas CCM models in a dual capacity, analogous to the cyanobacterial vCA PC1 and proteobacterial DAB. Under very-low CO_2, the protein forms a bimolecular complex with an ~60% identical protein, LCIC, that localizes to the periphery of the pyrenoid (Yamano *et al.* 2010; Toyokawa *et al.* 2020). The localization is mediated by the binding of LCIC to transmembrane proteins situated at the thylakoids (see below). There, the vCA moiety of LCIB is believed to recapture unfixed CO_2 leaking from the pyrenoid (Figure 7.5(B1)). Under air-level CO_2, the pyrenoid-peripheral localization is lost. This could shift the function towards assisting LCI1 and a chloroplast envelope CO_2 channel (CIA8?) in generating an inward concentration gradient of CO_2 (Figure 7.5(B2)). Mutants of LCIB can grow under very-low CO_2 but not under air-level CO_2, clearly suggesting an essential role in sustaining Rubisco activity when the preferred source of inorganic carbon for Chlamydomonas is CO_2. More experiments are needed to disentangle the role of LCIB and explain the different localization under very-low and air-level CO_2.

7.3.4. *Stromal HCO_3^- is catalytically dehydrated to CO_2 inside thylakoid lumen*

Whereas cyanobacteria prevent the premature conversion of intracellular HCO_3^- to CO_2 by sequestering the CA into carboxysomes, Chlamydomonas co-opted the thylakoid lumen space and a resident CA. At least one additional transporter is therefore needed to supply stromal bicarbonate to the CA (Figure 7.5(B1,2)). The best candidates for that role are BST1-3, Ca^{2+}-activated transmembrane proteins of the bestrophin family that localize to thylakoids (Mukherjee *et al.* 2019). Immunoprecipitation experiments showed that the LCIC component of the CO_2 recapture system docks on two of these bestrophins (Mackinder *et al.* 2017). Only when all three BST genes are silenced, and when mutants are grown in very-low CO_2, is growth fully impaired, leaving open the possibility that there is some redundancy in this final uptake step. The catalytically released CO_2 then freely diffuses out of the thylakoid lumen to the surrounding Rubisco matrix anchored onto these membranes.

Several authors believe that this luminal CA is important for maintaining PSII activity in removing protons from the oxygen-evolving site, by reforming HCO_3^- from CO_2 (Shutova *et al.* 2008; Terentyev *et al.* 2019). One way to reconcile this conflict (the same CA would need to run in opposite directions when cells are photosynthetically active) is by recognizing that the membranes associated with the pyrenoid have their distinct biochemical and ultrastructural identity (Figure 7.6(A,B)). Although continuous with and resulting from the merger of several thylakoids, pyrenoid tubules are depleted in PSII, enriched in PSI and form a central reticulated network (Engel *et al.* 2015).

7.3.5. *The* Chlamydomonas *pyrenoid has a complex architecture*

At the heart of the Chlamydomonas pyrenoid, a reticulated network of tubules forms a scaffold anchoring a single spherical aggregate of loosely packed Rubisco (Figure 7.5(A)). When cells are photosynthetically active and the CCM is fully induced, the Rubisco matrix is surrounded by a starch sheath assembled from a few plates roughly in the shape of spherical lunes (Meyer *et al.* 2016). The starch sheath is unique to chloroplasts of the "green" lineage and may play a role in minimizing CO_2 leakage. Most algal pyrenoids are reduced to a densely packed Rubisco matrix, usually anchored around membranes. Continuity of the pyrenoid membranes with thylakoids is not always visible in thin-section electron micrographs. A wider use of cryo-electron tomography and focused ion beam milling scanning electron microscopy could resolve this uncertainty.

7.3.6. *Rubisco and EPYC1 condense into a bimolecular complex*

EPYC1 (for Essential Pyrenoid Component) encodes an intrinsically disordered protein that localizes to the pyrenoid and links Rubisco into a densely packed matrix (Mackinder *et al.* 2016). The precise nature of the binding was resolved *in vitro*, by probing interactions between synthetic EPYC1 peptides and purified Rubisco, and by single-particle cryo-electron microscopy of purified Rubisco saturated with EPYC1 peptides (He *et al.* 2020). EPYC1 has five highly similar and evenly spaced repeats, and each repeat contains a motif that binds with one Rubisco small subunit, via multiple electrostatic and hydrophobic interactions. The tetragonal symmetry of Rubisco allows for eight binding sites with EPYC1. Chlamydomonas, α-cyanobacteria and β-cyanobacteria therefore use distinct regions on the Rubisco surface to assemble the CCM microcompartment (see comparative Figure 7.7, below).

The two-component Rubisco–EPYC1 system spontaneously condenses in solution, forming spherical aggregates and the entrapped Rubisco is catalytically competent (Freeman Rosenzweig *et al.* 2017; Wunder *et al.* 2018). But additional molecular factors are needed to anchor the Rubisco–EPYC1 matrix near the CA where CO_2 is being released, that is, around tubules.

7.3.7. *Rubisco is anchored to tubules and starch plates by proteins sharing a binding motif*

The number of proteins that constitute a mature pyrenoid in Chlamydomonas is not known precisely. Purified pyrenoids yield close to 200 different proteins when examined with mass spectrometry (Zhan *et al.* 2018). And Rubisco binds to dozens of proteins other than EPYC1 (Mackinder *et al.* 2017). A unifying principle emerged with the discovery of a multivalent motif shared by a small set of chloroplast proteins that are simultaneously found in the pyrenoid proteome and the Rubisco interactome. The motif is similar to the motif that allows EPYC1 to bind to Rubisco.

The Rubisco-binding capacity of this motif was validated *in vitro* and *in vivo*. Addition of the motif to a non-pyrenoid-related stromal protein retargeted the protein to the pyrenoid, while deletion of the motif on a pyrenoid protein cancelled the pyrenoid-specific localization. This established that the motif is sufficient and necessary for pyrenoid localization, and that Rubisco acts as a hub for pyrenoid assembly (Meyer *et al.* 2020b).

Two hitherto uncharacterized proteins with the motif have predicted transmembrane domains and localize to the tubules (Figure 7.6(C)). One of these is a fourth member of the BST family, already implicated in the channeling of HCO_3^- to the luminal CA. Two other motif-containing proteins localize to the interface of the Rubisco–EPYC1 matrix and the starch sheath (Figure 7.6(C)). Mutants of one of these proteins containing a starch-binding domain have multiple pyrenoids dispersed throughout the chloroplast (Itakura *et al.* 2019). There is now a pressing need to characterize all motif-containing proteins and to investigate if the unifying principle of this Rubisco-binding motif explains the architecture of other algal pyrenoids.

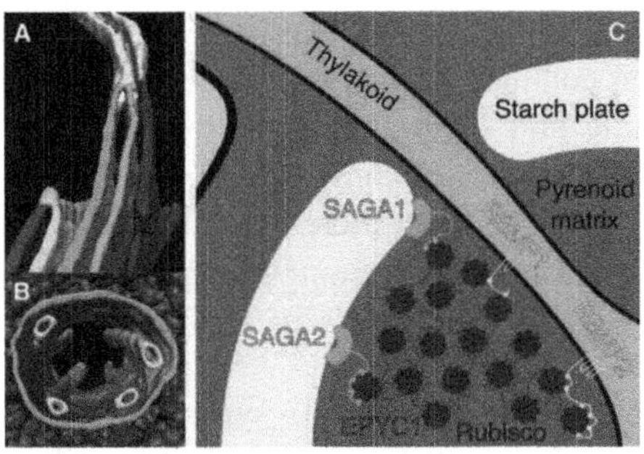

Figure 7.6. *A Rubisco-binding motif orchestrates the architecture of the Chlamydomonas pyrenoid around tubules. A, Thylakoid stacks transition to pyrenoid tubules. Colors indicate lumen connectivity (green and cyan merge to yellow) (image credit:* (Engel et al. 2015)*). B, Cross-section of a pyrenoid tubule (green), revealing a bundle of minitubules (orange). The lumen of each minitubule is continuous with the stromal thylakoid interspace. Minitubules might channel substrates and products between Rubisco and the rest of the Calvin–Benson–Bassham cycle in the stroma (by analogy with carboxysomal pores), but this has not been tested experimentally (image credit:* (Engel et al. 2015)*). C, Five pyrenoid proteins share a multivalent Rubisco-binding motif. Two uncharacterized Rubisco-binding membrane proteins (RBMP1 and RBMP2, in yellow) anchor the Rubisco–EPYC1 matrix (in gray and red) to the tubules. Two peripheral proteins with starch-binding domains (SAGA1* (Itakura et al. 2019) *and the uncharacterized homolog SAGA2 (in cyan)) mediate interactions between the Rubisco–EPYC1 matrix and the starch plates. For a color version of this figure, see www.iste.co.uk/maberly/photosynthesis.zip*

7.4. Introducing an aquatic CCM into crops could increase biomass production

Unsustainable economic and population growth is calling for innovative ideas. A new "green revolution" is needed to reduce agricultural land use without compromising the demands for food production. The engineering of an algal CCM into crop plants could help overcome this challenge. The addition of bicarbonate transporters to the chloroplast inner envelope is predicted to increase leaf CO_2 uptake by 10% and by up to 60% if Rubisco is confined to a microcompartment (McGrath and Long 2014). The associated increase in yield is expected to be upwards of 40%. The molecular toolkit at the disposal of synthetic biologists is now reasonably well characterized, as seen above. Below, key studies are highlighted, illustrating that aquatic CCM components could easily be expressed, transiently or stably, in chloroplasts of the model plants *Nicotiana tabacum* (tobacco) and *Arabidopsis thaliana.*

7.4.1. *Proto-carboxysomes and proto-pyrenoids assemble in chloroplasts*

Structurally stable carboxysomes self-assemble in tobacco chloroplasts with only three α-carboxysomal components: a major shell protein (that forms hexamers), a Rubisco linker and a cyanobacterial Form IA Rubisco (Long *et al.* 2018). Transgenic lines continue to express the native Rubisco (future experiments will look into repressing this co-expression) but only the cyanobacterial form is encapsulated. The ability to assemble carboxysomes without a pentameric protein is a remarkable result in its own right, highlighting the importance of a cargo in stabilizing the organelle. It will be interesting to see if the addition of an α-carboxysomal CA and corresponding linker is compatible with this minimalistic and probably leakage-prone shell. More likely, all components will be needed to mimic the selective permeability of a wild-type carboxysome shell. An earlier attempt, with β-carboxysomal components, had produced structures resembling carboxysomes without a discernible cargo (Lin *et al.* 2014).

Efforts to introduce an algal pyrenoid into a higher plant initially focused on generating pyrenoid-compatible Rubisco forms (Atkinson *et al.* 2017). Arabidopsis mutants missing several of the Rubisco small subunit genes (most land plants have multiple small subunit isoforms) were complemented with a chimeric variant mixing native Arabidopsis small subunit sequence and the regions needed in algae for condensing into a pyrenoid matrix (Meyer *et al.* 2012). The complemented lines restored near wild-type growth; an encouraging result given that Rubisco is notoriously resistant to mutagenesis without negative impact on the catalytic properties. Co-expression of full-length Chlamydomonas EPYC1 and the pyrenoid-competent Rubisco led to the formation of liquid–liquid phase-separated droplets *in vitro*, but no proto-pyrenoid was discernible *in planta* (Atkinson *et al.* 2019). This was remedied by replacing the EPYC1 N-terminal sequence with an Arabidopsis Rubisco small subunit transit peptide and incorporation into a high expression vector. Transgenic lines assembled a proto-pyrenoid in each chloroplast (Atkinson *et al.* 2020).

These experiments stress the importance of expressing a carboxysome or pyrenoid-competent Rubisco, that is, capable of interacting with matching cyanobacterial or eukaryotic linkers. Now that the binding mechanism between the different linkers and corresponding Rubisco forms are better understood (Figure 7.7), synthetic linkers compatible with land plant Rubisco could be explored. This would circumvent the need to modify or replace the native land plant Rubisco, which remains challenging. Presently, transgenic plants with synthetic carboxysomes require elevated CO_2 for maintenance, whereas chloroplasts with proto-pyrenoid currently provide no growth advantage.

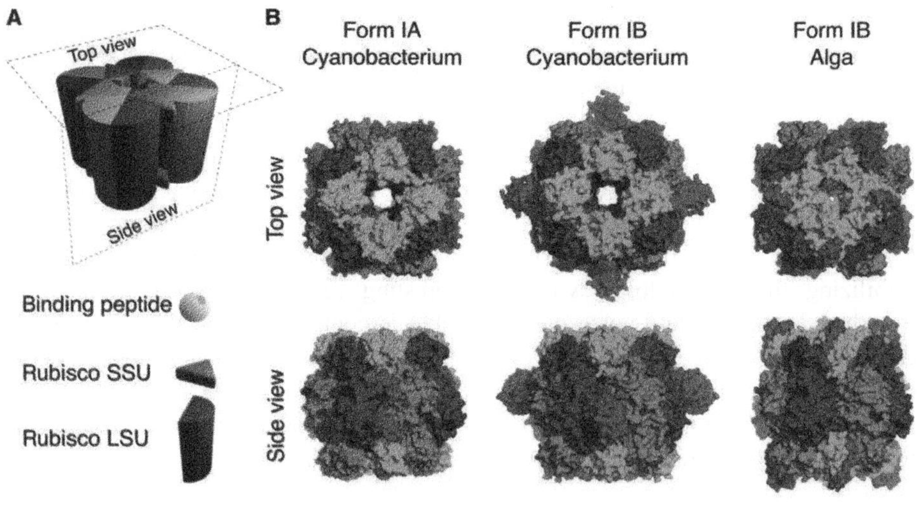

Figure 7.7. *Rubisco-linkage sites in cyanobacteria and Chlamydomonas. A, In most photosynthetic organisms, Rubisco is a hexadecameric assemblage of eight large subunits (four dimers, dark and light gray) and eight small subunits (cyan and green-cyan). B, Top and side views (as shown in A) of a Rubisco crystal with attached linker peptides. For a color version of this figure, see www.iste.co.uk/maberly/photosynthesis.zip*

Future efforts will need to concentrate on co-localizing a carbonic anhydrase with or near the Rubisco assemblages. Potential leakage points created by the endogenous CAs present in many cellular compartments will also need to be addressed (DiMario *et al.* 2017). Tapping into existing intracellular pools of HCO_3^- is predicted to provide an ~10% growth advantage, but the effects on pH homeostasis have not been properly evaluated. Some form of initial CO_2 capture by conversion to HCO_3^-, mimicking the first step of the C_4 and CAM mechanisms, might also be necessary.

7.4.2. Cyanobacterial and algal inorganic carbon transporters can be targeted to the chloroplast envelope

Two of the three cyanobacterial bicarbonate transporters (BicA and SbtA) have now been successfully expressed at the inner chloroplast envelope of tobacco and Arabidopsis with nuclear-encoded transgenes (Rolland *et al.* 2016; Uehara *et al.* 2016, 2020). No effect on growth has yet been reported.

Three Chlamydomonas inorganic carbon uptake systems (LCI1, HLA3 and LCIA) have also been successfully expressed individually in tobacco and Arabidopsis (Atkinson *et al.* 2016). Localization recapitulated the native algal localization and required no codon optimization. Retargeting to the chloroplast envelope of the plasmalemma transporters (to prioritize stromal accumulation of inorganic carbon) was not successful even after replacing the native N-terminus with the signaling peptide of an Arabidopsis chloroplast membrane ABC-type transporter. However, these experiments predated the discovery of the HLA3 and LCI1 supramolecular complexes, which is likely important for the correct assemblage and function of the two proteins. No adverse growth effect was measured in transgenic lines, but there was no benefit either, dispelling optimistic predictions from modeling. More work is needed to understand the modes of activation of the different transporters.

7.5. Conclusion

The past 10 years have witnessed an acceleration in the discovery rate of the molecular details of the mechanisms underpinning half of our planet's primary CO_2 assimilation. Major beneficiaries are synthetic biologists seeking alternatives to the C_4 rice project, to address food security issues and optimize anthropogenic land use. Bioengineers now have a toolkit of dozens of molecular candidates at their disposal. While current efforts focus on replicating cyanobacterial or algal inorganic carbon uptake systems, novel combinations mixing prokaryotic and eukaryotic components should also be attempted. Other beneficiaries are evolutionary and molecular biologists. The discovery of a common binding motif, pivotal to the Chlamydomonas pyrenoid, will inform new search strategies to identify the molecular basis of pyrenoids in ecologically important phytoplankton species. Ultimately, this will help us understand the multiple independent origins of aquatic CCMs and their associated Rubisco microcompartment, another noteworthy example of convergent evolution.

7.6. References

Amoroso, G., Sültemeyer, D., Thyssen, C., Fock, H.P. (1998). Uptake of HCO_3^- and CO_2 in cells and chloroplasts from the microalgae *Chlamydomonas reinhardtii* and *Dunaliella tertiolecta*. *Plant Physiology*, 116, 193–201.

Atkinson, N., Feike, D., Mackinder, L.C.M., Meyer, M.T., Griffiths, H., Jonikas, M.C., Smith, A.M., McCormick, A.J. (2016). Introducing an algal carbon-concentrating mechanism into higher plants: Location and incorporation of key components. *Plant Biotechnology Journal*, 14, 1302–1315.

Atkinson, N., Leitão, N., Orr, D.J., Meyer, M.T., Carmo-Silva, E., Griffiths, H., Smith, A.M., McCormick, A.J. (2017). Rubisco small subunits from the unicellular green alga Chlamydomonas complement Rubisco-deficient mutants of Arabidopsis. *New Phytologist*, 214, 655–667.

Atkinson, N., Velanis, C.N., Wunder, T., Clarke, D.J., Mueller-Cajar, O., McCormick, A.J. (2019). The pyrenoidal linker protein EPYC1 phase separates with hybrid Arabidopsis-Chlamydomonas Rubisco through interactions with the algal Rubisco small subunit. *Journal of Experimental Botany*, 70, 5271–5285.

Atkinson, N., Mao, Y., Chan, K.X., McCormick, A.J. (2020). Condensation of Rubisco into a proto-pyrenoid in higher plant chloroplasts. *Nature Communications*, 11, 6303.

Badger, M.R., Kaplan, A., Berry, J.A. (1978). A mechanism for concentrating CO_2 in *Chlamydomonas reinhardtii* and *Anabaena variabilis* and its role in photosynthetic CO_2 fixation. *Carnegie Institution Yearbook*, 77, 362–366.

Badger, M.R., Hanson, D., Price, G.D. (2002). Evolution and diversity of CO_2 concentrating mechanisms in cyanobacteria. *Functional Plant Biology*, 29, 161–173.

Badger, M.R., Price, G.D., Long, B.M., Woodger, F.J. (2006). The environmental plasticity and ecological genomics of the cyanobacterial CO_2 concentrating mechanism. *Journal of Experimental Botany*, 57, 249–265.

Bar-On, Y.M. and Milo, R. (2019). The global mass and average rate of Rubisco. *Proceedings of the National Academy of Sciences of the United States of America*, 116, 4738–4743.

Beardall, J. and Raven, J.A. (2020). Structural and biochemical features of carbon acquisition in algae. In *Photosynthesis in Algae: Biochemical and Physiological Mechanisms*, Larkum, A.W.D., Grossman, A.R., Raven, J.A. (eds). Springer, Cham.

Berg, I.A. (2011). Ecological aspects of the distribution of different autotrophic CO2 fixation pathways. *Applied and Environmental Microbiology*, 77, 1925–1936.

Burnap, R.L., Hagemann, M., Kaplan, A. (2015). Regulation of CO_2 concentrating mechanism in cyanobacteria. *Life*, 5, 348–371.

Cai, F., Menon, B.B., Cannon, G.C., Curry, K.J., Shively, J.M., Heinhorst, S. (2009). The pentameric vertex proteins are necessary for the icosahedral carboxysome shell to function as a CO_2 leakage barrier. *PLOS ONE*, 10, e7521.

Cai, F., Sutter, M., Cameron, J.C., Stanley, D.N., Kinney, J.N., Kerfeld, C.A. (2013). The structure of CcmP, a tandem bacterial microcompartment domain protein from the β-carboxysome, forms a subcompartment within a microcompartment. *Journal of Biological Chemistry*, 288, 16055–16063.

Cameron, J.C., Wilson, S.C., Bernstein, S.L., Kerfeld, C.A. (2013). Biogenesis of a bacterial organelle: The carboxysome assembly pathway. *Cell*, 155, 1131.

Chaijarasphong, T., Nichols, R.J., Kortright, K.E., Nixon, C.F., Teng, P.K., Oltrogge, L.M., Savage, D.F. (2016). Programmed ribosomal frameshifting mediates expression of the α-Carboxysome. *Journal of Molecular Biology*, 428, 153–164.

Chen, A.H., Robinson-Mosher, A., Savage, D.F., Silver, P.A., Polka, J.K. (2013). The bacterial carbon-fixing organelle is formed by shell envelopment of preassembled cargo. *PLOS ONE*, 8, e76127.

Desmarais, J.J., Flamholz, A.I., Blikstad, C., Dugan, E.J., Laughlin, T.G., Oltrogge, L.M., Chen, A.W., Wetmore, K., Diamond, S., Wang, J.Y., Savage, D.F. (2019). DABs are inorganic carbon pumps found throughout prokaryotic phyla. *Nature Microbiology*, 4, 2204–2215.

DiMario, R.J., Clayton, H., Mukherjee, A., Ludwig, M., Moroney, J.V. (2017). Plant carbonic anhydrases: Structures, locations, evolution, and physiological roles. *Molecular Plant*, 10, 30–46.

Dufresne, A., Salanoubat, M., Partensky, F., Artiguenave, F., Axmann, I.M., Barbe, V., Duprat, S., Galperin, M.Y., Koonin, E.V., Le Gall, F., Makarova, K.S., Ostrowski, M., Oztas, S., Robert, C., Rogozin, I.B., Scanlan, D.J., Tandeau de Marsac, N., Weissenbach, J., Wincker, P., Wolf, Y.I., Hess, W.R. (2003). Genome sequence of the cyanobacterium *Prochlorococcus marinus* SS120, a nearly minimal oxyphototrophic genome. *Proceedings of the National Academy of Sciences of the United States of America*, 100, 10020–10025.

Engel, B.D., Schaffer, M., Cuellar, L.K., Villa, E., Plitzko, J.M., Baumeister, W. (2015). Native architecture of the chlamydomonas chloroplast revealed by *in situ* cryo-electron tomography. *eLife*, 4, e04889.

Fang, W., Si, Y., Douglass, S., Casero, D., Merchant, S.S., Pellegrini, M., Ladunga, I., Liu, P., Spaldinga, M.H. (2012). Transcriptome-wide changes in *Chlamydomonas reinhardtii* gene expression regulated by carbon dioxide and the CO_2-concentrating mechanism regulator CIA5/CCM1. *Plant Cell*, 24, 1876–1893.

Faulkner, M., Rodriguez-Ramos, J., Dykes, G.F., Owen, S.V., Casella, S., Simpson, D.M., Beynon, R.J., Liu, L.N. (2017). Direct characterization of the native structure and mechanics of cyanobacterial carboxysomes. *Nanoscale*, 9, 10662–10673.

Faulkner, M., Szabó, I., Weetman, S.L., Sicard, F., Huber, R.G., Bond, P.J., Rosta, E., Liu, L.N. (2020). Molecular simulations unravel the molecular principles that mediate selective permeability of carboxysome shell protein. *Scientific Reports*, 10, 17501.

Field, C.B., Behrenfeld, M.J., Randerson, J.T., Falkowski, P. (1998). Primary production of the biosphere: Integrating terrestrial and oceanic components. *Science*, 281, 237–240.

Folea, I.M., Zhang, P., Nowaczyk, M.M., Ogawa, T., Aro, E.M., Boekema, E.J. (2008). Single particle analysis of thylakoid proteins from *Thermosynechococcus elongatus* and *Synechocystis* 6803: Localization of the CupA subunit of NDH-1. *FEBS Letters*, 582, 249–254.

Freeman Rosenzweig, E.S., Xu, B., Kuhn Cuellar, L., Martinez-Sanchez, A., Schaffer, M., Strauss, M., Cartwright, H.N., Ronceray, P., Plitzko, J.M., Förster, F., Wingreen, N.S., Engel, B.D., Mackinder, L.C.M., Jonikas, M.C. (2017). The eukaryotic CO_2-concentrating organelle is liquid-like and exhibits dynamic reorganization. *Cell*, 171, 148–162.

Gantt, E. and Conti, S.F. (1969). Ultrastructure of blue-green algae. *Journal of Bacteriology*, 97, 1486–1493.

Gao, H., Wang, Y., Fei, X., Wright, D.A., Spalding, M.H. (2015). Expression activation and functional analysis of HLA3, a putative inorganic carbon transporter in *Chlamydomonas reinhardtii*. *Plant Journal*, 82, 1–11.

Garcia-Alles, L.F., Root, K., Maveyraud, L., Aubry, N., Lesniewska, E., Mourey, L., Zenobi, R., Truan, G. (2019). Occurrence and stability of hetero-hexamer associations formed by β-carboxysome CcmK shell components. *PLOS ONE*, 14, e0223877.

Hagemann, M. and Kaplan, A. (2020). Is the structure of the CO_2-hydrating complex I compatible with the cyanobacterial CO_2-concentrating mechanism? *Plant Physiology*, 183, 460–463.

Hantke, M.F., Hasse, D., Maia, F.R.N.C., Ekeberg, T., John, K., Svenda, M., Loh, N.D., Martin, A.V., Timneanu, N., Larsson, D.S.D., van der Schot, G., Carlsson, G.H., Ingelman, M., Andreasson, J., Westphal, D., Liang, M., Stellato, F., DePonte, D.P., Hartmann, R., Kimmel, N., Kirian, R.A., Seibert, M.M., Mühlig, K., Schorb, S., Ferguson, K., Bostedt, C., Carron, S., Bozek, J.D., Rolles, D., Rudenko, A., Epp, S., Chapman, H.N., Barty, A., Hajdu, J., Andersson, I. (2014). High-throughput imaging of heterogeneous cell organelles with an X-ray laser. *Nature Photonics*, 8, 943–949.

Hatch, M.D. (2002). C_4 photosynthesis: Discovery and resolution. *Photosynthesis Research*, 78, 251–256.

He, S., Chou, H.T., Matthies, D., Wunder, T., Meyer, M.T., Atkinson, N., Martinez-Sanchez, A., Jeffrey, P.D., Port, S.A., Patena, W., He, G., Chen, V.K., Hughson, F.M., McCormick, A.J., Mueller-Cajar, O., Engel, B.D., Yu, Z., Jonikas, M.C. (2020). The structural basis of Rubisco phase separation in the pyrenoid. *Nature Plants*, 6, 1480–1490.

Iancu, C.V., Ding, H.J., Morris, D.M., Dias, D.P., Gonzales, A.D., Martino, A., Jensen, G.J. (2007). The structure of isolated *Synechococcus* strain WH8102 carboxysomes as revealed by electron cryotomography. *Journal of Molecular Biology*, 372, 764–773.

Iancu, C.V., Morris, D.M., Dou, Z., Heinhorst, S., Cannon, G.C., Jensen, G.J. (2010). Organization, structure, and assembly of α-carboxysomes determined by electron cryotomography of intact cells. *Journal of Molecular Biology*, 396, 105–117.

Itakura, A.K., Chan, K.X., Atkinson, N., Pallesen, L., Wang, L., Reeves, G., Patena, W., Caspari, O., Roth, R., Goodenough, U., McCormick, A.J., Griffiths, H., Jonikas, M.C. (2019). A Rubisco-binding protein is required for normal pyrenoid number and starch sheath morphology in *Chlamydomonas reinhardtii*. *Proceedings of the National Academy of Sciences of the United States of America*, 116, 18445–18454.

Jin, S., Sun, J., Wunder, T., Tang, D., Cousins, A.B., Sze, S.K., Mueller-Cajar, O., Gao, Y.G. (2016). Structural insights into the LCIB protein family reveals a new group of β-carbonic anhydrases. *Proceedings of the National Academy of Sciences of the United States of America*, 113, 14716–14721.

Kellogg, E.A. (2013). C_4 photosynthesis. *Current Biology*, 23, 594–599.

Kerfeld, C.A., Aussignargues, C., Zarzycki, J., Cai, F., Sutter, M. (2018). Bacterial microcompartments. *Nature Reviews Microbiology*, 16, 277–290.

Kinney, J.N., Salmeen, A., Cai, F., Kerfeld, C.A. (2012). Elucidating essential role of conserved carboxysomal protein CcmN reveals common feature of bacterial microcompartment assembly. *Journal of Biological Chemistry*, 287, 17729–17736.

Kono, A. and Spalding, M.H. (2020). LCI1, a *Chlamydomonas reinhardtii* plasma membrane protein, functions in active CO_2 uptake under low CO_2. *Plant Journal*, 102, 1127–1141.

Kono, A., Chou, T.H., Radhakrishnan, A., Bolla, J.R., Sankar, K., Shome, S., Su, C.C., Jernigan, R.L., Robinson, C.V., Yu, E.W., Spalding, M.H. (2020). Structure and function of LCI1: A plasma membrane CO_2 channel in the Chlamydomonas CO_2 concentrating mechanism. *Plant Journal*, 102, 1107–1126.

Koropatkin, N.M., Koppenaal, D.W., Pakrasi, H.B., Smith, T.J. (2007). The structure of a cyanobacterial bicarbonate transport protein, CmpA. *Journal of Biological Chemistry*, 282, 2606–2614.

Kowata, H., Tochigi, S., Takahashi, H., Kojima, S. (2017). Outer membrane permeability of cyanobacterium *Synechocystis* sp. strain PCC 6803. *Journal of Bacteriology*, 199, e00371-17.

Laughlin, T.G., Bayne, A.N., Trempe, J.F., Savage, D.F., Davies, K.M. (2019). Structure of the complex I-like molecule NDH of oxygenic photosynthesis. *Nature*, 566, 411–414.

Launay, H., Huang, W., Maberly, S.C., Gontero, B. (2020). Regulation of carbon metabolism by environmental conditions: A perspective from diatoms and other chromalveolates. *Frontiers in Plant Science*, 11, 1033.

Li, S., Roy, P., Travesset, A., Zandi, R. (2018). Why large icosahedral viruses need scaffolding proteins. *Proceedings of the National Academy of Sciences of the United States of America*, 115, 10971–10976.

Lin, M.T., Occhialini, A., Andralojc, P.J., Devonshire, J., Hines, K.M., Parry, M.A.J., Hanson, M.R. (2014). β-carboxysomal proteins assemble into highly organized structures in *Nicotiana* chloroplasts. *Plant Journal*, 79, 1–12.

Long, B.M., Badger, M.R., Whitney, S.M., Price, G.D. (2007). Analysis of carboxysomes from *Synechococcus* PCC7942 reveals multiple Rubisco complexes with carboxysomal proteins CcmM and CcaA. *Journal of Biological Chemistry*, 282, 29323–29335.

Long, B.M., Hee, W.Y., Sharwood, R.E., Rae, B.D., Kaines, S., Lim, Y.L., Nguyen, N.D., Massey, B., Bala, S., von Caemmerer, S., Badger, M.R., Price, G.D. (2018). Carboxysome encapsulation of the CO_2-fixing enzyme Rubisco in tobacco chloroplasts. *Nature Communications*, 9, 3570.

Maberly, S.C., Ball, L.A., Raven, J.A., Sültemeyer, D. (2009). Inorganic carbon acquisition by chrysophytes. *Journal of Phycology*, 45, 1052–1061.

Machingura, M.C., Bajsa-Hirschel, J., Laborde, S.M., Schwartzenburg, J.B., Mukherjee, B., Mukherjee, A., Pollock, S.V., Förster, B., Price, G.D., Moroney, J.V. (2017). Identification and characterization of a solute carrier, CIA8, involved in inorganic carbon acclimation in *Chlamydomonas reinhardtii*. *Journal of Experimental Botany*, 68, 3879–3890.

Mackinder, L.C.M. (2018). The Chlamydomonas CO_2-concentrating mechanism and its potential for engineering photosynthesis in plants. *New Phytologist*, 217, 54–61.

Mackinder, L.C.M., Meyer, M.T., Mettler-Altmann, T., Chen, V.K., Mitchell, M.C., Caspari, O., Freeman Rosenzweig, E.S., Pallesen, L., Reeves, G., Itakura, A., Roth, R., Sommer, F., Geimer, S., Mühlhaus, T., Schroda, M., Goodenough, U., Stitt, M., Griffiths, H., Jonikas, M.C. (2016). A repeat protein links Rubisco to form the eukaryotic carbon-concentrating organelle. *Proceedings of the National Academy of Sciences of the United States of America*, 113, 5958–5963.

Mackinder, L.C.M., Chen, C., Leib, R.D., Patena, W., Blum, S.R., Rodman, M., Ramundo, S., Adams, C.M., Jonikas, M.C. (2017). A spatial interactome reveals the protein organization of the algal CO_2-concentrating mechanism. *Cell*, 171, 133–147.

Mariscal, V., Moulin, P., Orsel, M., Miller, A.J., Fernández, E., Galván, A. (2006). Differential regulation of the Chlamydomonas Nar1 gene family by carbon and nitrogen. *Protist*, 157, 421–433.

McGinn, P.J., Price, G.D., Maleszka, R., Badger, M.R. (2003). Inorganic carbon limitation and light control the expression of transcripts related to the CO_2-concentrating mechanism in the cyanobacterium *Synechocystis* sp. strain PCC6803. *Plant Physiology*, 132, 218–229.

McGrath, J.M. and Long, S.P. (2014). Can the cyanobacterial carbon-concentrating mechanism increase photosynthesis in crop species? A theoretical analysis. *Plant Physiology*, 164, 2247–2261.

McGurn, L.D., Moazami-Goudarzi, M., White, S.A., Suwal, T., Brar, B., Tang, J.Q., Espie, G.S., Kimber, M.S. (2016). The structure, kinetics and interactions of the β-carboxysomal β-carbonic anhydrase, CcaA. *Biochemical Journal*, 473, 4559–4572.

Menon, B.B., Dou, Z., Heinhorst, S., Shively, J.M., Cannon, G.C. (2008). *Halothiobacillus neapolitanus* carboxysomes sequester heterologous and chimeric Rubisco species. *PLOS ONE*, e3570.

Meyer, M.T., Genkov, T., Skepper, J.N., Jouhet, J., Mitchell, M.C., Spreitzer, R.J., Griffiths, H. (2012). Rubisco small-subunit α-helices control pyrenoid formation in Chlamydomonas. *Proceedings of the National Academy of Sciences of the United States of America*, 109, 19474–19479.

Meyer, M.T., McCormick, A.J., Griffiths, H. (2016). Will an algal CO_2-concentrating mechanism work in higher plants? *Current Opinion in Plant Biology*, 31, 181–188.

Meyer, M.T., Goudet, M.M.M., Griffiths, H. (2020a). The algal pyrenoid. In *Photosynthesis in Algae: Biochemical and Physiological Mechanisms*, Larkum, A.W.D., Grossman, A.R., Raven, J.A. (eds). Springer, Cham.

Meyer, M.T., Itakura, A.K., Patena, W., Wang, L., He, S., Emrich-Mills, T., Lau, C.S., Yates, G., Mackinder, L.C.M., Jonikas, M.C. (2020b). Assembly of the algal CO_2-fixing organelle, the pyrenoid, is guided by a Rubisco-binding motif. *Science Advances*, 6, eabd2408.

Mukherjee, A., Lau, C.S., Walker, C.E., Rai, A.K., Prejean, C.I., Yates, G., Emrich-Mills, T., Lemoine, S.G., Vinyard, D.J., Mackinder, L.C.M., Moroney, J.V. (2019). Thylakoid localized bestrophin-like proteins are essential for the CO_2 concentrating mechanism of *Chlamydomonas reinhardtii. Proceedings of the National Academy of Sciences of the United States of America*, 116, 16915–16920.

Niederhuber, M.J., Lambert, T.J., Yapp, C., Silver, P.A., Polka, J.K. (2017). Superresolution microscopy of the β-carboxysome reveals a homogeneous matrix. *Molecular Biology of the Cell*, 28, 2734–2745.

Ohnishi, N., Fukuzawa, H., Mukherjee, B., Tsujikawa, T., Yanase, M., Nakano, H., Moroney, J.V. (2010). Expression of a low CO_2-inducible protein, LCI1, increases inorganic carbon uptake in the green alga *Chlamydomonas reinhardtii. Plant Cell*, 22, 3105–3117.

Oltrogge, L.M., Chaijarasphong, T., Chen, A.W., Bolin, E.R., Marqusee, S., Savage, D.F. (2020). Multivalent interactions between CsoS2 and Rubisco mediate α-carboxysome formation. *Nature Structural and Molecular Biology*, 27, 281–287.

Omata, T., Price, G.D., Badger, M.R., Okamura, M., Gohta, S., Ogawa, T. (1999). Identification of an ATP-binding cassette transporter involved in bicarbonate uptake in the cyanobacterium *Synechococcus* sp. strain PCC 7942. *Proceedings of the National Academy of Sciences of the United States of America*, 96, 13571–13576.

Peña, K.L., Castel, S.E., De Araujo, C., Espie, G.S., Kimber, M.S. (2010). Structural basis of the oxidative activation of the carboxysomal γ-carbonic anhydrase, CcmM. *Proceedings of the National Academy of Sciences of the United States of America*, 107, 2455–2460.

Price, G.D. and Badger, M.R. (1989). Expression of human carbonic anhydrase in the cyanobacterium *Synechococcus* PCC7942 creates a high CO_2-requiring phenotype: Evidence for a central role for carboxysomes in the CO_2 concentrating mechanism. *Plant Physiology*, 91, 505–513.

Price, G.D., Woodger, F.J., Badger, M.R., Howitt, S.M., Tucker, L. (2004). Identification of a SulP-type bicarbonate transporter in marine cyanobacteria. *Proceedings of the National Academy of Sciences of the United States of America*, 101, 18228–18233.

Rolland, V., Badger, M.R., Price, G.D. (2016). Redirecting the cyanobacterial bicarbonate transporters BicA and SbtA to the chloroplast envelope: Soluble and membrane cargos need different chloroplast targeting signals in plants. *Frontiers in Plant Science*, 7, 185.

Ryan, P., Forrester, T.J.B., Wroblewski, C., Kenney, T.M.G., Kitova, E.N., Klassen, J.S., Kimber, M.S. (2019). The small RbcS-like domains of the β-carboxysome structural protein CcmM bind RubisCO at a site distinct from that binding the RbcS subunit. *Journal of Biological Chemistry*, 294, 2593–2603.

Sandrini, G., Matthijs, H.C.P., Verspagen, J.M.H., Muyzer, G., Huisman, J. (2014). Genetic diversity of inorganic carbon uptake systems causes variation in CO_2 response of the cyanobacterium *Microcystis*. *ISME Journal*, 8, 589–600.

Sasso, S., Stibor, H., Mittag, M., Grossman, A.R. (2018). The natural history of model organisms from molecular manipulation of domesticated *Chlamydomonas reinhardtii* to survival in nature. *eLife*, 7, e39233.

Schmid, M.F., Paredes, A.M., Khant, H.A., Soyer, F., Aldrich, H.C., Chiu, W., Shively, J.M. (2006). Structure of *Halothiobacillus neapolitanus* carboxysomes by cryo-electron tomography. *Journal of Molecular Biology*, 364, 526–535.

Schuller, J.M., Saura, P., Thiemann, J., Schuller, S.K., Gamiz-Hernandez, A.P., Kurisu, G., Nowaczyk, M.M., Kaila, V.R.I. (2020). Redox-coupled proton pumping drives carbon concentration in the photosynthetic complex I. *Nature Communications*, 11, 494.

Scott, K.M., Harmer, T.L., Gemmell, B.J., Kramer, A.M., Sutter, M., Kerfeld, C.A., Barber, K.S., Bari, S., Boling, J.W., Campbell, C.P., Gallard-Gongora, J.F., Jackson, J.K., Lobos, A., Mounger, J.M., Radulovic, P.W., Sanson, J.M., Schmid, S., Takieddine, C., Warlick, K.F., Whittaker, R. (2020). Ubiquity and functional uniformity in CO_2 concentrating mechanisms in multiple phyla of bacteria is suggested by a diversity and prevalence of genes encoding candidate dissolved inorganic carbon transporters. *FEMS Microbiology Letters*, 367, fnaa106.

Shibata, M., Ohkawa, H., Kaneko, T., Fukuzawa, H., Tabata, S., Kaplan, A., Ogawa, T. (2001). Distinct constitutive and low-CO_2-induced CO_2 uptake systems in cyanobacteria: Genes involved and their phylogenetic relationship with homologous genes in other organisms. *Proceedings of the National Academy of Sciences of the United States of America*, 98, 11789–11794.

Shibata, M., Katoh, H., Sonoda, M., Ohkawa, H., Shimoyama, M., Fukuzawa, H., Kaplan, A., Ogawa, T. (2002). Genes essential to sodium-dependent bicarbonate transport in cyanobacteria: Function and phylogenetic analysis. *Journal of Biological Chemistry*, 277, 18658–18664.

Shively, J.M., Ball, F., Brown, D.H., Saunders, R.E. (1973). Functional organelles in prokaryotes: Polyhedral inclusions (carboxysomes) of *Thiobacillus neapolitanus*. *Science*, 182, 584–586.

Shutova, T., Kenneweg, H., Buchta, J., Nikitina, J., Terentyev, V., Chernyshov, S., Andersson, B., Allakhverdiev, S.I., Klimov, V.V., Dau, H., Junge, W., Samuelsson, G. (2008). The photosystem II-associated Cah3 in Chlamydomonas enhances the O_2 evolution rate by proton removal. *EMBO Journal*, 27, 782–791.

So, A.K., Espie, G.S., Williams, E.B., Shively, J.M., Heinhorst, S., Cannon, G.C. (2004). A novel evolutionary lineage of carbonic anhydrase (ε Class) is a component of the carboxysome shell. *Journal of Bacteriology*, 186, 623–630.

Sommer, M., Sutter, M., Gupta, S., Kirst, H., Turmo, A., Lechno-Yossef, S., Burton, R.L., Saechao, C., Sloan, N.B., Cheng, X., Chan, L.J.G., Petzold, C.J., Fuentes-Cabrera, M., Ralston, C.Y., Kerfeld, C.A. (2019). Heterohexamers formed by CcmK3 and CcmK4 increase the complexity of beta carboxysome shells. *Plant Physiology*, 179, 156–167.

Steemann Nielsen, E. (1952). The use of radio-active carbon (C^{14}) for measuring organic production in the sea. *ICES Journal of Marine Science*, 18, 117–140.

Sutter, M., Greber, B., Aussignargues, C., Kerfeld, C.A. (2017). Assembly principles and structure of a 6.5 MDa bacterial microcompartment shell. *Science*, 356, 1293–1297.

Sutter, M., Laughlin, T.G., Sloan, N.B., Serwas, D., Davies, K.M., Kerfeld, C.A. (2019). Structure of a synthetic β-carboxysome shell. *Plant Physiology*, 181, 1050–1058.

Terentyev, V.V., Shukshina, A.K., Shitov, A.V. (2019). Carbonic anhydrase CAH3 supports the activity of photosystem II under increased pH. *Biochimica et Biophysica Acta – Bioenergetics*, 1860, 582–590.

Toyokawa, C., Yamano, T., Fukuzawa, H. (2020). Pyrenoid starch sheath is required for LCIB localization and the CO_2-concentrating mechanism in green algae. *Plant Physiology*, 182, 1883–1893.

Twarock, R. and Luque, A. (2019). Structural puzzles in virology solved with an overarching icosahedral design principle. *Nature Communications*, 10, 4414.

Uehara, S., Adachi, F., Ito-Inaba, Y., Inaba, T. (2016). Specific and efficient targeting of cyanobacterial bicarbonate transporters to the inner envelope membrane of chloroplasts in Arabidopsis. *Frontiers in Plant Science*, 7, 16.

Uehara, S., Sei, A., Sada, M., Ito-Inaba, Y., Inaba, T. (2020). Installation of authentic BicA and SbtA proteins to the chloroplast envelope membrane is achieved by the proteolytic cleavage of chimeric proteins in Arabidopsis. *Scientific Reports*, 10, 2353.

Wang, Y. and Spalding, M.H. (2014). Acclimation to very low CO_2: Contribution of limiting CO_2 inducible proteins, LCIB and LCIA, to inorganic carbon uptake in *Chlamydomonas reinhardtii*. *Plant Physiology*, 166, 2040–2050.

Wang, C., Sun, B., Zhang, X., Huang, X., Zhang, M., Guo, H., Chen, X., Huang, F., Chen, T., Mi, H., Yu, F., Liu, L.N., Zhang, P. (2019a). Structural mechanism of the active bicarbonate transporter from cyanobacteria. *Nature Plants*, 5, 1184–1193.

Wang, H., Yan, X., Aigner, H., Bracher, A., Nguyen, N.D., Hee, W.Y., Long, B.M., Price, G.D., Hartl, F.U., Hayer-Hartl, M. (2019b). Rubisco condensate formation by CcmM in β-carboxysome biogenesis. *Nature*, 566, 131–135.

Watson, G.M.F. and Tabita, F.R. (1997). Microbial ribulose 1,5-bisphosphate carboxylase/oxygenase: A molecule for phylogenetic and enzymological investigation. *FEMS Microbiology Letters*, 146, 13–22.

Wunder, T., Cheng, S.L.H., Lai, S.K., Li, H.Y., Mueller-Cajar, O. (2018). The phase separation underlying the pyrenoid-based microalgal Rubisco supercharger. *Nature Communications*, 9, 5076.

Xu, M., Bernát, G., Singh, A., Mi, H., Rögner, M., Pakrasi, H.B., Ogawa, T. (2008). Properties of mutants of *Synechocystis* sp. strain PCC 6803 lacking inorganic carbon sequestration systems. *Plant and Cell Physiology*, 49, 1672–1677.

Yamano, T., Tsujikawa, T., Hatano, K., Ozawa, S.I., Takahashi, Y., Fukuzawa, H. (2010). Light and low-CO_2-dependent LCIB-LCIC complex localization in the chloroplast supports the carbon-concentrating mechanism in *Chlamydomonas reinhardtii*. *Plant and Cell Physiology*, 51, 1453–1468.

Yamano, T., Sato, E., Iguchi, H., Fukuda, Y., Fukuzawa, H. (2015). Characterization of cooperative bicarbonate uptake into chloroplast stroma in the green alga *Chlamydomonas reinhardtii*. *Proceedings of the National Academy of Sciences of the United States of America*, 112, 7315–7320.

Zhan, Y., Marchand, C.H., Maes, A., Mauries, A., Sun, Y., Dhaliwal, J.S., Uniacke, J., Arragain, S., Jiang, H., Gold, N.D., Martin, V.J.J., Lemaire, S.D., Zerges, W. (2018). Pyrenoid functions revealed by proteomics in Chlamydomonas reinhardtii. *PLOS ONE*, 13, e0185039.

8

Environmental Variability and Its Control of Productivity

Alessandra NORICI[1], Caterina GEROTTO[1], John BEARDALL[2,3] and John A. RAVEN[4,5,6]

[1] *Laboratorio di Fisiologia delle Alghe e delle Piante, Dipartimento di Scienze della Vita e dell'Ambiente, Università Politecnica delle Marche, Ancona, Italy*

[2] *State Key Laboratory of Marine Environmental Science & College of Ocean and Earth Sciences, Xiamen University, China*

[3] *School of Biological Sciences, Monash University, Clayton, Victoria, Australia*

[4] *Division of Plant Science, University of Dundee at the James Hutton Institute, Invergowrie, United Kingdom*

[5] *University of Technology, Climate Change Cluster, Faculty of Science, Sydney, Australia*

[6] *School of Biological Sciences, University of Western Australia, Crawley, Australia*

8.1. Introduction

Aquatic oxygenic photosynthesis occurs in incredibly diverse groups of cyanobacteria, eukaryotic phytoplankton, macroalgae and macrophytes (see Chapters 2 and 3). It is modulated by abiotic (e.g., nutrient availability, light, temperature) and biotic (e.g., interspecific competition, prey–predator interactions, infection with pathogenic organisms) factors linked to the environment. Here, focusing on the cyanobacteria and microalgae which globally are responsible for the bulk of aquatic photosynthesis, we address the main abiotic factors controlling biosynthetic processes within cells and therefore the elemental and macromolecular composition of photoautotrophs in balance with the environment.

Blue Planet, Red and Green Photosynthesis,
coordinated by Stephen Christopher MABERLY and Brigitte GONTERO.

It is noteworthy that photosynthetic organisms have evolved in a constantly changing world. The origin of oxygenic photosynthesis was most likely closer to the origin of life, in the early Archean, than suggested in previous studies, as a late innovation occurred in the most recent common ancestor of Cyanobacteria more than 2.4 billion years ago (Cardona *et al.* 2019; Oliver *et al.* 2021). Since then, physico-chemical transformations undergone by the oceans have shaped (at least in part) the evolutionary trajectories that lead to contemporary primary production. The present-day large-scale human effects on Earth (the Anthropocene) have further challenged phototrophs: every day, huge amounts of CO_2 are emitted into the atmosphere and dissolve in the oceans, changing ocean chemistry, lowering pH, warming ocean waters, shoaling and steepening the thermocline and limiting nutrient delivery from deep waters to surface layers. Moreover, top predators are declining in numbers and nutrient inputs into coastal waters continue to rise due to human activities. Unravelling how environment controls the functions of phototrophs is essential to our ability to interpret biological responses to present and future conditions and to preserve highly threatened ecosystems (Armbrust 2009).

8.2. Macro- and micronutrients in aquatic environments during Earth's history and their biological functions

In addition to C (already addressed in Chapters 4–7), O and H, a number of other elements are essential to cell structure and function. In accordance with their concentration in waters, these are often referred to as macronutrients (e.g. N, P, S), present in micro- to nanomolar concentrations, and micronutrients (e.g. Fe, Zn, Cu, Mn, Mo) ranging from nano- to picomolar levels (and are therefore called trace elements). Their chemical form and availability change according to the specific aquatic environment and have varied during the history of Earth. Less mentioned elements in eco-physiological studies or books are Cl, Mg, Ca and K. Although required to perform oxygenic photosynthesis, the ionic forms of these elements have been, and are generally, ubiquitous and in quantities unlikely to exhibit depletions able to limit algal productivity (Raven 2020, 2021). Rare events of biomass yield limitation by K^+ have been recorded in freshwaters during dense blooms (Talling 2010). Interestingly, Cl^-, which is primarily involved at low concentration as a cofactor of the photosystem II oxygen-evolving complex (PSII-OEC) and in its assembly, may also play a role in HCO_3^- entry in marine photosynthetic organisms (Raven 2020).

Here, we explore the main nutrient requirements in relation to their capacity to limit photosynthesis and cell growth in nature.

8.2.1. *Nitrogen*

Nitrogen (N), in the reduced form of NH_4^+, is incorporated into carbon skeletons to synthesize amino acids (thus proteins), nucleic acids and other organic molecules (e.g. chlorophylls). N sources available to photosynthetic organisms changed over geological time with respect to the redox state of the aquatic environment and the evolving microbiota.

As long as pO_2 remained low in the oceans, the N source for assimilation into organic matter may have been relatively abundant in the form of NH_4^+ (Anbar and Knoll 2002). The situation likely changed due to oxidation events occurring late in the Proterozoic era which increased pO_2 (the Great Oxidation Event being the first, ca. 2,400 Ma; Holland 2006). Nitrification and denitrification became important on a global scale, and NH_4^+ was converted in surface seawater to NO_3^-, NO_2^-, N_2 and N_2O.

It is important to note that the different chemical forms of inorganic N, depending on their oxidation state, require a different number of reducing steps and thus a range of energy costs for their assimilation by phototrophs. At the same time, since multiple metal ions are used as cofactors by the enzymes of the N assimilation pathway, metal cell quotas (especially Fe and Mo requirements, as explained below) change accordingly. Growth and photosynthesis are strongly affected by the chemical form, as well as concentration, of N available in the environment.

8.2.1.1. *Dissolved N_2*

The most abundant nitrogen form, N_2, is chemically stable and poorly reactive because of the strength of the N–N triple bond. However, N_2 is biologically fixed in an energy-consuming process by some heterotrophic archaea and bacteria, as well as some anoxygenic photosynthetic bacteria and cyanobacteria, when other forms of inorganic nitrogen are unavailable (Raymond *et al.* 2004). Marine N_2-fixing cyanobacteria thus provide a critical source of new and biologically available N and support a large fraction of total biological productivity in the tropical and subtropical oceans.

$$N_2 + 16\ ATP + 8\ e^- + 8\ H^+ \rightarrow 2\ NH_3 + H_2 + 16\ ADP + 16\ P_i$$

The reaction is catalyzed by an enzymatic complex termed nitrogenase which consists of two component metalloproteins, the molybdenum–iron (MoFe-) protein and the iron (Fe-) protein, where the Fe-protein is a nucleotide-binding and electron-donor component, and the MoFe-enzyme contains the substrate-reducing site (Peters and Szilagyi 2011). In other nitrogenases, Mo is replaced by vanadium (V) or Fe (Eady 1996).

The fixation of N_2 is incompatible with the presence of O_2, which damages nitrogenase, with re-synthesis of nitrogenase needed for continued N_2 fixation. N_2-fixers, also called diazotrophs, separate N_2 fixation spatially or temporally from photosynthesis. In the first case, N_2 fixation is restricted to specialized cells termed heterocysts (e.g. in the filamentous cyanobacteria *Anabaena*) in which microaerobic conditions can be maintained as they have characteristics such as the possession of cell wall structures that minimize O_2 diffusion into the cell while allowing sufficient N_2 entry. They also lack PSII and thus oxygen evolution. Within the heterocyst, glutamate is synthesized from the reduced nitrogen (NH_4^+) obtained from N_2 fixation and carbon skeletons that are imported from the adjacent vegetative (photosynthetic) cells. When a temporal separation is accomplished in non-heterocystous cyanobacteria, the organisms usually perform photosynthesis during the day and N_2 fixation at night (Fay 1992; Dixon *et al.* 2004), as in the unicellular N_2 fixer *Crocosphaera* (Shi *et al.* 2010). Interestingly, since diazotrophy requires two to five times more Fe than for phytoplankton growing on NH_4^+ and only slightly higher than for phytoplankton growing on NO_3^- (Sañudo-Wilhelmy *et al.* 2001), the marine *Crocosphaera watsonii* daily synthesizes and degrades metalloenzymes belonging to the metabolisms of N_2 fixation and photosynthesis according to the diel cycle. Such a strategy, although expensive, allows an ~40% lower requirement for external Fe and represents a competitive adaptation (see glossary) to low Fe open oceans (Saito *et al.* 2011). *Trichodesmium* is an exception to the spatial or temporal division.

Trichodesmium is an abundant, filamentous, colony-forming cyanobacterium residing in oligotrophic tropical and subtropical oceans. It is a globally important non-heterocystous diazotroph since it fixes 60–80 Tg new N annually, accounting for almost 40% of global marine N_2 fixation (Carpenter *et al.* 2008). It can exist in the water column as single filaments (trichomes) and as large colonies (~1 mm in diameter). Even single filaments of *Trichodesmium* are capable of fixing N_2 in the light (Milligan *et al.* 2007) without forming colonies, possibly through the formation of microaerobic zones via the light-dependent reduction of oxygen by a flavodiiron-protein-dependent Mehler-like reaction (Zhang *et al.* 2009; see also section 8.4.4.2). Colonies consist of aggregates of up to several hundred trichomes and form fusiform colonies when aligned in parallel (called "tufts"), or spherical colonies when aligned radially (called "puffs"). To sustain the Fe demand of nitrogenase, *Trichodesmium* strongly relies on aeolian sources of Fe such as airborne dust deposited on the ocean surface. *Trichodesmium* can sense Fe and selectively interact with Fe-rich dust particles (Kessler *et al.* 2020). This ability assists

Trichodesmium to obtain Fe from dust and facilitate its growth and subsequent contribution to nutrient cycling and productivity in the ocean. Other factors (such as P) may control N_2 fixation once the Fe quota is met (i.e. under the high-Fe conditions of the central region of the Atlantic Ocean, which is reached by the aeolian dust deposition; Sañudo-Wilhelmy *et al.* 2001) or co-limit it (Mills *et al.* 2004). Recently, N_2-fixing cyanobacteria have been found not only in tropical and subtropical waters but also in nutrient-enriched coastal (i.e. North America) and low-temperature higher latitude (i.e. Arctic Ocean) waters where uncultured UCYN-A cyanobacteria live in symbiosis with algal hosts (e.g. *Braarudosphaera*) and fix N_2 during the day (Mulholland *et al.* 2012; Harding *et al.* 2018; Zeher and Capone 2020).

8.2.1.2. *Inorganic nitrogen is also present as NO_3^-, NO_2^- and NH_4^+*

In the present oceans and freshwaters, nitrate (NO_3^-) is predominant: <0.2 μM at the surface and up to 40 μM at depth in the ocean. Thus, upwelling is a major source of N for the photic zone. Aquatic photoautotrophs have mostly evolved high-affinity nitrogen transporters (HAN/NiT) that actively acquire external NO_3^-/NO_2^- (Fernandez and Galvan 2007). Nitrate reduction to NH_4^+ involves nitrate reductase (NR), catalyzing NO_3^- reduction to NO_2^-, and nitrite reductase (NiR), reducing NO_2^- to NH_4^+. Both enzymes are cytosolic in cyanobacteria, while NiR is plastidial in eukaryotes. NR is a metalloenzyme that forms homodimers or homotetramers. Each monomer of NR consists of three domains binding a Mo atom, heme (Cyt_{557}) and FAD as prosthetic groups that drive electrons from NAD(P)H. NiR has an iron–sulfur cluster and a siroheme, typical of enzymes that catalyze reactions involving six electrons (also the sulfite reductase, involved in sulfur assimilation and highly similar to NiR):

$$NO_3^- + 8\,e^- + 10\,H^+ \rightarrow NH_4^+ + 3H_2O$$

The reducing power is in the form of NAD(P)H and ferredoxin and ultimately originates from photosynthesis. Consistent with this is the observation that the redox status of the plastoquinone pool controls the activity and expression of *Chlamydomonas reinhardtii* NR (Giordano *et al.* 2005). NiR is constitutively expressed and highly active, avoiding the accumulation of NO_2^-, which is very toxic to cells. In *C. reinhardtii*, the genes involved in NO_3^- reduction (*nar* genes) are grouped into two "clusters" on chromosome IX. The signal triggering the transcription of *nar* genes appears to involve the *nit2* gene which acts as the last effector of a signal transduction cascade that activates *nar* genes and requires no NH_4^+ to be present in the cell, though the presence of NO_3^- is not crucial (Galvàn and Fernandez 2001).

In contrast, NH_4^+ is the source of inorganic N with the lowest energy assimilation cost. In the present ocean, the average concentration of NH_4^+ is below 0.1 μM. The NH_4^+ form is in equilibrium with the unprotonated NH_3 form, with their respective concentrations depending on the environmental pH; at pH < 9.25 (pKa at 25°C), the dominant form is NH_4^+; above that pH, the dominant form is NH_3. While the uncharged NH_3 is significantly lipid-soluble and can cross the plasmamembranes (the permeability is 6–7 μm s^{-1}; Raven *et al.* 2008), NH_4^+, whose hydration volume is similar to that of K^+, can enter the cell through low-affinity K^+ channels, according to the electrochemical gradient (Boyd and Gradmann 1999; Hoopen *et al.* 2010; Giordano and Raven 2014). However, this is not important in seawater where the K^+ concentration of 9.7 mM outcompetes ecologically relevant concentrations of NH_4^+. Furthermore, flowering plants have NH_4^+-insensitive high-affinity K^+ transporters (Sante-Maria *et al.* 2000); such transporters may also occur in algae. In addition, NH_4^+ is taken up by specific ammonium transporters (AMT; Fernandez *et al.* 2009). High concentrations of NH_3 inhibit algal photosynthesis and growth (Azov and Golman 1982), possibly by uncoupling the proton motive force across the thylakoid membranes. For this reason, reduced N is immediately assimilated in algae principally via the enzymes glutamine synthetase (GS) and glutamine-oxoglutarate aminotransferase (GOGAT). Maintenance of this route requires the transfer of 2-oxoglutaric acid from the tricarboxylic acid cycle (TCA) to generate glutamate (Miflin and Lea 1982; Muro-Pastor *et al.* 2005). As a result, NH_4^+ is incorporated into organic matter (Weger *et al.* 1998). The TCA is therefore used as a source of carbon skeletons but, since it has no net production of metabolic intermediates, the cycle must be supplied with substrates by anaplerotic reactions to support its biosynthetic role. Anaplerosis consists of β-carboxylation of a 3C compound (phosphoenolpyruvate [PEP] or pyruvate [Pyr]) to produce a 4C intermediate of the TCA cycle (oxaloacetate [OAA] and malate). The β-carboxylation accounts for 5% of the organic matter produced by phytoplankton (Norici and Giordano 2002). In the presence of high concentrations of NH_4^+, a high anaplerotic rate may require inorganic C supply at sustainable rates through the help of a mitochondrial carbonic anhydrase (mtCA) (Giordano *et al.* 2003; see also Rai *et al.* 2021). Indeed, the expression of mtCA is governed by the C-to-N ratio (Giordano *et al.* 2003).

8.2.1.3. *Organic sources of dissolved nitrogen*

Organic N forms available to phytoplankton include urea (0.1–0.5 μM) and amino acids (0.01–0.03 μM), mostly found in open oceans because of excretion by bacteria and zooplankton or release by lysed algal cells. In many inland waters, higher concentrations of dissolved organic nitrogen (DON) are found, even exceeding those of inorganic nitrogen. They derive from natural and anthropogenic inputs in catchment areas. The use of dissolved organic nitrogen is negatively correlated with the total nitrogen and ammonia concentration, being a potential nutrient resource in freshwaters when the inorganic forms are limited (Mackay *et al.* 2020).

8.2.2. *Phosphorus*

Most phosphorus (P) in the biosphere comes from the weathering of PO_4^{3-}-containing terrestrial minerals, mostly apatite ($Ca_5(PO_4)_3OH$). The degree of elevation of continents relative to sea level has greatly influenced, on a geological scale, the availability of P reaching the aquatic biosphere (i.e. fresh waters and oceans) via river runoff and wind. The increased phosphate flux since the Paleoproterozoic into shallow oceans favored photosynthetic primary production and the appearance of biogenic phosphatic stromatolites (Bekker and Holland 2012). In the last century, there has been an increasing anthropogenic influence on the P biogeochemical cycle, as phosphate rocks are increasingly exploited for the industrial production of P-rich fertilizers, and this mined phosphorus is then introduced into soil and partially transferred to river systems (Baturin 2003). The average P content in the ocean is 70–72 μg/l, and it is unevenly distributed along the water column with the lowest concentration in the surface (1–40 μg/l) (Baturin 2003), as is the case for most nutrients (see section 8.5).

In water, the concentration of free inorganic P is usually low (nanomolar or even sub-nanomolar) and depends on the redox state of the environment. Mostly found as $HPO_4^{2-}/H_2PO_4^-$, dissolved inorganic P may form complexes with oxides and hydroxides of Fe^{3+} (i.e. Fe_2O_3) or Al^{3+} but does not bind with the corresponding compounds of Fe^{2+} (FeO, FeS). The amount of metal complexes is highly variable and depends on the nature of the surrounding rocks (if basic such as basalts, the concentration of oxides is higher) and the intensity of weathering (e.g. in relation to the intensity of rains). The Fe concentration and Fe redox state are therefore major regulators of phosphate availability. Lower Fe concentrations, and lower O_2 concentrations, mean a lower concentration of Fe^{3+}, less complexation of phosphate and hence more alga-available P.

Photoautotrophs can also access the dissolved organic phosphorus (DOP) returning to the environment through excretion and as when organisms die and decay. Another source of soluble organic P comes from the discharge of industrial and domestic wastewaters. Monophosphate esters (C–O–P bond) are externally hydrolyzed by phosphatases, with inorganic phosphate uptake; phosphonates (C–P bond) can be taken up by cyanobacteria, and inorganic phosphate is released by phosphonatase, a C–P lyase (Dyhrman *et al.* 2006). DOP are increasingly utilized when very low surface phosphate concentrations are present, as in oligotrophic tropical and subtropical areas (Torres-Valdes *et al.* 2009). The same is true in rivers and inland water basins (Mackay *et al.* 2020).

In sediments, P is mainly composed of particulate organic P (POP) (~40%) and particulate inorganic P (PIP) (~25%) coming from remineralized P, which reacts with calcium ions to form insoluble calcium phosphate compounds, specifically carbonate fluorapatite (Paytan and McLaughlin 2007). Once P-rich sediments are accumulated, P must be incorporated into rocks, re-emerge and be subject to further weathering before it can again be available as inorganic P to organisms. Its recycling rate and residence time in oceanic waters are about 20,000 yr (Baturin 2003), so slow because no important gaseous forms contain P.

Unlike N and S, which are assimilated in their most reduced form, bioavailable PO_4^{3-}, once within the cell, is rapidly assimilated by algae in its most oxidized form. PO_4^{3-} is a functional component of many molecules in cell metabolism. Inside the cell, PO_4^{3-} is incorporated into ATP via photophosphorylation and oxidative phosphorylation; a range of enzymes, mainly kinases and phosphatases, then control its entry and exit into and from the various intracellular organic pools. In some biochemical pathways, pyrophosphate $P_2O_7^{4-}$ is used instead (Carnal and Black 1983; Perez-Castinero *et al.* 2001; Cade-Mendin and Payton 2010; Sanz-Luque *et al.* 2020).

Major cellular reservoirs of inorganic P include nucleic acids and phospholipids as well as phosphoproteins, metabolites (e.g. ATP, NAD(P)H, glyceraldehyde 3-phosphate (GAP), PEP, hexoses phosphate) in catabolic and anabolic pathways and signaling molecules (Rees and Raven 2021). Moreover, as a response to fluctuations of PO_4^{3-} in the natural environment, some algae are capable of luxury uptake, that is, they accumulate intracellular P beyond their growth needs (8–16 times the minimum cell-quota found in previously P-starved organisms; Morel 1987) in a PO_4^{3-}-replete medium. These species may have a competitive advantage over less efficient organisms in PO_4^{3-} acquisition (see glossary) and storage once PO_4^{3-} becomes limiting (Orchard *et al.* 2010; Sanz-Luque *et al.* 2020). They polymerize PO_4^{3-} into polyphosphate (polyP) chains that are composed of three to hundreds of PO_4^{3-} groups linked by phosphoanhydride bonds that have a high free energy of hydrolysis (Sanz-Luque *et al.* 2020). PolyP chains are stored in granules called polyP bodies (PPBs) which are often found in acidocalcisomes in eukaryotic phototrophs (although they can be present also in the nucleus, mitochondria, cytoplasm, cell wall and endoplasmic reticulum) (Ruiz *et al.* 2001; Yagisawa *et al.* 2009; Aksoy *et al.* 2014) or in the center of the cell and near carboxysomes in cyanobacteria (Liberton *et al.* 2011). PolyP chains provide a reservoir of chemical bond energy for driving biological processes as well as other roles such as acting to buffer changes in cellular PO_4^{3-}, as chaperone in protein folding, binding and sequestering cations, thereby contributing to heavy metal tolerance (Sanz-Luque *et al.* 2020; Rees and Raven 2021, and references therein).

In oligotrophic regions of the ocean where inorganic P is scarce, eukaryotic photoautotrophs can reduce their cellular P requirements by 50% through decreases in the content of phospholipids and RNA or substituting phospholipids with non-P membrane lipids (i.e. N-containing betaine lipids); such mechanisms allow phytoplankton to maintain growth in P-limited conditions albeit at the expense of decreased growth rates (Van Mooy *et al.* 2009).

8.2.3. *Sulfur*

The Sulfur (S) cell quota in algae is similar to that of P (Giordano 2013). S-containing compounds include the amino acids cysteine and methionine, sulfolipids, gluthathione, vitamins and cofactors such as CoA (Takahashi *et al.* 2011; Giordano and Prioretti 2016). It plays fundamental functional roles, being a key element in the redox regulation of enzymes and in electron transport as Fe–S clusters embedded in several electron transporters such as Photosystem I (PSI), Cytb_6/Fe_{nh} and ferredoxin (Fdx).

Sulfur is acquired by photosynthetic cells as sulfate (SO_4^{2-}), the most oxidized (+6) form of S, thanks to membrane transporters, but it is assimilated (see glossary) into most organic compounds as sulfide (S^{2-}, oxidation number -2). There are also organic sulfate esters (S +6) in cell walls of some algae, and also sulfoquinovose diglyceride (S +4) in thylakoid membranes (Mizusawa and Wada 2012; Le Costaouëc *et al.* 2017). SO_4^{2-} acquisition relies on different types of transporters. SULTR transporters catalyze a cotransport H^+/SO_4^{2-} taking advantage of the proton gradient across membranes, and are widespread among photosynthetic organisms, while another family of Na^+/SO_4^{2-} transporters, SLT, has been identified in freshwater green algae such as *C. reinhardtii* and *Volvox carteri* (and in non-vascular plants). In cyanobacteria and in the chloroplasts of green algae, the SO_4^{2-} transport is catalyzed by ATP-dependent ABC transporters (Giordano and Prioretti 2016).

Sulfur assimilation into most organic molecules is an energy-demanding reductive process. First, SO_4^{2-} is activated by ATP sulfurylase (ATPS), producing adenosine 5'-phosphosulfate (APS). This step lowers the standard redox potential from -454 mV of SO_4^{2-}/SO_3^{2-} to -60 mV of APS/SO_3^{2-} (Giordano and Raven 2014), allowing the subsequent action of APS reductase (APR) which produces sulfite, SO_3^{2-}, using reduced glutathione as an electron donor (Giordano and Prioretti 2016).

Sulfite is further reduced to S^{2-} by the action of sulfite reductase (SiR), a ferredoxin-dependent enzyme (Giordano and Prioretti 2016). S^{2-} is then incorporated into the S-containing amino acid cysteine by the action of two enzymes, serine acetyl transferase (SAT) and o-acetyl-serine thiol lyase (OAS-TL). Cysteine can then be further processed to produce other S-containing compounds such as glutathione and methionine. S assimilation by photosynthetic organisms then allows organic S-containing molecules to be available for other living organisms through the food web, for example, metazoans assimilate reduced S from Cys and Met amino acids ingested with their diet (Patron *et al.* 2008). Knowledge on S assimilation in photosynthetic cells is mainly related to higher plant species, while information on algae is still incomplete and under investigation. The above-described sulfate reduction takes place in chloroplasts (Takahashi *et al.* 2011; Giordano and Prioretti 2016), with the exception of *Euglena gracilis* where those reactions occur in mitochondria (Saidha *et al.* 1988). For some enzymes, however, multiple isoforms might occur with different localization within the cell. For example, some species possess two ATPS isoforms, one plastidial and a second one cytosolic (Takahashi *et al.* 2011; Giordano and Prioretti 2016). Similarly, multiple isoforms of SAT and OAS-TL can exist, which can be localized to chloroplasts, mitochondria and/or the cytosol, characterized by uneven activity in species where all isoforms are present (as in the vascular plant *Arabidopsis*) (Wirtz and Hell 2006). Available data allow us to hypothesize differences in the regulatory point for S assimilation between higher plants and algae, with APR being more important in the former and ATPS in the latter, particularly in marine species where ATPS appears to be redox-regulated (Prioretti *et al.* 2014).

In addition to the reduction pathway, APS can be phosphorylated by APS kinase (APK) generating 3′-phosphoadenosine 5′-phosphosulfate (PAPS) to obtain sulfated compounds. Moreover, methionine is the precursor of dimethylsulfoniopropionate (DMSP). DMSP is suggested to play a number of roles in microalgal cells. They range from being an osmolyte alternative to glycine betaine (from which it differs being an S-containing rather than an N-containing molecule), a cryoprotectant and an anti-grazing molecule (Norici *et al.* 2005; Takahashi *et al.* 2011; Giordano and Prioretti 2016). Further, a product of DMSP enzymatic degradation, dimethylsulfide (DMS), is released in the atmosphere, contributing to the biogeochemical cycle of S in nature. The production of DMSP is, generally, related to the cell C:S ratio in microalgae. The main DMSP producers in oceans, mostly the dinoflagellates and coccolithophores, are in fact characterized on average by a lower C:S ratio compared to green algae (Norici *et al.* 2005; Bullock *et al.* 2017).

Freshwater and marine algae face different S availability in their habitats. In freshwater environments, SO_4^{2-} can occur in limiting concentrations for algal cells (from 10 µM to 1 mM; Giordano *et al.* 2005; Ratti *et al.* 2011). Conversely, today's oceanic SO_4^{2-} concentration is about 28 mM. However, the oceanic SO_4^{2-} concentration has changed during the history of the Earth, with concentrations inferred to be about 1–5 mM in the Proterozoic ocean, increasing to about 13–27 mM in the late Paleozoic–early Mesozoic oceans (Ratti *et al.* 2011) and reaching today, likely, the highest concentration of 28 mM. The dominant marine phytoplankton species have also changed during these eras, with green lineage species dominant in the past, when SO_4^{2-} concentration was low, and red lineage species dominant in today's high $[SO_4^{2-}]$ oceans. A correlation between SO_4^{2-} concentration and species dominance has thus been hypothesized in the so-called "sulfate facilitation hypothesis" (Ratti *et al.* 2011), based on the average S cell quota of species belonging to the green and red algal lineages, which is lower in the former. Therefore, the increase in oceanic SO_4^{2-} concentrations may have contributed to the rise to dominance of the red lineage phytoplankton species.

8.2.4. *Silicon*

Silicon (Si) is used in biomineralization processes by several ochristan phototrophs (Bacillariophyceae, Parmophyceae, Chrysophyceae, Synurophyceae and silicoflagellates), and also by some Prymnesiophyceae and Prasinophyceae (Raven and Giordano 2009; Durak *et al.* 2016; Medlin 2016; Langer *et al.* 2021). Among the algae cited above, diatoms are responsible for up to 40% of the world's aquatic photosynthesis (Nelson *et al.* 1995) and Si is a major and essential nutrient for these organisms; therefore, Si availability can shape the community of primary producers and influence diatom primary production.

Hydrated amorphous silica is the major component of the diatom cell wall, the frustule. Fitness advantages that such a cell wall provides in terms of fast sinking (buoyancy control), mechanical defense, nutrient uptake, UV-B screen and PAR (Photosynthetic Active Radiation, 400–700 nm) waveguide will not be discussed here but the reader is referred to Raven and Waite (2004) and Ellegaard *et al.* (2016) for further information on these aspects of diatom frustules.

The dissolved Si mostly comes from natural processes such as weathering of silicate rocks. Undissociated silicic acid $Si(OH)_4$, which accounts for about 97% of the dissolved Si in seawater, is acquired actively by cells; the ionic form $SiO(OH)_3^-$, which represents the remaining 3% (Stumm and Morgan 1981) has, in some cases, also been shown to be taken up (Martin-Jezequel *et al.* (2000) and references therein). An energy-dependent sodium/silicic acid symporter has been identified in marine diatoms and the corresponding enzyme in freshwater diatoms may use both sodium and potassium (Thamatrakoln *et al.* 2008). Silicic acid transporter (SIT) genes have been isolated for the first time in *Cylindrotheca fusiformis*; the expression of transporter genes differed during frustule synthesis and suggested specific roles in the overall control of Si transport (Hildebrand *et al.* 1997, 1998; Thamatrakoln *et al.* 2008). Once acquired, soluble Si is maintained at supersaturated concentration and deposited in specialized vesicles without being diverted to internal pools for storage (Sullivan 1979; Martin-Jezequel *et al.* 2000 and references therein). The silicon deposition vesicles (SDV) are surrounded by a lipid bilayer associated with proteins, named the silicalemma (Martin-Jezequel *et al.* 2000; Tesson *et al.* 2017; Hildebrand *et al.* 2018). The intralumenal pH of the SDV is acidic to favor silica polymerization, and the growing valves are enclosed within the silicalemma, which may also contribute to membrane components such as proteins. After the valve is completed, exocytosis occurs and parts of the SDV are likely embedded in the frustule, which is protected by an organic scaffold from silica dissolution (since seawater is undersaturated and slightly alkaline). Frustulins, proteins associated with the organic scaffold rather than to the cell wall, are not necessarily derived from SDV proteins; species-specific silaffins and long-chain polyamines isolated from cell wall silica are instead active in the SDV at low pH (Kroger *et al.* 2000). Newly discovered proteins associated with biosilica in the SDV are silacidins (Wenzl *et al.* 2008) and silicanin-1 (Kotzsch *et al.* 2017), both likely involved in the morphological formation of the frustule (Hildebrand *et al.* 2018; Görlich *et al.* 2019). Si uptake and silicification are fueled by aerobic respiration (via oxidative phosphorylation) (Martin-Jezequel *et al.* 2000).

The Si cell quota differs in different species (partly due to cell size) and is directly affected by growth and indirectly by environmental factors (Martin-Jezequel *et al.* 2000). In fact, when nutrient availability is low such as during N or P limitation, then the diatom Si pool is larger and consequently the frustule is thicker (De Tommasi *et al.* 2017). Metal ions ($Fe^{2+/3+}$, Zn^{2+}) can also influence Si uptake and pools in a species-specific manner (De La Rocha *et al.* 2000). The effect of light regimes on soluble Si pools and silicification is still controversial: low light might increase silicification (Taylor 1985; Martin-Jezequel *et al.* 2000) or not (Su *et al.* 2018). Interestingly, in areas and seasons of Si-limited oceans (i.e. Southern Ocean; Pinkerton *et al.* 2021), reduced levels of Si uptake and silicification have been observed, while diatoms maintained nearly maximal division rates (Olsen and Paasche 1986). In contrast, freshwater diatoms are generally much more silicified than are marine diatoms, including during the bloom seasons (Conley and Kilham 1989). It is noteworthy that in the generally brackish Baltic Sea, where the extensive long-term damming of rivers by human activities has reduced the natural supply of dissolved Si to the sea, many of the dominant marine diatoms are less silicified and consequently less sensitive to low concentrations of dissolved Si (Spilling *et al.* 2018).

Macroevolutionary patterns in diatom frustules show a similar adaptation towards smaller and less silicified forms in response to ocean Si depletion over geological time (Finkel *et al.* 2005, 2010). Indeed, fossil records and molecular evidence suggest that when diatoms first appeared in the Triassic period (perhaps as early as 250 Myr ago), they were not silicified; the earliest fossils come from the Jurassic, some 190 Myr ago (Sims *et al.* 2006), showing centric heavily silicified cells. A gradual evolution from the centric to the less silicified pennate symmetry then occurred (Medlin 2016) as the available dissolved Si concentration decreased (Finkel and Kotrc 2010). In fact, in the absence of biosilicifiers, Si concentration was determined by inorganic reactions with respect to amorphous silica solubility and/or with other Si mineral complexes. With the appearance of sponges and radiolarians in the late Proterozoic, biomineralization caused a drawdown of dissolved Si concentrations (Conley *et al.* 2017). Dissolved Si was further reduced in oceans by the diatom radiation during the Cenozoic era (Finkel and Kotrc 2010; Conley *et al.* 2017) changing from more than 1 mM (solubility is limited to around 2 mM) in the Precambrian time to present concentrations of 2–100 μM (Garcia *et al.* 2006; Conley *et al.* 2017). Growth and Si uptake kinetics saturate at concentrations lower than 60 μM (Finkel and Kotrc 2010).

8.2.5. *Iron, copper, manganese, zinc, molybdenum, nickel*

The trace metal nutrients Fe, Cu, Mn, Zn, Mo and Ni are essential cofactors/catalysts for indispensable functions, like photosynthesis, respiration, N assimilation and CO_2-concentrating mechanisms. For example, Fe atoms are found in PSII, PSI, Cyt b_6f, Cytc_6, Fdx as well as in the N-assimilating enzymes, Cu in plastocyanin and cytochrome oxidase, Mn in PSII (OEC) and iota-CA, Zn in CAs and some SITs and Mo in the nitrogenase of N fixation and in nitrate reductase (Blaby-Haas and Merchant (2017) and literature therein; Jensen *et al.* 2019). It is also worth mentioning the Ni-form of superoxide dismutase (SOD), besides the Cu/Zn and Fe/Mn forms (Dupont *et al.* 2008; Morel *et al.* 2020). Urease is another Ni metalloenzyme, predominantly used by phytoplankton to hydrolyze urea (Rees and Bekheet 1982).

Phototrophs have evolved responses to fluctuations in supply of these elements. For instance, metal replacement has been reported in CAs where Cd, Co, Fe or Mn can be used as a cofactor instead of Zn (Lane and Morel 2000a, 2000b; Jensen *et al.* 2020; Morel *et al.* 2020). These divalent metals may compete for the uptake, and a deficiency in a given metal (Zn) can lead to enhanced uptake of others (Cd or Co) (Morel *et al.* (2020) and references therein).

On the other hand, if accumulated in excess within the cell, the same metals can be toxic and react in the presence of oxygen resulting in ROS production and oxidative stress. Cellular metal concentrations thus must be finely regulated to strictly meet metabolic demand when external concentrations vary. For example, when Fe demand exceeds its supply, phototrophs express non-reductive Fe uptake through unique high-affinity transporters of dissolved Fe (transferrin-like proteins) and siderophore-bound Fe in addition to reduction–oxidation–transport mechanisms (Allen *et al.* 2007; Gao *et al.* 2021).

Algae can also release stored Fe from acidocalcisomes and ferritins (Blaby-Haas and Merchant 2017). Ferritin assembles into a multimeric complex capable of storing up to 4,500 Fe(III) atoms as an iron oxide mineral by oxidizing iron at ferroxidase centers and releasing reduced iron on cellular demand. Ferritin is involved in Fe buffering: for example, in acetate-grown cells of *Chlamydomonas reinhardtii,* ferritin mediates an Fe-sparing/recycling process involving degradation of the Fe-rich PSI complex, ferredoxin and Cyt$b_6f$$Fe_{nh}$, in favor of the mitochondrial respiratory chain with a reduced Fe quota (Busch *et al.* 2008; Page *et al.* 2012). Diatoms are known to modify their photosynthetic apparatus in order to require less iron, by decreasing the PSI:PSII ratio or by replacing cytochrome c_6 (Fe-containing) with plastocyanin (Cu-containing) (Strzepek and Harrison 2004; Peers *et al.* 2006; Gao *et al.* 2021). Substitution of Fdx by flavodoxin, using flavin instead of iron to carry electrons, has also been observed in diatoms under Fe-limited conditions (Setif *et al.* 2001). Conversely, when Fe supply exceeds demand, increased efflux and sequestering of the ions in the acidic vacuoles occur (Blaby-Haas and Merchant 2017). Compartmentalization is used in these conditions to trap toxic concentrations of metals (seen also in the case of Cd) or to avoid mis-metalation from competitive ions (i.e. Cu during Zn starvation) (Waldron *et al.* 2009).

Primary productivity in 30–40% of the world's contemporary oceans is limited by the availability of iron, particularly in open-ocean regions of the Southern Ocean, equatorial Pacific Ocean and north Pacific Ocean (Moore *et al.* 2004). These high-nutrient, low-chlorophyll (HNLC) regions are characterized by extremely low concentrations of iron and high concentrations of other essential nutrients, such as nitrate and phosphate. Low Fe availability limits primary production by hampering N assimilation and therefore growth (Figure 8.1) (Pinkerton *et al.* 2021). Iron-enrichment experiments carried out to date confirm that iron fertilization does increase carbon fixation while phytoplankton community shifts from one dominated by small cyanobacteria to one dominated by pennate diatoms (Boyd *et al.* 2007).

The ability for enhanced iron storage and protection against oxidative stress provided by ferritin in pennate diatoms was probably responsible for their numerical dominance in the blooms following iron fertilization (Marchetti *et al.* 2009).

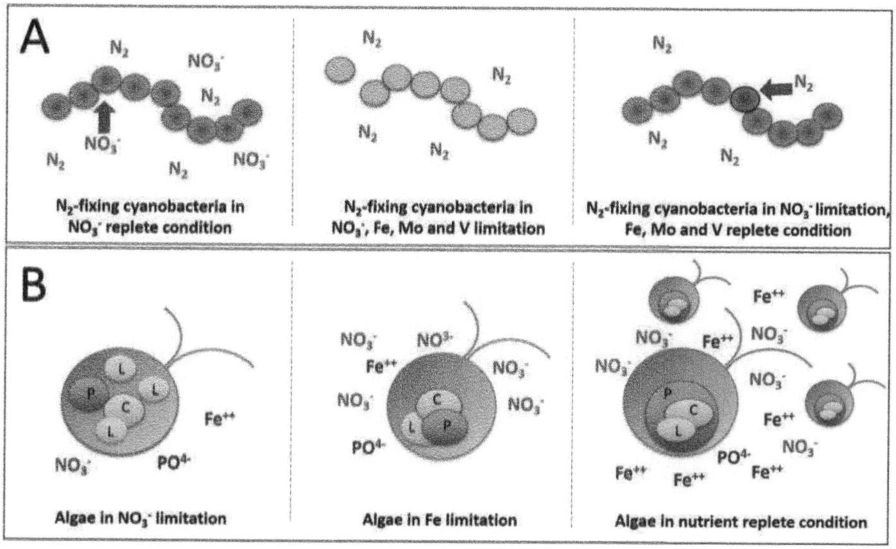

Figure 8.1. *A schematic representation showing how nutrients may impact primary production and macromolecular composition in phytoplankton. A) diazotrophic cyanobacteria in oligotrophic waters may be limited by trace metals for the synthesis and catalysis of the N_2-fixing enzyme, nitrogenase; the blue arrow highlights the N-form taken up; B) algae under N limitation have a smaller protein pool (P) and often more lipids (L) than algae in replete conditions; algae under Fe limitation have less proteins than algae in the replete condition since Fe is required for N-assimilating enzymes. P, proteins; L, lipids; C, carbohydrates. For a color version of this figure, see www.iste.co.uk/maberly/photosynthesis.zip*

Iron has affected the primary production since the Proterozoic (2,500–542 Ma) due to a shift in the redox state of the oceans which changed the bioavailable Fe^{2+} to bio-unavailable Fe^{3+} (Anbar and Knoll 2002). In the anoxic waters of the Archean Eon and the Paleoproterozoic, Fe^{2+} was abundant (50 µM) and formed banded iron formations (BIFs) (Holland 1973) (Figure 8.2). BIFs are chemical sediment deposits that primarily consist of Fe-bearing minerals and silica. When oceans became sulfidic in the mid Proterozoic (after roughly 1,800 Ma), deep waters were still anoxic and became more reducing rather than more oxidizing despite the rise of dissolved oxygen in surface waters (Canfield 1998) (Figure 8.2). In sulfidic conditions, Fe, involved in biological N_2 fixation and nitrate assimilation, was removed from solution in H_2S-bearing waters and deposited as Fe-sulfides instead of BIFs. The redox state of the oceans was therefore directly coupled to N bioavailability (Figure 8.2). In anoxic and sulfidic oceans, Mo was also scarce because of rapid removal and limited N metabolism. Enhanced ocean oxygenation and oxidative weathering made Mo anions (MoO_4^{2-}) more abundant at the surface (Figure 8.2). In contemporary oxic oceans, Fe concentration is more than three orders of magnitude lower as compared to the concentration in anoxic times and Mo is the most abundant transition metal in the oceans, with a concentration of 105 nM (Figure 8.2).

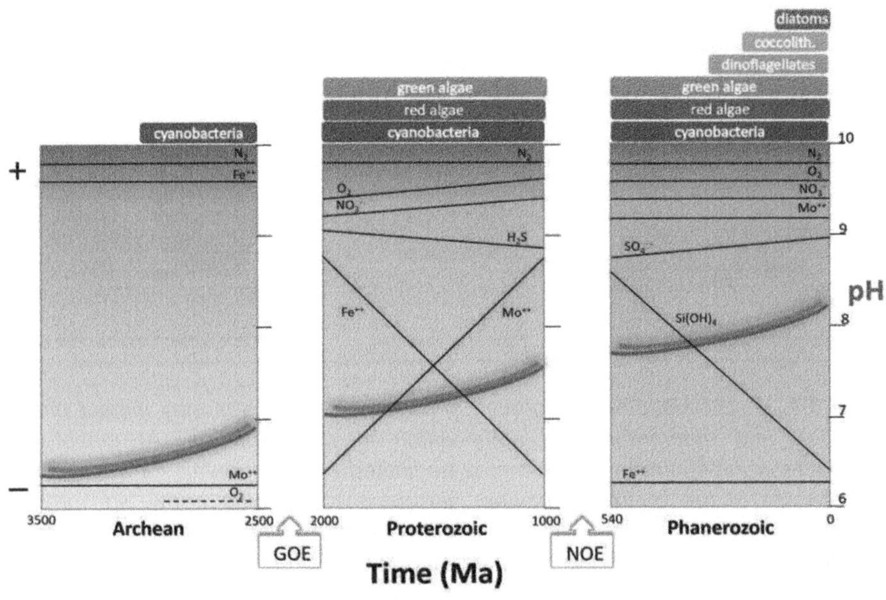

Figure 8.2. *A schematic diagram showing changes in nutrients and pH through geologic eras. Ocean chemistry in three specific ranges of time (Archean from 3,500 to 2,500 Ma, Proterozoic from 2,000 to 1,000 Ma, Phanerozoic from 540 to 0 Ma) is depicted. Nutrient concentrations are not given in absolute values but presented as trends. Above each time period, the phytoplankton radiation is highlighted. GOE, Great Oxygenation Event; NOE, Neoproterozoic Oxygenation Event; Ma, million years ago. For a color version of this figure, see www.iste.co.uk/maberly/photosynthesis.zip*

8.3. The ultimate element limiting productivity and cell stoichiometry

Back in the 19th century, Liebig's law of the minimum states that agricultural yields are limited at any given time by a single resource (1841) and half-a-century later, Blackman (1905) showed that photosynthetic growth is limited by a single factor (i.e. temperature, light, "supply of materials"), following the same principle of the minimum. During recent decades, however, research has proved that more often in oligotrophic oceans, phytoplankton growth is limited by multiple resources simultaneously (Arrigo 2005). Resource co-limitation has been particularly invoked at the entire community level. The best-known example of community co-limitation is found in Fe-replete oligotrophic waters where N_2-fixing cyanobacteria are abundant, and N and P are scarce. Notably, (i) addition of NO_3^- to these waters stimulates non-N_2 fixers in the population capable of assimilating dissolved organic P using extracellular alkaline phosphatase and/or able to substitute P-containing molecules with N-containing ones (see section 8.2.2). Moreover, the P requirement might be higher in N_2 fixers due to a higher mRNA and rRNA quota needed for nitrogenase synthesis and resynthesis after damage. (ii) Addition of P stimulates cyanobacteria such as *Trichodesmium* (Arrigo (2005) and references therein); N_2 fixers would in fact support their N requirement through N_2 fixation and be more competitive in comparison with non-N_2 fixers.

In oligotrophic waters, the growth of N_2 fixers is especially favored when the NO_3^- upwelling to the surface, after O_2 decomposition of sinking organic matter in bottom waters, is reduced by microbial denitrification and anammox (Mulder *et al.* 1995). Since N demand in the euphotic zone (see glossary) exceeds NO_3^- fluxes from depth, N_2 fixation fuels new production and helps determine the amount of C and N that can be consumed by other organisms or exported to depth (Capone *et al.* 2005). In addition to N, P is also found in low concentrations over much of the open ocean, limiting marine phytoplankton growth rates (Tyrrell 1999; Elser *et al.* 2007; Krom *et al.* 2010). More often, P constrains primary production in freshwaters (Elser *et al.* 2007). This is partly due to the anthropogenic input of fertilizers: runoff into rivers contains more N than P since P is much less mobile in soils. Moreover, increased N emissions into the atmosphere have caused increased N deposition in lakes over Europe and North America and the higher concentrations of inorganic N have stimulated phytoplankton biomass (Bergstrom and Jansson 2006). Observations also suggest that anthropogenic N deposition has pushed inland water to greater P-limitation than before the industrial age (Bergstrom and Jansson 2006). Despite the unfavorable conditions for diazotrophy in rivers or freshwater basins, the N_2 fixation rate of heterotrophic diazotrophs (bacteria/archaea) associated with free-floating polysaccharide-based (high C:N ratio) aggregates has been measured at several times higher compared to rates reported from many oligotrophic marine environments (Geisler *et al.* 2020). Co-limitation of P and N has also been documented in inland waters, especially in small shallow lakes (Maberly *et al.* 2020).

Nutrient availability is said to exert a bottom–up control on phytoplankton cell growth and productivity (Moore *et al.* 2013). Nutrient limitation, including co-limitation by several nutrients, results in competition among phytoplankton species that have evolved a range of strategies to optimize nutrient acquisition driven by adaptation to environmental limitation. Some examples are described in the previous paragraphs.

Phytoplankton community structure is also influenced by top–down controls including predation by grazers and viral infection. However, recent discoveries regarding viral proteins acting to amend host nutrient limitation have merged bottom–up and top–down regulations of phytoplankton communities (Monier *et al.* 2017). Indeed, marine viruses harbor host-derived genes encoding a diverse range of putative functions. These "auxiliary metabolic genes" (AMGs) are thought to modulate host function to improve fitness of the virus and, in some cases and temporarily, of the host. Viruses can interfere with the hosts' nutrient uptake capabilities by encoding their own transporters (i.e. phosphate transporters, NH_4^+ transporters) (York *et al.* 2017). For example, the virus OtV6 infects *Ostreococcus tauri*, which is the smallest known free-living non-symbiotic eukaryote on Earth. Among the host-derived genes contained in the viral genome, there is a functional ammonium transporter which is closely related to the Amt/Mep/Rh superfamily of proteins (Monier *et al.* 2017). Its expression changes how the host accesses diverse N sources during infection.

Nutrient availability in surface waters is closely linked to the composition in elements assimilated by phytoplankton. This concept is expressed in oceanography by the Redfield ratio (Redfield 1934). He found that plankton have an average atomic C:N:P stoichiometry of 106:16:1 (Redfield 1958); many laboratory and field studies which followed such pioneering work have revised and extended such stoichiometry to other elements (Geider and La Roche 2002; Ho *et al.* 2003; Quigg *et al.* 2003). Quigg and coauthors (2011) proposed a ratio of $(C_{132}N_{18}P_1S_{0.99})_{1000}Fe_{18}Mn_{2.8}Zn_{1.34}Cu_{0.50}Co_{0.11}Cd_{0.11}$. In addition, Ho *et al.* (2003) report a mmol Mo to mol P ratio of 0.03, both under nutrient-replete conditions.

Variations in major and trace element composition across phytoplankton reflect the evolutionary history of the phyla and acclimation to environmental (biogeochemical) conditions. By evolutionary history, we consider the host and plastid endosymbioses, resulting in different genetic constraints, and the selection pressures associated with environmental conditions that characterized aquatic chemistry through time. Evolutionary history appears to be more important in determining phytoplankton composition in nutrient-rich environments, while acclimation strategies are driving factors under nutrient limitation, as found in oligotrophic gyres.

The elemental stoichiometry of phototrophs is linked to their macromolecular composition, that is, cellular architecture and biochemistry, at the phylum level (Finkel *et al.* 2016). Nevertheless, in nutrient-sufficient conditions, the median macromolecular composition of exponentially growing microalgae is roughly 32% protein, 17% lipid, 15% carbohydrate, 17% ash, 6% RNA, 1% chlorophyll *a* and 1% DNA expressed as percent dry weight (Finkel *et al.* 2016). The major contributors to cell composition are cell wall structure (i.e. silicified in diatoms, calcified in some haptophytes, organic in cyanobacteria and other eukaryotes), light-harvesting apparatus (see the next section) and storage pools (i.e. TAG droplets, starch), underlining how functions shape organic matter composition in primary producers (Figure 8.1(B)). Notably, each macromolecular pool is characterized by a different cost in terms of chemical energy (ATP equivalents) invested to allocate C into cellular matter with the costs for proteins>lipids>carbohydrates (Montechiaro *et al.* 2006). Energetic constraints to biomass composition become relevant when energy availability limits growth.

8.4. Light variability and effect on photosynthesis

8.4.1. *Light-harvesting and photosynthetic electron transport*

Photosynthetic organisms harvest sunlight energy via pigment–protein complexes localized in thylakoid membranes. Upon energy transfer between light-harvesting pigments and eventual transfer of excitation energy to Photosystem II (PSII) and PSI reaction centers (chlorophyll special pairs), charge separation occurs, driving photosynthetic electron transport through the multi-subunit thylakoid protein complexes of PSII, Cytb_6/Fe_{nh} and PSI and several soluble electron transporters (Figure 8.3). After charge separation at the PSII reaction center (P680), electrons move to plastoquinone molecules. Plastoquinol (reduced plastoquinone) then diffuses in the thylakoid membrane and donates electrons to the membrane-embedded Cytb_6/Fe_{nh} complex. Concomitantly, the oxidized PSII reaction center chlorophyll ($P680^+$) receives an electron released from water at the lumen-exposed Mn_4CaO_5 oxygen evolving complex (OEC), generating, after four rounds of charge separation, O_2 as a byproduct. Cytb_6/Fe_{nh} transfers the electrons to the lumen-soluble carrier plastocyanin (PC) or cytochrome c_6. After light absorption and charge separation at PSI, electrons move to ferredoxin (Fdx), the ferredoxin-NADPH reductase (FNR) and finally reduce $NADP^+$ to NADPH. The oxidized PSI reaction center chlorophyll ($P700^+$) is reduced by the electron coming from PC or Cytc_6. Altogether, these components drive linear electron flow (LEF) from water to NADPH. In parallel to electron transport, protons accumulate in the thylakoid lumen. This generates an electrochemical potential, the so-called proton motive force (pmf), across the thylakoid membrane. The pmf comprises both a H^+ concentration gradient (ΔpH) and an electric gradient (Δψ) and drives ATP synthesis. Products of the "light reactions" of photosynthesis, ATP and reducing power are then used in the Calvin–Benson–Bassham (CBB) cycle to fix CO_2 into organic compounds and, more generally, to sustain cell metabolism and growth (see Chapters 1 and 5).

Photosystem reaction centers (RC), which enclose the chlorophyll special pairs (P680 for PSII and P700 for PSI) and the other PS core electron transporters, are widely conserved among oxygenic photosynthetic organisms. Common to oxygenic photosynthetic autotrophs of different phylogeny is also the use of chl *a* as the main photosynthetic pigment, with few exceptions. Recent reports showed that, rarely, chl *d* or chl *f* are also present in oxygenic photosynthetic organisms. The photosynthetic apparatus of the cyanobacterium *Acaryochloris marina* is dominated by the far-red-absorbing chl *d*, although chl *a* still occupies some of the key electron transfer positions (Renger and Schlodder 2008). Another cyanobacterium, *Chroococcidiopsis thermalis*, when grown under far-red illumination, contained both chl *d* and *f* in addition to chl *a*, which was still the most abundant (Nürnberg *et al.* 2018).

Besides chl *a*, the specific accessory pigments and light-harvesting antennae system have changed during the evolution of photosynthetic organisms (Allen *et al.* 2011; Caffarri *et al.* 2014). Cyanobacterial light harvesting involves the phycobilisomes, large thylakoid–peripheral protein complexes which comprise the subunits allophycocyanin, phycocyanin and phycoerythrin binding the linear tetrapyrrole pigments phycocyanobilin (the former two) and phycoerythrobilin (the latter). Phycobilisomes as the light-harvesting system are also found in red algae, glaucophytes and crytophytes, the latter displaying a unique phycobilisomes localized on the thylakoid luminal side (Spear-Bernstein and Miller 1989; Greenwold *et al.* 2019). All other groups of eukaryotic algae instead contain the light-harvesting complexes (LHC) as pigment-binding protein complexes embedded in the thylakoid membrane. LHCs are present in different isoforms and can accommodate chl *a*, chl *b* or chl *c*, and several carotenoids, according to the species. Algae belonging to the green lineage contain chl *a* and *b*, as do land plants, while groups originating from the red lineage, such as diatoms, include chl *a* and chl *c* as accessory chl. The specific LHCs are often named according to their associations with photosystems or their main pigments. Thus, green algae are characterized by the presence of LHCA and LHCB antennae, chl *a*/*b*-binding proteins bound to PSI and PSII, respectively. Species from the red lineage instead have LHCF and LHCR isoforms, which are preferentially bound to PSII and PSI. In diatoms, the main LHCF isoforms are also known as FCP (fucoxanthin–chlorophyll proteins) while

VCP (violaxanthin–chlorophyll protein) is the main LHC subunit in the Ochrophyta *Nannochloropsis* (Sukenik *et al.* 2000; Koziol *et al.* 2007; Engelken *et al.* 2010; Neilson and Durnford 2010; Büchel 2020).

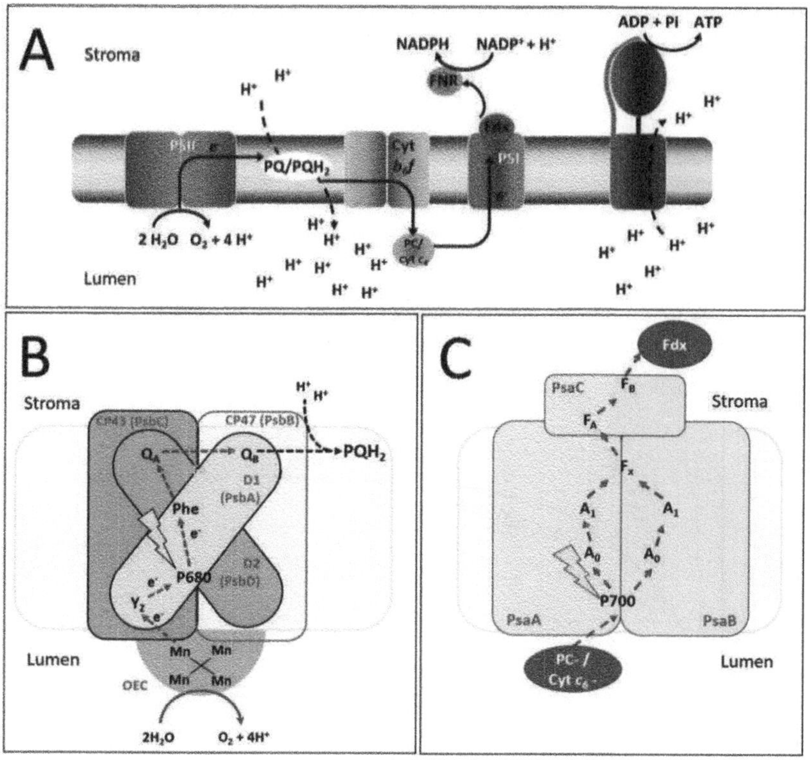

Figure 8.3. *Photosynthetic electron transport chain. A) Schematic view of the linear electron transport chain (LET). Light energy absorbed at PSII and PSI powers charge separation and electron transport. PSII starts the electron transfer to plastoquinone, $Cytb_6fFe_{nh}$, PC and/or $Cytc_6$ and PSI. The oxidized PSII special pair ($P680^+$) is reduced by water splitting at the oxygen evolving complex (OEC). PSI reduces Fdx, then FNR, and finally NADPH is produced. Electron transport is coupled with proton translocation into the lumen, generating the proton motive force (pmf) used by the ATP synthase to produce ATP. B,C) Simplified view of PSII and PSI reaction centers (RC). Only the main PS subunits involved in coordinating the various PS core electron transporters are depicted. Phe, pheophytin; Q_A, Q_B, quinones; Yz, tyrosine; A_0, chlorophyll; A_1, phylloquinone; F_X, F_A, F_B, FeS clusters. Red dashed lines: after light energy absorption, electrons move from the special pair to electron acceptors. Blue dashed lines: electrons move from electron donors to reduce the oxidized special pair. For a color version of this figure, see www.iste.co.uk/maberly/photosynthesis.zip*

The fraction of the incident sunlight spectrum capable of being absorbed by photosynthetic pigments is called PAR and ranges from the more energetic blue wavelengths to red light (400–700 nm), with different pigments being characterized by different absorption spectra. This results in species (and growth-condition)-specific absorption spectra. In an aquatic ecosystem, such a diversified pigment composition can allow multiple species to co-exist in the same environment, as light absorption by each species well complements light utilization by the others (Stomp *et al.* 2004).

8.4.2. *Photosynthesis versus irradiance (P vs. E) curves*

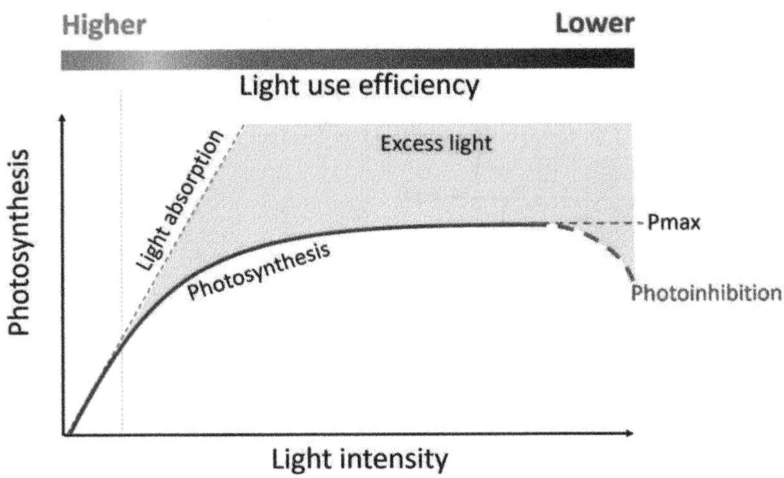

Figure 8.4. *Photosynthetic versus irradiance curve (P vs. E). The relationship between light intensity and photosynthesis is schematically depicted. Light use efficiency is higher when light intensity is low, as most of the light captured is used to drive electron transport and C fixation. Conversely, when light intensity exceeds the photosynthetic capacity of the photosynthetic cells (Pmax, which itself is dependent on the specific environmental conditions cells are facing), photosynthesis reaches saturation and excess light energy must be safely dissipated. If light irradiation exceeds cell photoprotection capacity, photoinhibition (i.e. a decrease in photosynthesis) might also occur. For a color version of this figure, see www.iste.co.uk/maberly/photosynthesis.zip*

Figure 8.4 exemplifies the relationship between photosynthesis and light irradiance. Photon absorption by photosynthetic cells proportionally increases with the irradiance (dashed line in Figure 8.4). In contrast, the photosynthesis curve (measured either as electron transport rate (ETR), net CO_2 assimilation rate or O_2 evolution) as a function of light intensity reaches saturation (Pmax). In more detail, at low light (the green segment in Figure 8.4), photosynthesis is light-limited, that is, photosynthesis proportionally increases to irradiance because the light intensity is not saturating the photosynthetic apparatus and the energy captured is efficiently used to drive electron transport. The initial slope of the curve (alpha) is a measure of the incident photon use efficiency and includes the fraction of light energy absorbed and quantum yield of photosynthesis. For higher light intensities, instead, photosynthesis saturates, with Pmax indicating the maximal rate of photosynthesis at saturating light (Erickson *et al.* 2015). The consequences of exposure to excess light energy, which eventually might lead to photoinhibition, that is, a decrease in photosynthesis, are discussed in section 8.4.4 below. Alpha and Pmax may vary according to the species and may be modulated according to the environmental conditions (e.g. CO_2 availability).

8.4.3. *Aquatic ecosystems: temporal and depth variations of light*

In the natural environment, light is highly variable, both in its intensity and quality. Natural light changes range from seasonal and daily variations in average solar irradiance to abrupt fluctuations in light intensity due, for example, to cloud movement in the sky. Aquatic species, in addition, face light variations because of waves, of cells moving within the water column, of absorption by the water itself and by optically active substances. At the surface, light intensity is altered by waves, which exert a focusing/defocusing effect of incoming light, creating local and sudden modifications in PAR. Further, light is exponentially attenuated in its intensity as it penetrates the water column, and this is accompanied by changes in the spectral distribution of the light at depth (change in light quality) (Falkowski and Raven 2007; Stomp *et al.* 2007; Depauw *et al.* 2012). Water itself absorbs in the red and infrared portions of the spectrum. This has the consequence that light penetrating into water is increasingly enriched in blue and green wavelengths. Moving deeper in the water column, the light spectrum is further enriched in the green fraction, as an effect of light absorption by photosynthetic organisms themselves. An additional component of the solar spectrum reaching the ocean surface is ultraviolet radiation (wavelengths below 400 nm). UVB radiation in particular is potentially damaging to a range of biological processes, including photosynthesis, and the effects of UV radiation and its interactions with other environmental changes are described by Gao *et al.* (see Chapter 9).

Notably, the underwater light spectrum is diverse across different aquatic ecosystems depending on the concentration of dissolved organic matter (also called gilvin), inanimate particles (also called tripton) and phytoplankton cells themselves (Stomp *et al.* 2007). Gilvin and some forms of tripton strongly absorb in the blue region of the spectrum. The phytoplankton community is influenced accordingly, as species with different pigments are favored in competition with others and become dominant in different waters. For example, in clear oligotrophic waters, phytoplankton and organic matter concentration are low, the underwater light spectrum is enriched in blue wavelengths and the genus *Prochlorococcus* dominates the community in the deep part of the euphotic zone of oceans due to its ability to absorb such wavelengths more efficiently thanks to divinyl-chl pigments (Ito and Tanaka 2011). In coastal waters like the Baltic Sea where green light is enriched due to high gilvin and tripton concentrations, red-colored cyanobacteria dominate and use phycoerythrin to absorb such photons (Stomp *et al.* 2007).

Water transparency is often expressed as Secchi disc readings or as the diffuse attenuation coefficient (K) (Kirk 1994). The light attenuation coefficient is useful to allow the calculation of PAR at a specific depth, for example, the daily integrated PAR at the sea floor. It is also an indicator for the environmental state over time. For instance, light attenuation increases due to increasing amounts of gilvin and tripton as a consequence of eutrophication (nutrient enrichment of water bodies leading to massive phototrophic growth).

8.4.4. *Physiological processes associated with exposure to variable light intensities (acclimation and regulation of photosynthesis)*

The temporal and spatial variabilities in light intensity and in its spectrum affect photosynthetic efficiency and productivity. Low light can constrain productivity and growth, while excess light energy also poses multiple challenges to photosynthetic cells. Further, microalgae, particularly those living in the upper mixed layer (see glossary), are exposed to frequent and sudden variation in light exposure, due to their vertical movement within the water column. According to the particular depth of the euphotic zone and upper mixed layer, microalgae might live in limiting light conditions for most of the time, yet, due to vertical movement, they might suddenly face a burst of high light. Different mechanisms are thus activated in response of the "average" environmental conditions as well as their sudden changes. They have different time scales and deal with acclimation and regulation of photosynthesis, respectively (see glossary). On the short term (seconds or minutes), photosynthetic cells regulate their light absorption and photosynthetic electron transport by activating/deactivating regulatory proteins. Such processes are addressed in section 8.4.4.2. On a longer time scale, instead, cells can actively modify their proteome to acclimate to the (average) environmental conditions.

8.4.4.1. *Acclimation of photosynthesis*

In order to optimize photosynthesis in low-light irradiance, cells enhance the light-harvesting capacity by modulating the number and/or the size of the "photosynthetic units" (PSU) (Figure 8.5(A)). A photosynthetic unit comprises both PS core and related antenna subunits, and the relative ratio of those two components can be kept invariable (change in the number of PSU) or modified (change in the size of PSU). Overall, the effect is to increase the amount of chlorophyll per cell to maximize light energy harvesting and to support cell growth and metabolism. If the number of PSU is changed, both the light-harvesting proteins and the PS core complexes are accumulated in a greater amount. In the second case, cells are modulating their "antenna size" (photosynthetic unit size), that is, the number of antennae subunits serving each PS core (Falkowski and Owens 1980; Fisher *et al.* 1989).

If photosynthetic cells are acclimated to high light (Figure 8.5(B)), conversely, a fraction of antenna protein is usually degraded, while the capacity to fix CO_2 is increased, for example, by inducing the expression of Calvin–Benson–Bassham cycle enzymes, to make use of the products of light reactions under saturating light intensity (Fisher *et al.* 1989; Simionato *et al.* 2011). Pigment composition is also modified. Degradation of antenna proteins decreases the chlorophyll content per cell, whereas photoprotective carotenoids are usually accumulated to a greater extent compared to lower-light conditions (higher carotenoid to chlorophyll ratio), to protect cell components from oxidative stress (Wilhelm *et al.* 2014; Erickson *et al.* 2015; Virtanen *et al.* 2021).

8.4.4.2. *Regulation of photosynthetic light reactions*

Photosynthetic organisms also evolved several processes to regulate the flow of excitation energy and electron transport (Figure 8.5(C)) to cope with variable environmental conditions. This is particularly relevant when cells are exposed to a sudden increase in light energy due to surface waves or water column mixing, as described above. The light energy that exceeds photosynthetic electron transport capacity needs to be safely dissipated in order to avoid oxidative damage to the photosynthetic apparatus itself and, more generally, to all cell components.

Despite falling into specific categories (i.e. heat dissipation, state transitions, alternative electron routes), the key proteins required for the activation of such processes have adapted during evolution. Although the diversity of response is presently only emerging, current knowledge suggests each algal species to be likely characterized by a particular combination, regulation and amplitude of the different mechanisms, as exemplified below.

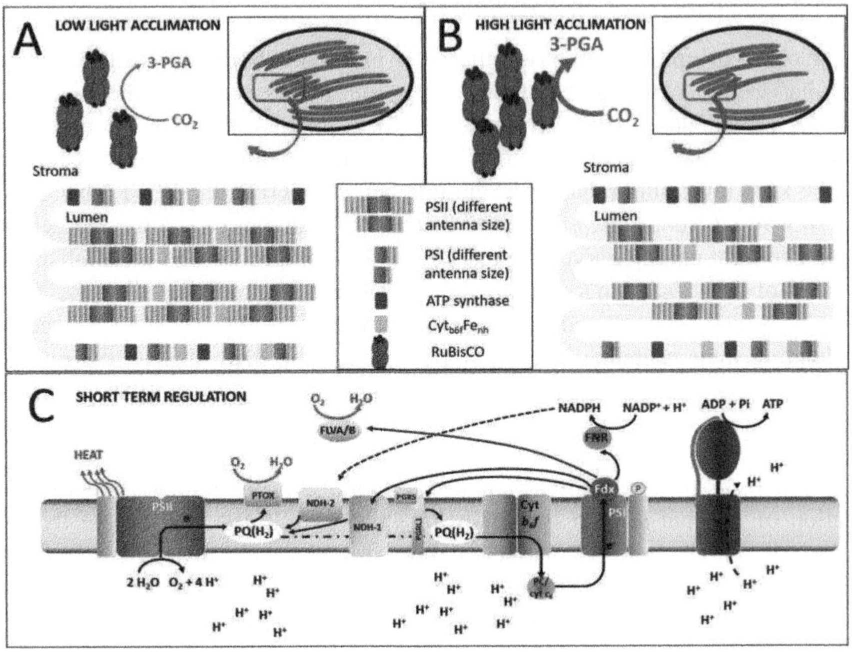

Figure 8.5. *Schematic view of the short-term regulation and long-term acclimation of the photosynthetic apparatus. A,B) Long-term acclimation of photosynthesis. The picture shows common changes of chloroplast ultrastructure and photosystem accumulation in the thylakoids upon prolonged low-light exposure (A) or prolonged high-light irradiation (B). In low-light conditions, usually, thylakoid membrane numbers within the chloroplast are increased, together with the number and/or size (i.e. number of associated antennae to PS RC) of photosystems with respect to high-light exposure. Rubisco content usually also increases in HL to make use of the available energy for C fixation. The cartoons of the different protein complexes depicted are not to scale. The figure depicts a situation in which all mentioned strategies are induced. The actual strategies detected (i.e. either modulation of thylakoid membrane numbers or PS number and/or of PS size) is species-specific. C) Short-term regulation of photosynthesis. The cartoon depicts different proteins involved in the regulation of the light phase of photosynthesis. They range from proteins involved in the dissipation of exceeding energy absorbed as heat at PSII to various proteins working as alternative electron donors or acceptors (PTOX, NDH-1, NDH-2, PGRL1/PGR5, FLV, all in light blue), to antenna proteins that can undergo reversible phosphorylation to balance the energy distribution between PSII and PSI. Note that proteins identified in different model species are presented together in the picture, although they are present in various combinations in the different phylogenetic groups. Refer to the main text and references therein for species-specific features. For a color version of this figure, see www.iste.co.uk/maberly/photosynthesis.zip*

The fastest photoprotection mechanism, induced within seconds/minutes upon an increase in light intensity, activates the dissipation of exceeding energy in the form of heat – a process also known as non-photochemical quenching (NPQ). In cyanobacteria, the orange carotenoid protein (OCP) is a photoactive protein which induces heat dissipation within the phycobilisomes (Muzzopappa and Kirilovsky 2020). In eukaryotic algae, heat dissipation of excess energy generally depends on three components: (i) increased pH gradient across the thylakoid membranes, which is achieved in high irradiance by chloride entry into, or potassium exit from, the lumen, in parallel to proton accumulation into the lumen, leading the *pmf* (see section 8.4.4.1 above) to become dominated by the pH component at the expense of the electrical potential difference (Raven 2020); (ii) the accumulation of specific LHC(-like) isoforms (named LHCSR or LHCX); and (iii) the reversible accumulation of specific carotenoids. Many algae possess the xanthophyll cycle, in which violaxanthin is reversibly converted in zeaxanthin, the latter playing a crucial role in photoprotection. In some phylogenetic groups, like diatoms, the carotenoids involved in xanthophyll cycle are instead diadinoxanthin/diatoxanthin.

Most information on microalgal heat dissipation photoprotective process comes from the characterization of a few model species, particularly the green alga *C. reinhardtii* and the diatoms *Phaeodactylum tricornutum* and *Thalassiosira pseudonana*. In *C. reinhardtii*, the model freshwater green alga, dissipation of excess energy absorbed as heat requires the accumulation of LHCSR protein (whose expression itself is induced by high-light acclimation), which acts as a pH sensor, thanks to protonatable residues exposed in the thylakoid lumen, and as an energy quencher, through carotenoids, particularly lutein (Peers *et al.* 2009; Bonente *et al.* 2011). Activation of this process thus takes advantage of the pH component of the *pmf*: the more electrons are transported in the thylakoid electron transport chain, the more H^+ are translocated into the thylakoid lumen (see section 8.4.4.1), activating heat dissipation of excess light energy absorbed. For this reason, this mechanism is often referred to as "feedback de-excitation". Although it is not strictly required for the initial heat dissipation activation, zeaxanthin accumulation also contributes to *C. reinhardtii* photoprotection and depends on the activity of a recently discovered atypical violaxanthin de-epoxidase (VDE) enzyme (Niyogi *et al.* 1997; Li *et al.* 2016). Recently, also PSBS, the thylakoid protein which has acquired a key role for heat dissipation of excess energy in land plants, was found to be transiently expressed in *C. reinhardtii*. Although the exact physiological role of PSBS in *C. reinhardtii* is still unclear, it is believed to facilitate the acclimation to strong light rather than being directly involved in the heat dissipation mechanism (Redekop *et al.* 2020). In the model pennate diatom *P. tricornutum*, and more generally in diatoms, the key mechanism is the pH-dependent regulation of the xanthophyll cycle, leading to accumulation of the carotenoid diatoxanthin (Dt) in excess light conditions. The LHCX1 protein, constitutively accumulated in contrast to the previous example of *C. reinhardtii* LHCSR, acts as an enhancer of NPQ but requires the accumulation of Dt to induce quenching (Bailleul *et al.* 2010; Zhu and Green 2010; Goss and Lepetit 2015). The specific contribution of LHCSR/LHCX proteins and of xanthophyll cycle in photoprotection is thus different among species (Goss and Lepetit 2015). Such differences are most likely not unique to the given examples of model species. Consistently, in the Prasinophyta *Mantoniella squamata*, a partial xanthophyll cycle has been characterized. Instead of accumulating zeaxanthin, such species accumulates antheraxanthin, which, however, fully substitutes zeaxanthin in its quenching role (Goss *et al.* 1998).

Available studies on several other species, although lacking detailed molecular characterization of the underlying molecular mechanisms, showed a great variety of NPQ induction kinetics and amplitude. Some species are able to activate a strong NPQ after a switch from dark to light, while in other cases, NPQ induction is slow and with low amplitude. Further, some species require pre-acclimation to strong light to induce NPQ, suggesting that key proteins are expressed only after high-light acclimation, as is the case for *C. reinhardtii*. Such a diverse behavior strongly suggests NPQ adapted according to the phylogenetic position and ecological niche occupied by each species (Gerotto and Morosinotto 2013; Goss and Lepetit 2015; Minagawa and Tokutsu 2015; Fisher *et al.* 2020; Lacour *et al.* 2020).

Induction of heat dissipation allows cells to modulate the amount of energy reaching the reaction centers. Other mechanisms instead allow electron transport in the thylakoid membranes to be regulated. Several proteins are involved in alternative electron flow routes, which donate or accept electrons from the LEF modulating ATP/NADPH ratio and/or avoiding over-reduction of LEF and photodamage. They include cyclic electron flow (CEF) at PSI, flavodiiron (FDP or FLV) proteins, plastid terminal oxidase (PTOX) and, in specific cases, hydrogen production (Peltier *et al.* 2010; Allahverdiyeva *et al.* 2015b). In the photosystem PSI CEF, electrons received by Fdx can be re-injected into the electron transport chain generating PQH_2, favoring ATP production over NADPH generation. At least two routes for Fdx – PQH_2 oxidoreductase activity exist. One is dependent on type I or type II NADH dehydrogenase. Type I NDH (NDH-1) is a multi-subunit thylakoid membrane complex identified, for example, in cyanobacteria. Despite the name, they are now believed to receive electrons from Fdx and not NAD(P)H. Type II NDH (NDH-2) is a single-subunit enzyme which substitutes NDH-1 activity in many microalgae, which lost *ndh* genes (Peltier *et al.* 2016). The second CEF pathway instead relies on PGR5/PGRL1, although the exact function of these proteins in CEF is still under extensive investigation (Nawrocki *et al.* 2019). FLV proteins instead work as an electron sink downstream of PSI performing a Mehler-like reaction (FLV1/3 in cyanobacteria and FLVA/B in eukaryotes). They thus catalyze a water–water cycle in which electrons extracted from water at PSII are used by FLVs to photoreduce O_2 to H_2O (Allahverdiyeva *et al.* 2013, 2015a). PTOX enzymes can accept electrons from plastoquinol, reducing molecular oxygen to water. The PTOX reaction might occur also in the dark, and participates in chlororespiration together with NDH (Nawrocki *et al.* 2015; Peltier *et al.* 2016).

A different process, state transitions, allows the redistribution of energy between PSII and PSI to balance their activities and thus optimize photosynthetic electron transport. In cyanobacteria, multiple underlying molecular models have been proposed, from phycobilisomes movement to spillover mechanisms (Calzadilla and Kirilovsky 2020). In green microalgae (and plants), state transitions rely on reversible phosphorylation of major LHCII subunits to adjust the energy distribution between the two photosystems. This process is particularly relevant in *C. reinhardtii*, in which up to 80% of LHCII can detach from PSII, contributing to energy quenching or migrating to PSI (Depège *et al.* 2003; Allorent *et al.* 2013; Nagy *et al.* 2014; Ünlü *et al.* 2014). In contrast, the presence of such a mechanism in diatoms is still under debate (Wilhelm *et al.* 2014; Xu *et al.* 2017).

Even if all these processes are activated to regulate photosynthetic electron transport according to the external conditions, cells cannot completely avoid damage to the photosynthetic apparatus during strong illumination and so-called photoinhibition (i.e. a decrease of photosynthetic efficiency; see also Figure 8.4) occurs. The PSII subunit D1 is the primary target of photodamage but an effective repair mechanism for PSII, in which a damaged D1 subunit is replaced with a newly synthesized one, has evolved in photosynthetic cells (Mulo *et al.* 2012). However, when the rate of damage exceeds the rate of repair, damaged PSII reaction centers are accumulated and photoinhibition occurs, reducing the photosynthetic efficiency and the productivity and growth of photosynthetic cells.

8.5. Photosynthesis and primary production in the water column

When applied to the water column, the following conditions can be recognized, based on the availability of light and nutrient resources (Figure 8.6).

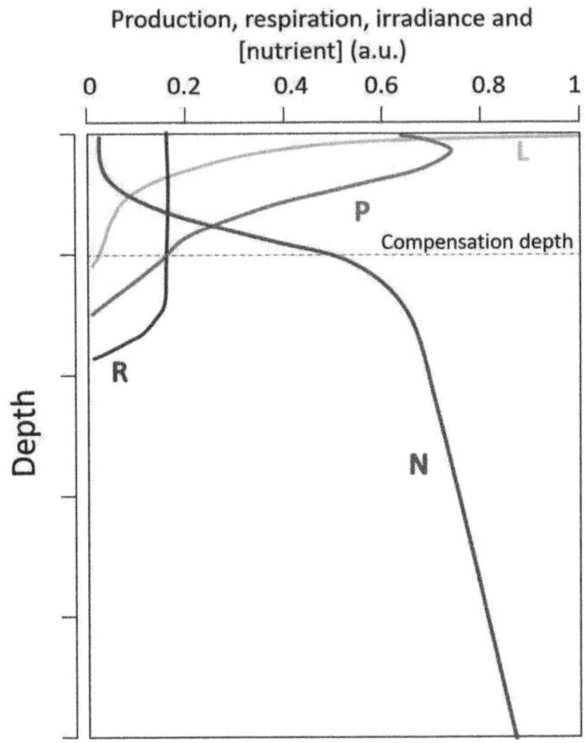

Figure 8.6. *Vertical profile of cellular photosynthesis (P), cellular respiration (R), attenuation of irradiance (L) and nutrient concentration (N) in a water column. The position in the water column corresponding to 1% of the surface irradiance is often taken as the limit of the euphotic zone (130–200 m) and the compensation depth (where P=R over 24 h). For a color version of this figure, see www.iste.co.uk/maberly/photosynthesis.zip*

At the surface, irradiation exceeds photosynthesis saturation and photoinhibition frequently occurs (see section 8.4.2). Photosynthetic rates and productivity are thus reduced. Below the surface, deeper in the euphotic zone, the environmental conditions in terms of both light and nutrient availability are the most favorable within the water column for photosynthetic cells and allow a peak in photosynthesis and in net primary production (NPP, see glossary).

Phytoplankton respiration is light independent and assumed to be 10% of the maximal photosynthesis (Ryther 1954). The lower limit of the euphotic zone (compensation depth) matches with the depth at which the gross daily photosynthesis is equal to phytoplankton daily respiration. Although compensation depth is often taken as corresponding to a light intensity of roughly 1% of the surface irradiance, net photosynthesis (see glossary) is frequently possible at light levels well below 1% surface light (Raven *et al.* 2000). Further down in the water column, photosynthesis becomes energy-limited and negative. As shown in Figure 8.6, after the peak within the euphotic zone, the photosynthesis curve shows a monotonic decay with depth.

As observed in previous sections, nutrient availability affects the spatial and temporal distribution of primary production in the euphotic zone. In temperate regions, the upper waters gain nutrients from the deep ocean during the winter thanks to convective mixing through the water column. Inorganic nutrients are, in fact, more concentrated at depth where the organic forms are oxidized to the inorganic ones and assimilation by phototrophs does not occur. With the arrival of the spring temperatures, the water column stratifies, separating the upper mixed layer (see glossary) from the deep waters, and phytoplankton blooms consume the nutrients contained in the upper mixed layer. The vertical nutrient profile shows the concentration gradient resulting from phototrophic uptake (Figure 8.6).

Polar seas have a homogeneous temperature through the water column, and tropical oceans exhibit a thermocline that is consistent across the year. Therefore, warm waters move towards the poles on the surface of the oceans and a corresponding flux of cold waters move towards the equator at depth. The so-called thermohaline circulation creates clockwise gyres of water circulation in the northern hemisphere and counterclockwise gyres in the southern hemisphere. At the mid-latitude oceans, solar heating induces thermal stratification of the water column according to the season, as already described. Because of the gyres, upwelling circulation is generated along continent boundaries in tropical regions enriching the coasts of nutrients. The rate of nutrient supply from depth to the euphotic zone ultimately controls phototrophic biomass in the euphotic zone.

For surface waters above the thermocline, inputs of DON and DOP may sustain primary production in the tropical and subtropical oligotrophic oceans (Torres-Valdes *et al.* 2009). DON accounts for more than 90% of the total dissolved N in waters above the thermocline (> 5 μM) and it is less important within the thermocline due to greater concentrations of nitrate. Again, DOP provides the dominant contribution to the total dissolved P (more than 75% in the upper 50 m) with values higher than 0.2 μM and minima of 0.05 μM in tropic and subtropical gyres (Torres-Valdes *et al.* 2009).

NPP can be remotely inferred by satellite images of the water body (ocean) color. Phytoplankton blooms affect the color of the water by altering light reflectance spectrum, blue–green (440–550 nm) spectral band ratios, spectral band difference between the bloom area and the nearby bloom-free area, thanks to generic- (chl *a*) and species-specific algal pigment absorption, such as phycobiliproteins for cyanobacteria, fucoxanthin for diatoms and peridinin for dinoflagellates (Blondeau-Patissier *et al.* 2014). In addition to the blue and green bands, the red (620–700 nm) and near-infrared (>700 nm) bands have also been used to derive chl *a* concentration from remote-sensing data due to the reduced influence from light absorption of non-algal suspended particles and colored dissolved organic matter on the red and near-infrared bands. Currently, there are several algorithms for deriving chl *a* concentration and phytoplankton biomass from the remote-sensing reflectance; each one has advantages and disadvantages (Blondeau-Patissier *et al.* 2014). Moreover, satellite-based data are limited by cloud coverage, ice (ice algae grow within or below sea ice which stops satellite from seeing the colors) and depth. Earth system models to simulate depth or ice-integrated estimates of NPP are being developed to implement satellite observations of phytoplankton biomass (mostly using chlorophyll as proxy) (Pinkerton *et al.* 2021).

8.6. Glossary

Acquisition, assimilation. **Acquisition:** process of uptake of a nutrient. **Assimilation**: process of incorporating the acquired nutrient into organic compounds.

Euphotic zone: part of the water column in which the net primary production takes place. The below limit of the euphotic zone equals the *compensation depth*, that is, the depth at which the daily gross photosynthetic C fixation (see glossary) and the respiratory losses are balanced. Compensation depth is often taken as the depth at which light intensity is 1% with respect to the surface. Euphotic zone can be influenced by multiple factors, like light intensity itself, temperature, nutrient availability and their chemical forms.

Gross photosynthesis: the light-dependent C fixation, that is, the transfer of electrons from water to CO_2 as terminal acceptor, powered by light energy. It is directly proportional to the photosynthetic electron transport.

Net photosynthesis: the difference between gross photosynthesis (light-dependent CO_2 fixation) and respiratory losses (electron flow from organic C to O_2), so the net C fixation.

***Net primary production* (NPP):** the photosynthetically fixed C available for other trophic levels (total rate of organic carbon production by phototrophs minus their respiration).

Regulation, acclimation, adaptation. **Regulation** (of photosynthesis light reactions): short-term mechanisms activated or switched-off on seconds/minutes time scale. These changes take advantage of proteins already present in the chloroplast and thylakoid membranes, which are feedback-activated or deactivated according to the actual electron transport, for example, by sensing the luminal pH or by the redox poise in the chloroplast stroma. **Acclimation** is a longer-term process (e.g. day, seasons), which also include changes in the proteome (i.e. synthesis or degradation of certain proteins). **Adaptation** requires genome-associated changes.

***Upper mixed layer*:** upper water portion which is easily mixed. It does not necessarily correspond to the euphotic zone.

8.7. Acknowledgments

This chapter is dedicated to Mario Giordano for his teaching and research commitment especially on the physiological responses of algae to environmental factors.

8.8. References

Aksoy, M., Pootakham, W., Grossman, A.R. (2014). Critical function of a *Chlamydomonas reinhardtii* putative polyphosphate polymerase subunit during nutrient deprivation. *Plant Cell*, 26, 4214–4229.

Allahverdiyeva, Y., Mustila, H., Ermakova, M., Bersanini, L., Richaud, P., Ajlani, G., Battchikova, N., Cournac, L., Aro, E.-M. (2013). Flavodiiron proteins Flv1 and Flv3 enable cyanobacterial growth and photosynthesis under fluctuating light. *Proceedings of the National Academy of Sciences of the United States of America*, 110, 4111–4116.

Allahverdiyeva, Y., Isojärvi, J., Zhang, P., Aro, E.-M. (2015a). Cyanobacterial oxygenic photosynthesis is protected by flavodiiron proteins. *Life*, 5, 716–743.

Allahverdiyeva, Y., Suorsa, M., Tikkanen, M., Aro, E.-M. (2015b). Photoprotection of photosystems in fluctuating light intensities. *Journal of Experimental Botany*, 66, 2427–2436.

Allen, M.D., del Campo, J.A., Kropat, J., Merchant, S.S. (2007). FEA1, FEA2, and FRE1, encoding two homologous secreted proteins and a candidate ferrireductase, are expressed coordinately with FOX1 and FTR1 in iron-deficient *Chlamydomonas reinhardtii*. *Eukaryotic Cell*, 6, 1841–1852.

Allen, J.F., de Paula, W.B.M., Puthiyaveetil, S., Nield, J. (2011). A structural phylogenetic map for chloroplast photosynthesis. *Trends in Plant Science*, 16, 645–655.

Allorent, G., Tokutsu, R., Roach, T., Peers, G., Cardol, P., Girard-Bascou, J., Seignerin-Berny, D., Petroutsos, D., Kuntz, M., Breyton, C., Franck, F., Wollman, F.-A., Niyogi, K.K., Krieger-liszkay, A., Minagawa, J., Finazzi, G. (2013). A dual strategy to cope with high light in *Chlamydomonas reinhardtii*. *Plant Cell*, 25, 545–557.

Anbar, A.D. and Knoll, A.H. (2002). Proterozoic ocean chemistry and evolution: A bioinorganic bridge? *Science*, 297, 1137–1142.

Armbrust, E.V. (2009). The life of diatoms in the world's oceans. *Nature*, 459, 185–192.

Arrigo, K. (2005). Marine microorganisms and global nutrient cycles. *Nature*, 437, 349–355.

Azov, Y. and Golmer, J.C. (1982). Free ammonia inhibition of algal photosynthesis in intensive cultures. *Applied and Environmental Microbiology*, 43, 735–739.

Bailleul, B., Rogato, A., de Martino, A., Coesel, S., Cardol, P., Bowler, C., Falciatore, A., Finazzi, G. (2010). An atypical member of the light-harvesting complex stress-related protein family modulates diatom responses to light. *Proceedings of the National Academy of Sciences of the United States of America*, 107, 18214–18219.

Baturin, G.N. (2003). Phosphorus cycle in the ocean. *Lithology and Mineral Resources*, 38, 101–119.

Bekker, A. and Holland, H.D. (2012). Oxygen overshoot and recovery during the early Paleoproterozoic. *Earth and Planetary Science Letters*, 317, 295–304.

Bergstrom, A.K. and Jansson, M (2006). Atmospheric nitrogen deposition has caused nitrogen enrichment and eutrophication of lakes in the northern hemisphere. *Global Change Biology*, 12, 635–643.

Blaby-Haas, C.E. and Merchant, S.S. (2017). Regulating cellular trace metal economy in algae. *Current Opinion in Plant Biology*, 39, 88–96.

Blackman, F.F. (1905). Optima and limiting factors. *Annals of Botany*, 19, 281–295.

Blondeau-Patissier, D., Gower, J.F.R., Dekker, A.G., Phinn, S.R., Brando, V.E. (2014). A review of ocean color remote sensing methods and statistical techniques for the detection, mapping and analysis of phytoplankton blooms in coastal and open oceans. *Progress in Oceanography*, 123, 123–144.

Bonente, G., Ballottari, M., Truong, T.B., Morosinotto, T., Ahn, T.K., Fleming, G.R., Niyogi, K.K., Bassi, R. (2011). Analysis of LhcSR3, a protein essential for feedback de-excitation in the green alga *Chlamydomonas reinhardtii*. *PLoS Biology*, 9, e1000577.

Boyd, C.M. and Gradmann, D. (1999). Electrophysiology of the marine diatom *Coscinodiscus wailesii* III. Uptake of nitrate and ammonium. *Journal of Experimental Botany*, 50, 461–467.

Boyd, P.W., Jickells, T., Law, C.S., Blain, S., Boyle, E.A., Buesseler, K.O., Coale, K.H., Cullen, J.J., de Baar, H.J.W., Follows, M., Harvey, M., Lanchelot, C., Levasseur, M., Owens, N.P.J., Pollard, R., Rivkin, R.B., Sarmiento, J., Schoemann, V., Smetacek, V., Takeda, S., Tsuda, A., Turner, S., Watson, A.J. (2007). Mesoscale iron enrichment experiments 1993–2005: Synthesis and future directions. *Science*, 215, 612–617.

Büchel, C. (2020). Light harvesting complexes in chlorophyll c-containing algae. *Biochimica et Biophysica Acta – Bioenergetics*, 1861, 148027.

Bullock, H.A., Luo, H., Whitman, W.B. (2017). Evolution of dimethylsulfoniopropionate metabolism in marine phytoplankton and bacteria. *Frontiers in Microbiology*, 8, 637.

Busch, A., Rimbauld, B., Naumann, B., Rensch, S., Hippler, M. (2008). Ferritin is required for rapid remodeling of the photosynthetic apparatus and minimizes photo-oxidative stress in response to iron availability in *Chlamydomonas reinhardtii*. *Plant Journal*, 55(2), 201–211.

Cade-Menun, B.J. and Payton, A. (2010). Nutrient temperature and light stress alter phosphate and carbon forms in culture-grown algae. *Marine Chemistry*, 121, 27–36.

Caffarri, S., Tibiletti, T., Jennings, R.C., Santabarbara, S. (2014). A comparison between plant photosystem I and photosystem II architecture and functioning. *Current Protein & Peptide Science*, 15, 296–331.

Calzadilla, P.I. and Kirilovsky, D. (2020). Revisiting cyanobacterial state transitions. *Photochemical & Photobiological Sciences*, 19, 585–603.

Canfield, D.E. (1998). A new model for Proterozoic ocean chemistry. *Nature*, 396, 450–453.

Capone, D.G., Burns, J.A., Michaels, A.F., Montoya, J.P., Subramaniam, A., Carpenter, E.J. (2005). Nitrogen fixation by *Trichodesmium* spp.: An important source of new nitrogen to the tropical and subtropical North Atlantic Ocean. *Global Biogeochemical Cycles*, 19, GB2024.

Cardona, T., Sanchez-Baracaldo, P., Rutherford, A.W., Larkum, A.W. (2018). Early Archaean origin of photosystem II. *Geobiology*, 17, 127–150.

Carnal, N.W. and Black, C.C. (1963). Phosphofructokinase activities in photosynthetic organisms. The occurrence of pyrophosphate-dependent 6-phosphofructokinase in plants and algae. *Plant Physiology*, 71, 150–155.

Carpenter, E.J. and Capone, D.G. (2008). Nitrogen fixation in the marine environment. In *Nitrogen in the Marine Environment*, Capone, D.G., Bronk, D.A., Mulholland, M.R., Carpenter, E.J. (eds). Academic Press, Cambridge.

Conley, D.J. and Kilham, S.S. (1989). Differences in silica content between marine and freshwater diatoms. *Limnology and Oceanography*, 34(1), 205–213.

Conley, D.J., Frings, P.J., Fontorbe, G., Clymans, W., Stadmark, J., Hendry, K.R., Marron, A.O., De La Rocha, C.L. (2017). Biosilicification drives a decline of dissolved Si in the oceans through geologic time. *Frontiers in Marine Science*, 4, 397.

De La Rocha, C.L., Hutchins, D.A., Brzezinski, M.A. (2000). Effects of iron and zinc deficiency on elemental composition and silica production by diatoms. *Marine Ecology Progress Series*, 195, 71–79.

De Tommasi, E., Gielis, J., Rogato, A. (2017). Diatom frustule morphogenesis and function: A multidisciplinary survey. *Marine Genomics*, 35, 1–18.

Depauw, F.A., Rogato, A., D'Alcalà, M.R., Falciatore, A. (2012). Exploring the molecular basis of responses to light in marine diatoms. *Journal of Experimental Botany*, 63, 1575–1591.

Depège, N., Bellafiore, S., Rochaix, J.-D. (2003). Role of chloroplast protein kinase Stt7 in LHCII phosphorylation and state transition in Chlamydomonas. *Science*, 299, 1572–1575.

Dixon, R. and Kahn, D. (2004). Genetic regulation of biological nitrogen fixation. *Nature Reviews Microbiology*, 2, 621–631.

Dupont, C.L., Neupane, K., Shearer, J., Palenik, B. (2008). Diversity, function and evolution of genes coding for putative Ni-containing superoxide dismutases. *Environmental Microbiology*, 10, 1831–1843.

Durak, G., Taylor, A., Walker, C., Probert, I., de Vargas, C., Audic, S., Schroeder, D., Brownlee, C., Wheeler, G.L. (2016). A role for diatom-like silicon transporters in calcifying coccolithophores. *Nature Communication*, 7, 10543.

Dyhrman, S., Chappell, P., Haley, S., Moffett, J.W., Orchard, E.D., Waterbury, J.B., Webb, E.A. (2006). Phosphonate utilization by the globally important marine diazotroph *Trichodesmium*. *Nature*, 439, 68–71.

Eady, R.R. (1996). Structure–function relationships of alternative nitrogenases. *Chemical Reviews*, 96, 3013–3030.

Ellegaard, M., Lenau, T., Lundholm, N., Maibohm, C., Mørk Friis, S.M., Rottwitt, K., Su, Y. (2016). The fascinating diatom frustule – Can it play a role for attenuation of UV radiation? *Journal of Applied Phycology*, 28, 3295–3306.

Elser, J.J., Bracken, M.E.S., Cleland, E.E., Gruner, D.S., Harpole, W.S., Hillebrand, H., Ngai, J.T., Seabloom, E.W., Shurin, J.B., Smith, J.E. (2007). Global analysis of nitrogen and phosphorus limitation of primary producers in freshwater, marine and terrestrial ecosystems. *Ecology Letters*, 10, 1135–1142.

Engelken, J., Brinkmann, H., Adamska, I. (2010). Taxonomic distribution and origins of the extended LHC (light-harvesting complex) antenna protein superfamily. *BMC Evolutionary Biology*, 10, 233.

Erickson, E., Wakao, S., Niyogi, K.K. (2015). Light stress and photoprotection in *Chlamydomonas reinhardtii*. *Plant Journal*, 82, 449–465.

Falkowski, P.G. and Owens, T.G. (1980). Light-shade adaptation. Two strategies in marine phytoplankton. *Plant Physiology*, 66, 592–595.

Falkowski, P.G. and Raven, J.A. (2007). *Aquatic Photosynthesis*. Blackwell Publishing Ltd, Hoboken, NJ.

Fay, P. (1992). Oxygen relations of nitrogen fixation in cyanobacteria. *Microbialogical Reviews*, 56, 340–373.

Fernandez, E. and Galvan, A. (2007). Inorganic nitrogen assimilation in *Chlamydomonas*. *Journal of Experimental Botany*, 58, 2279–2287.

Fernandez, E., Llamas, A., Galvàn, A. (2009). Nitrogen assimilation and its regulation. In *The Chlamydomonas Source Book*, Stern, D.B. and Harris E.H. (eds), 2nd edition. Elsevier, Amsterdam.

Finkel, Z.V. and Kotrc, B. (2010). Silica use through time: Macroevolutionary change in the morphology of the diatom frustule. *Geomicrobiology Journal*, 27, 596–608.

Finkel, Z.V., Katz, M., Wright, J., Schofieid, O., Falkowski, P.G. (2005). Climatically driven macroevolutionary patterns in the size of marine diatoms over the Cenozoic. *Proceedings of the National Academy of Sciences of the United States of America*, 102, 8927–8932.

Finkel, Z.V., Follows, M.J., Liefer, J.D., Brown, C.M., Benner, I., Irwin, A.J. (2016). Phylogenetic diversity in the macromolecular composition of microalgae. *PLOS One*, 11, e0155977.

Fisher, T., Shurtz-Swirski, R., Gepstein, S., Dubinsky, Z. (1989). Changes in the levels of Ribulose-l,5-bisphosphate Carboxylase/Oxygenase(Rubisco) in Tetraedron minimum (Chlorophyta) during light and shade adaptation. *Plant and Cell Physiology*, 30, 221–228.

Fisher, N.L., Campbell, D.A., Hughes, D.J., Kuzhiumparambil, U., Halsey, K.H., Ralph, P.J., Suggett, D.J. (2020). Divergence of photosynthetic strategies amongst marine diatoms. *PLOS One*, 15, e0244252.

Galvàn, A. and Fernandez, E. (2001). Eukaryotic nitrite and nitrate transporters. *Cellular and Molecular Life Science*, 58, 225–233.

Gao, X., Bowler, C., Kazamia, E. (2021). Iron metabolism strategies in diatoms. *Journal of Experimental Botany*, 72(6), 2165–2180.

Garcia, H.E., Locarnini, R.A., Boyer, T.P., Antonov, J.I. (2006). Nutrients (phosphate, nitrate, silicate). In *World Ocean Atlas 2005*, Levitus, S. (ed.). U.S. Government Printing Office, Washington, DC.

Geider, R. and La Roche, J. (2002). Redfield revisited: Variability of C:N:P in marine microalgae and its biochemical basis. *European Journal of Phycology*, 37(1), 1–17.

Geisler, E., Bogler, A., Bar-Zeev, E., Rahav, E. (2020). Heterotrophic nitrogen fixation at the hyper-eutrophic Qishon River and estuary system. *Frontiers in Microbiology*, 11, 1370.

Gerotto, C. and Morosinotto, T. (2013). Evolution of photoprotection mechanisms upon land colonization: Evidence of PSBS-dependent NPQ in late Streptophyte algae. *Physiologia Plantarum*, 149, 583–598.

Giordano, M. (2013). Homeostasis: An underestimated focal point of ecology and evolution. *Plant Science*, 211, 92–101.

Giordano, M. and Prioretti, L. (2016). Sulphur and algae: Metabolism, ecology and evolution. In *The Physiology of Microalgae*, Borowitzka, M., Beardall, J., Raven, J. (eds). Springer, Cham.

Giordano, M. and Raven, J.A. (2014). Nitrogen and sulfur assimilation in plants and algae. *Aquatic Botany*, 118, 45–61.

Giordano, M., Norici, A., Forssen, M., Eriksson, M., Raven, J.A. (2003). An anaplerotic role for mitochondrial carbonic anhydrase in *Chlamydomonas reinhardtii*. *Plant Physiology*, 132, 2126–2134.

Giordano, M., Norici, A., Hell, R. (2005a). Sulfur and phytoplankton: Acquisition, metabolism and impact on the environment. *New Phytologist*, 166, 371–382.

Giordano, M., Chen, Y-B., Koblizek, M., Falkowski, P.G. (2005b). Regulation of nitrate reductase in *Chlamydomonas reinhardtii* by the redox state of the plastoquinone pool. *European Journal of Phycology*, 40(4), 345–352.

Görlich, S., Pawolski, D., Zlotnikov, I., Kröger, N. (2019). Control of biosilica morphology and mechanical performance by the conserved diatom gene Silicanin-1. *Communications Biology*, 2, 245.

Goss, R. and Lepetit, B. (2015). Biodiversity of NPQ. *Journal of Plant Physiology*, 172, 13–32.

Goss, R., Bohme, K., Wilhelm, C. (1998). The xanthophyll cycle of *Mantoniella squamata* converts violaxanthin into antheraxanthin but not to zeaxanthin: Consequences for the mechanism of enhanced non-photochemical energy dissipation. *Planta*, 205, 613–621.

Greenwold, M.J., Cunningham, B.R., Lachenmyer, E.M., Pullman, J.M., Richardson, T.L., Dudycha, J.L. (2019). Diversification of light capture ability was accompanied by the evolution of phycobiliproteins in cryptophyte algae. *Proceedings of the Royal Society B*, 286, 20190655.

Harding, K., Turk-Kubo, K.A., Sipler, R.E., Mills, M.M., Bronk, D.A., Zehr, J.P. (2018). Symbiotic unicellular cyanobacteria fix nitrogen in the Arctic Ocean. *Proceedings of the National Academy of Sciences of the United States of America*, 115, 13371–13375.

Hildebrand, M., Volcani, B.E., Gassmann, W., Schroeder, J.I. (1997). A gene family of silicon transporters. *Nature*, 385, 688–689.

Hildebrand, M., Dahlin, K., Volcani, B.E. (1998). Characterization of a silicon transporter gene family in *Cylindrotheca fusiformis*: Sequences, expression analysis, and identification of homologs in other diatoms. *Molecular and General Genetics*, 260, 480–486.

Hildebrand, M., Lerch, S.J.L., Shrestha, R.P. (2018). Understanding diatom cell wall silicification – Moving forward. *Frontiers in Marine Science*, 5, 125.

Ho, T.-Y., Quigg, A., Finkel, Z.V., Milligan, A.J., Wyman, K., Falkowski, P.G. (2003). The elemental composition of some marine phytoplankton. *Journal of Phycology*, 39, 1145–1159.

Holland, H.D. (1973). The oceans: A possible source of iron in iron-formations. *Economic Geology*, 68, 1169–1172.

Holland, H.D. (2006). The oxygenation of the atmosphere and oceans. *Philosophical Transactions of the Royal Society of London B, Biological Sciences*, 361, 903–915.

Hoopen, F.T., Cuin, T.A., Pedas, P., Hegelund, J.N., Shabala, S., Schjoerring, J.K., Jahn, T.P. (2010). Competition between uptake of ammonium and potassium in barley and *Arabidopsis* roots: Molecular mechanisms and physiological consequences. *Journal of Experimental Botany*, 61, 2303–2315.

Ito, H. and Tanaka, A. (2011). Evolution of a divinyl chlorophyll-based photosystem in *Prochlorococcus*. *Proceedings of the National Academy of Sciences of the United States of America*, 108, 18014–18019.

Jensen, E.L., Clement, R., Kosta, A., Maberly, S.C., Gontero, B. (2019). A new widespread subclass of carbonic anhydrase in marine phytoplankton. *The ISME Journal*, 13, 2094–2106.

Jensen, E.L., Maberly, S.C., Gontero, B. (2020). Insights on the functions and ecophysiological relevance of the diverse carbonic anhydrases in microalgae. *International Journal of Molecular Sciences*, 21, 2922.

Kessler, N., Armoza-Zvuloni, R., Wang, S., Basu, S., Weber, P.K., Stuart, R.K., Shaked, Y. (2020). Selective collection of iron-rich dust particles by natural *Trichodesmium* colonies. *The ISME Journal*, 14, 91–103.

Kirk, J.T.O. (1994). *Light and Photosynthesis in Aquatic Ecosystems*, 3rd edition. Cambridge University Press, Cambridge.

Kotzsch, A., Groeger, P., Pawolski, D., Bomans, P.H.H., Sommerdijk, N.A.J.M., Schlierf, M., Kröger, N. (2017). Silicanin-1 is a conserved diatom membrane protein involved in silica biomineralization. *BMC Biology*, 15, 65.

Koziol, A.G., Borza, T., Ishida, K.-I., Keeling, P., Lee, R.W., Durnford, D.G. (2007). Tracing the evolution of the light-harvesting antennae in chlorophyll a/b-containing organisms. *Plant Physiology*, 143, 1802–1816.

Kroger, N., Deutzmann, R., Bergsdorf, C., Sumper, M. (2000). Species-specific polyamines from diatoms control silica morphology. *Proceedings of the National Academy of Sciences of the United States of America*, 97, 14133–14138.

Krom, M.D., Emeis, K.C., Van Cappellen, P. (2010). Why is the eastern Mediterranean phosphorus limited? *Progress in Oceanography*, 85, 236–244.

Lacour, T., Babin, M., Lavaud, J. (2020). Diversity in xanthophyll cycle pigments content and related nonphotochemical quenching (NPQ) among microalgae: Implications for growth strategy and ecology. *Journal of Phycology*, 56, 245–263.

Lane, T.W. and Morel, F.M.M. (2000a). A biological function for cadmium in marine diatoms. *Proceedings of the National Academy of Sciences of the United States of America*, 97, 4627–4631.

Lane, T.W. and Morel, F.M.M. (2000b). Regulation of carbonic anhydrase expression by zinc, cobalt, and carbon dioxide in the marine diatom *Thalassiosira weissflogii*. *Plant Physiology*, 123, 345–352.

Langer, G., Taylor, A.R., Walker, C.E., Meyer, E.M., Ben Joseph, O., Gal, A., Harper, G.M., Probert, I., Brownlee, C., Wheeler, G.L. (2021). Role of silicon in the development of complex crystal shapes in coccolithophores. *New Phytologist*, 231, 1845–1857.

Le Costaouëc, T., Unamunzaga, C., Mantecon, L., Helbert, W. (2017). New structural insights into the cell-wall polysaccharide of the diatom *Phaeodactylum tricornutum*. *Algal Research*, 26, 172–179.

Li, Z., Peers, G., Dent, R.M., Bai, Y., Yang, S.Y., Apel, W., Leonelli, L., Niyogi, K.K. (2016). Evolution of an atypical de-epoxidase for photoprotection in the green lineage. *Nature Plants*, 2, 16140.

Liberton, M., Austin, J., Berg, R.H., Pakrasi, H.B. (2011). Unique thylakoid membrane architecture of a unicellular N_2-fixing cyanobacterium revealed by electron tomography. *Plant Physiology*, 155, 1656–1666.

Liebig, J. (1841). *Organic Chemistry in Its Applications to Agriculture and Physiology*. J. Owen, Cambridge.

Maberly, S.C., Pitt, J.A., Davies, P.S., Carvalho, L. (2020). Nitrogen and phosphorus limitation and the management of small productive lakes. *Inland Waters*, 10, 159–172.

Mackay, E., Feuchtmayr, H, De Ville, M., Thackeray, S.J., Callaghan, N., Marshall, M., Rhodes, G., Yates, C., Johnes, P., Maberly, S.C. (2020). Dissolved organic nutrient uptake by riverine phytoplankton varies along a gradient of nutrient enrichment. *Science of the Total Environment*, 722, 137837.

Marchetti, A, Parker, M.S., Moccia, L.P., Lin, E.O., Arrieta A.L., Ribalet, F., Murphy, M.E.P., Maldonado M.T., Armbrust, E.V. (2009). Ferritin is used for iron storage in bloom-forming marine pennate diatoms. *Nature*, 457, 467–470.

Martin-Jezequel, V., Hildebrand, M., Brzezinski, M.A. (2000). Silicon metabolism in diatoms: Implications for growth. *Journal of Phycology*, 36, 821–840.

Medlin, LK. (2016). Evolution of the diatoms: Major steps in their evolution and a review of the supporting molecular and morphological evidence. *Phycologia*, 55, 79–103.

Miflin, B.J. and Lea, P.J. (1982). Ammonia assimilation and amino acid metabolism. In *Nucleic Acids and Proteins in Plants I*, Boulter, D. and Parthier, B. (eds). Springer-Verlag, Berlin.

Milligan, A.J., Berman-Frank, I., Gerchman Y., Dismukes, G.C., Falkowski, P.G. (2007). Light-dependent oxygen consumption in nitrogen-fixing cyanobacteria plays a key role in nitrogenase protection. *Journal of Phycology*, 43, 845–852.

Mills, M.M., Ridame, C., Davies, M., La Roche, J., Geider, R.J. (2004). Co-limitation of nitrogen fixation by iron and phosphorus in the Eastern North Atlantic. *Nature*, 429, 292–294.

Minagawa, J. and Tokutsu, R. (2015). Dynamic regulation of photosynthesis in *Chlamydomonas reinhardtii*. *Plant Journal*, 82, 413–428.

Mizusawa, N. and Wada, H. (2012). The role of lipids in photosystem II. *Biochimica et Biophysica Acta*, 1817, 194–208.

Monier, A., Chambouvet, A., Milner, D.S., Attah, V., Terrado, R., Lovejoy, C., Moreau, H., Santoro, A.E., Derelle, E., Richards, T.A. (2017). Host-derived viral transporter protein for nitrogen uptake in infected marine phytoplankton. *Proceedings of the National Academy of Sciences of the United States of America*, 114, E7689–E7498.

Montechiaro, F., Hirschmugl, C.J., Raven, J.A., Giordano, M. (2006). Homeostasis of cell composition during prolonged darkness. *Plant, Cell and Environment*, 29, 2198–2204.

Moore, J.K., Diney, S.C., Lindsay, K. (2004). Upper ocean ecosystem dynamics and iron cycling in a global three-dimensional model. *Global Biogeochemical Cycles*, 18, GB4028.

Moore, C., Mills, M., Arrigo, K., Berman-Frank, I., Bopp, L., Boyd, P.W., Galbraith, E.D., Geider, R.J., Guieu, C., Jaccard, S.L., Jickells, T.D., La Roche, J., Lenton, T.M., Mahowald, N.M., Marañón, E., Marinov, I., Moore, J.K., Nakatsuka, T., Oschlies, A., Saito, M.A., Thingstad, T.F., Tsuda, A., Ulloa, O. (2013). Processes and patterns of oceanic nutrient limitation. *Nature Geoscience*, 6, 701–710.

Morel, F.M.M. (1987). Kinetics of nutrient uptake and growth in phytoplankton. *Journal of Phycology*, 23, 137–150.

Morel, F.M.M., Lam, P.J., Saito, M.A. (2020). Trace metal substitution in marine phytoplankton. *Annual Review of Earth and Planetary Sciences*, 48, 19.1–19.27.

Mulder, A., van de Graaf, A.A., Robertson, L.A., Kuenen, J.G. (1995). Anaerobic ammonium oxidation discovered in a denitrifying fluidized bed reactor. *FEMS Microbiology Ecology*, 16, 177–184.

Mulholland, M.R., Bernhardt, P.W., Blanco-Garcia, J.L., Mannino, A., Hyde, K., Mondragon, E., Turk, K., Moisander, P.H., Zehr, J.P. (2012). Rates of dinitrogen fixation and the abundance of diazotrophs in North American coastal waters between Cape Hatteras and Georges Bank. *Limnology and Oceanography*, 57, 1067–1083.

Mulo, P., Sakurai, I., Aro, E.-M. (2012). Strategies for psbA gene expression in cyanobacteria, green algae and higher plants: From transcription to PSII repair. *Biochimica et Biophysica Acta – Bioenergetics*, 1817, 247–257.

Muro-Pastor, M.I., Reyes, J.C., Florencio, F.J. (2005). Ammonium assimilation in cyanobacteria. *Photosynthesis Research*, 83, 135–150.

Muzzopappa, F. and Kirilovsky, D. (2020). Changing color for photoprotection: The orange carotenoid protein. *Trends in Plant Science*, 25, 92–104.

Nagy, G., Ünnep, R., Zsiros, O., Tokutsu, R., Takizawa, K., Porcar, L., Moyet, L., Petroutsos, D., Garab, G., Finazzi, G., Minagawa, J. (2014). Chloroplast remodeling during state transitions in *Chlamydomonas reinhardtii* as revealed by noninvasive techniques *in vivo*. *Proceedings of the National Academy of Sciences of the United States of America*, 111, 5042–5047.

Nawrocki, W.J., Tuorasse, N.J., Taly, A., Rappaport, F., Wollman, F.-A. (2015). The plastid terminal oxidase: Its elusive function points to multiple contributions to plastid physiology. *Annual Review of Plant Biology*, 66, 49–74.

Nawrocki, W.J., Bailleul, B., Picot, D., Cardol, P., Rappaport, F., Wollman, F.-A., Joliot, P. (2019). The mechanism of cyclic electron flow. *Biochimica et Biophysica Acta – Bioenergetics*, 1860, 433–438.

Neilson, J.A.D. and Durnford, D.G. (2010). Structural and functional diversification of the light-harvesting complexes in photosynthetic eukaryotes. *Photosynthesis Research*, 106, 57–71.

Nelson, D.M., Tréguer, P., Brzezinski, M.A., Leynaert, A., Quéguiner, B. (1995). Production and dissolution of biogenic silica in the ocean: Revised global estimates, comparison with regional data and relationship to biogenic sedimentation. *Global Biogeochemical Cycles*, 9, 359.

Niyogi, K.K., Bjorkman, O., Grossman, A.R. (1997). Chlamydomonas xanthophyll cycle mutants identified by video imaging of chlorophyll fluorescence quenching. *Plant Cell*, 9, 1369–1380.

Norici, A. and Giordano, M. (2002). Anaplerosis in microalgae. *Recent Research and Development in Plant Physiology*, 3, 153–164.

Norici, A., Hell, R., Giordano, M. (2005). Sulfur and primary production in aquatic environments: An ecological perspective. *Photosynthesis Research*, 86, 409–417.

Nürnberg, D.J., Morton, J., Santabarbara, S., Telfer, A., Joliot, P., Antonaru, L.A., Ruban, A.V., Cardona, T., Krausz, E., Boussac, A., Fantuzzi, A., Rutherford, A.W. (2018). Photochemistry beyond the red limit in chlorophyll f-containing photosystems. *Science*, 360, 1210–1213.

Oliver, T., Sanchez-Baracaldo, P., Larkum, A.W., Rutherford, A.W., Cardona, T. (2021). Time-resolved comparative molecular evolution of oxygenic photosynthesis. *Biochimica et Biophysica Acta – Bioenergetics*, 1862, 148400.

Orchard, E.D., Benitez-Nelson, C.R., Pellechia, P.J., Lomas, M.W., Dyhrman, S.T. (2010). Polyphosphate in *Trichodesmium* from the low phosphorus Sargasso Sea. *Limnology and Oceanography*, 55, 2161–2169.

Page, M.D., Allen, M.D., Kropat, J., Urzica, E.I., Karpowicz, S.J., Hsieh, S.I., Loo, J.A., Merchant, S.S. (2012). Fe sparing and Fe recycling contribute to increased superoxide dismutase capacity in iron-starved *Chlamydomonas reinhardtii*. *Plant Cell*, 24, 2649–2665.

Patron, N.J., Durnford, D.G., Kopriva, S. (2008). Sulfate assimilation in eukaryotes: Fusions, relocations and lateral transfers. *BMC Evolutionary Biology*, 8, 39.

Paytan, A. and McLaughlin, K. (2007). The oceanic phosphorus cycle. *Chemical Reviews*, 107(2), 563–576.

Peers, G. and Price, N.M. (2006). Copper-containing plastocyanin used for electron transport by an oceanic diatom. *Nature*, 441, 341–344.

Peers, G., Truong, T.B., Ostendorf, E., Busch, A., Elrad, D., Grossman, A.R., Hippler, M., Niyogi, K.K. (2009). An ancient light-harvesting protein is critical for the regulation of algal photosynthesis. *Nature*, 462, 518–521.

Peltier, G., Tolleter, D., Billon, E., Cournac, L. (2010). Auxiliary electron transport pathways in chloroplasts of microalgae. *Photosynthesis Research*, 106, 19–31.

Peltier, G., Aro, E.-M., Shikanai, T. (2016). NDH-1 and NDH-2 plastoquinone reductases in oxygenic photosynthesis. *Annual Review of Plant Biology*, 67, 55–80.

Pérez-Castiñero, J.R., Gómez-García, R., López-Marqués, L.L., Losada, M., Serrano, A. (2001). Enxymatic systems of inorganic pyrophosphate bioenergetics in photosynthetic and heterotrophic protists: Remnants or metabolic cornerstones? *International Microbology*, 4, 135–142.

Peters, J.W. and Szilagyi, R.K. (2011). Exploring new frontiers of nitrogenase structure and mechanism. *Current Opinion in Chemical Biology*, 10, 101–108.

Pinkerton, M.H., Boyd, P.W., Deppeler, S., Hayward, A., Höfer, J., Moreau, S. (2021). Evidence for the impact of climate change on primary producers in the southern ocean. *Frontiers in Ecology and Evolution*, 9, 1–19.

Prioretti, L., Gontero, B., Hell, R., Giordano, M. (2014). Diversity and regulation of ATP sulfurylase in photosynthetic organisms. *Frontiers in Plant Science*, 5, 597.

Quigg, A., Finkel, Z.V., Irwin, A.J., Rosenthal, Y., Ho, T.-Y., Reinfelder, J.R., Schofield, O., Morel, F.M.M., Falkowski, P.G. (2003). The evolutionary inheritance of elemental stoichiometry in marine phytoplankton. *Nature*, 425, 291–294.

Quigg, A., Irwin, A.J., Finkel, Z.V. (2011). Evolutionary inheritance of elemental stoichiometry in phytoplankton. *Proceedings of the Royal Society B*, 278, 526–534.

Rai, A.K., Chen, T., Moroney, J.V. (2021). Mitochondrial carbon anhydrase are needed for optimal photosynthesis at low CO_2 concentrations in *Chlamydomonas*. *Plant Physiology*, kiab351.

Ratti, S., Knoll, A.H., Giordano, M. (2011). Did sulfate availability facilitate the evolutionary expansion of chlorophyll a+c phytoplankton in the oceans? *Geobiology*, 9, 301–312.

Raven, J.A. (2020). Chloride involvement in the synthesis, functioning and repair of the photosynthetic apparatus *in vivo*. *New Phytologist*, 227, 334–342.

Raven, J.A. (2021). Origin of the roles of potassium in biology. *BioEssays*, 43, 2000302.

Raven, J.A. and Giordano, M. (2009). Biomineralization by photosynthetic organisms: Evidence of coevolution of the organisms and their environment? *Geobiology*, 7, 140–154.

Raven, J.A. and Waite, A.M. (2004). The evolution of silicification in diatoms: Inescapable sinking and sinking as escape? *New Phytologist*, 162, 45–61.

Raven, J.A., Kübler, J., Beardall, J. (2000). Put out the light, and then put out the light. *Journal of the Marine Biological Association of the United Kingdom*, 80(1), 1–25.

Raven, J.A., Giordano, M., Beardall, J. (2008). Insights into the evolution of CCMs from comparisons with other resource acquisition processes. *Physiologia Plantarum*, 133, 4–14.

Raymond, J., Siefert, J.L., Staples, C.R., Blankenship, R.E. (2004). The natural history of nitrogen fixation. *Molecular Biology and Evolution*, 21, 541–554.

Redekop, P., Rothhausen, N., Rothhausen, N., Melzer, M., Mosebach, L., Dülger, E., Bovdilova, A., Caffarri, S., Hippler, M., Jahns, P. (2020). PsbS contributes to photoprotection in *Chlamydomonas reinhardtii* independently of energy dissipation. *Biochimica et Biophysica Acta – Bioenergetics*, 1861, 148183.

Redfield, A.C. (1934). On the proportions of organic derivatives in sea water and their relation to the composition of plankton. In *James Johnstone Memorial Volume*, Daniel, R.J. (ed.). Liverpool University Press, Liverpool.

Redfield A.C. (1958). The biological control of chemical factors in the environment. *American Scientist*, 46(3), 230A, 205–221.

Rees, T.A.V. and Bekheet, I.A. (1982). The role of nickel in urea assimilation by algae. *Planta*, 156, 385–387.

Rees, T.A.V. and Raven, J.A. (2021). The maximum growth rate hypothesis is correct for eukaryotic photosynthetic organisms, but not cyanobacteria. *New Phytologist*, 230, 601–611.

Renger, T. and Schlodder, E. (2008). The primary electron donor of photosystem II of the cyanobacterium *Acaryochloris marina* is a chlorophyll d and the water oxidation is driven by a chlorophyll a/chlorophyll d heterodimer. *The Journal of Physical Chemistry B*, 112, 7351–7354.

Ruiz, F.A., Marchesini, N., Seufferheld, M., Govindjee, Docampo, R. (2001). The polyphosphate bodies of *Chlamydomonas reinhardtii* possess a proton-pumping pyrophosphatase and are similar to acidocalcisomes. *Journal of Biological Chemistry*, 276, 46196–46203.

Ryther, J.H. (1954). The ratio of photosynthesis to respiration in marine plankton algae and its effect upon the measurement of productivity. *Deep-Sea Research*, 2, 134–139.

Saidha, T., Na, S.Q., Li, J.Y., Schiff, J.A. (1988). A sulphate metabolizing centre in *Euglena* mitochondria. *The Biochemical. Journal*, 253, 533–539.

Saito, M.A., Bertrand, E.M., Dutkiewicz, S., Bulygin, V.V., Moran, D.M., Monteiro, F.M., Follows, M.J., Valois, F.W., Waterbury, J.B. (2011). Iron conservation by reduction of metalloenzyme inventories in the marine diazotroph *Crocosphaera watsonii*. *Proceedings of the National Academy of Sciences of the United States of America*, 108, 2184–2189.

Sante-Maria, G.E., Danna, C.H., Czibener, C. (2000). High-affinity potassium transport in barley roots. Ammonia-sensitive and-insensitive pathways. *Plant Physiology*, 123, 297–306.

Sañudo-Wilhelmy, S.A., Kustka, A.B., Gobler, C.J., Hutchins, D.A., Yang, M., Lwiza, K., Burns, J., Capone, D.G., Raven, J.A., Carpenter, E.J. (2001). Phosphorus limitation of nitrogen fixation by *Trichodesmium* in the central Atlantic Ocean. *Nature*, 411, 66–69.

Sanz-Luque, E., Bhaya, D., Grossman, A.R. (2020). Polyphosphate: A multifunctional metabolite in cyanobacteria and algae. *Frontiers in Plant Science*, 11, 938.

Sétif, P. (2001). Ferredoxin and flavodoxin reduction by photosystem I. *Biochimica et Biophysica Acta – Bioenergetics*, 1507, 161–179.

Shi, T., Ilikchyan, I., Rabouille, S., Zehr, J.P. (2010). Genome-wide analysis of diel gene expression in the unicellular N^2-fixing cyanobacterium *Crocosphaera watsonii*, WH 8501. *The ISME Journal*, 4, 621–632.

Simionato, D., Sforza, E., Corteggiani Carpinelli, E., Bertucco, A., Giacometti, G.M., Morosinotto, T. (2011). Acclimation of *Nannochloropsis gaditana* to different illumination regimes: Effects on lipids accumulation. *Bioresource and Technology*, 102, 6026–6032.

Sims, P.A., Mann, D.G., Medlin, L.K (2006). Evolution of the diatoms: Insights from fossil, biological and molecular data. *Phycologia*, 45, 361–402.

Spear-Bernstein, L. and Miller, K.R. (1989). Unique location of the phycobiliprotein light-harvesting pigment in the Cryptophyceae. *Journal of Phycology*, 25, 412– 419.

Spilling, K., Olli, K., Lehtoranta, J., Kremp, A., Tedesco, L., Tamelander, T., Klais, R., Peltonen, H., Tamminen, T. (2018). Shifting diatom–dinoflagellate dominance during spring bloom in the Baltic Sea and its potential effects on biogeochemical cycling. *Frontiers in Marine Science*, 5, 327.

Stomp, M., Huisman, J., De Jongh, F., Veraart, A.J., Gerla, D., Rijkeboer, M., Ibelings, B.W., Wollenzien, U.I.A., Stal, L.J. (2004). Adaptive divergence in pigment composition promotes phytoplankton biodiversity. *Nature*, 432, 104–107.

Stomp, M., Huisman, J., Stal, L.J., Matthijs, H.C.P. (2007). Colorful niches of phototrophic microorganisms shaped by vibrations of the water molecule. *The ISME Journal*, 1, 271–282.

Strzepek, R.F. and Harrison, P.J. (2004). Photosynthetic architecture differs in coastal and oceanic diatoms. *Nature*, 431, 689–692.

Stumm, W. and Morgan, J.J. (1981). *Aquatic Chemistry*, 2nd edition. Wiley, New York.

Su, Y., Lundholm, N., Ellegaard, M. (2018). The effect of different light regimes on diatom frustule silicon concentration. *Algal Research*, 29, 36–40.

Sukenik, A., Livne, A., Apt, K.E., Grossman, A.R. (2000). Characterization of a gene encoding the light-harvesting violaxanthin-chlorophyll protein of *Nannochloropsis* sp. (Eustigmatophyceae). *Journal of Phycology*, 36, 563–570.

Sullivan, C.W. (1979). Diatom mineralization of silic acid IV. Kinetics of soluble Si pool formation in exponentially growing and synchronized *Navicula pelliculosa*. *Journal of Phycology*, 15, 210–216.

Takahashi, H., Kopriva, S., Giordano, M., Saito, K., Hell, R. (2011). Sulfur assimilation in photosynthetic organisms: Molecular functions and regulations of transporters and assimilatory enzymes. *Annual Review of Plant Biology*, 62, 157–184.

Talling, J.F. (2010). Potassium – A non-limiting nutrient in fresh waters? *Freshwater Reviews*, 3(2), 97–104.

Taylor, N.J. (1985). Silica incorporation in the diatom *Coscinodiscus granii* as affected by light intensity. *British Phycological Journal*, 20, 365–374.

Tesson, B., Lerch, S.J.L., Hildebrand, M. (2017). Characterization of a new protein family associated with the silica deposition vesicle membrane enables genetic manipulation of diatom silica. *Scientific Reports*, 7, 13457.

Thamatrakoln, K. and Hildebrand, M. (2008). Silicon uptake in diatoms revisited: A model for saturable and non-saturable uptake kinetics and the role of silicon transporters. *Plant Physiology*, 146, 1397–1407.

Torres-Valdés, S., Roussenov V.M., Sanders R., Reynolds S., Pan X., Mather R., Landolfi A., Wolff G.A., Achterberg E.P., Williams R.G. (2009). Distribution of dissolved organic nutrients and their effect on export production over the Atlantic Ocean. *Global Biogeochemical Cycles*, 23, GB4019.

Tyrrell, T. (1999). The relative influences of nitrogen and phosphorus on oceanic primary production. *Nature*, 400, 525–531.

Ünlü, C., Drop, B., Croce, R., van Amerongen, H. (2014). State transitions in *Chlamydomonas reinhardtii* strongly modulate the functional size of photosystem II but not of photosystem I. *Proceedings of the National Academy of Sciences of the United States of America*, 111, 3460–3465.

Van Mooy, B., Fredricks, H., Pedler, B., Dyhrman, S.T., Karl D.M., Koblížek M., Lomas, M.W., Mincer, T.J., Moore, L.R., Moutin, T., Rappé, M.S., Webb, E.A. (2009). Phytoplankton in the ocean use non-phosphorus lipids in response to phosphorus scarcity. *Nature*, 458, 69–72.

Virtanen, O., Khorobrykh, S., Tyystjärvi, E. (2021). Acclimation of *Chlamydomonas reinhardtii* to extremely strong light. *Photosynthesis Research*, 147, 91–106.

Waldron, K.J. and Robinson, N.J. (2009). How do bacterial cells ensure that metalloproteins get the correct metal? *Nature Review Microbiology*, 7, 25–35.

Weger, H.G., Birch, D.G., Elrifi, L.R., Turpin, D.H. (1988). Ammonium assimilation requires mitochondrial respiration in the light. A study with the green alga *Selenastrum minutum*. *Plant Physiology*, 86, 688–692.

Wenzl, S., Hett, R., Richthammer, P., Sumper, M. (2008). Silacidins: Highly acidic phosphopeptides from diatom shells assist in silica precipitation *in vitro*. *Angewandte Chemie International Edition English*, 47, 1729–1732.

Wilhelm, C., Jungandreas, A., Jakob, T., Goss, R. (2014). Light acclimation in diatoms: From phenomenology to mechanisms. *Marine Genomics*, 16, 5–15.

Wirtz, M. and Hell, R. (2006). Functional analysis of the cysteine synthase protein complex from plants: Structural, biochemical and regulatory properties. *Journal of Plant Physiology*, 163, 273–286.

Xu, K., Grant-burt, J.L., Donaher, N., Campbell, D.A. (2017). Connectivity among Photosystem II centers in phytoplankters: Patterns and responses. *Biochimica et Biophysica Acta – Bioenergetics*, 1858, 459–474.

Yagisawa, F., Nishida, K., Yoshida, M., Ohnuma, M., Shimada, T., Fujiwara, T., Yoshida, Y., Misumi, O., Kuroiwa, H., Kuroiwa, T. (2009). Identification of novel proteins in isolated polyphosphate vacuoles in the primitive red alga *Cyanidioschyzon merolae*. *Plant Journal*, 60, 882–893.

York, A. (2017). Algal virus boosts nitrogen uptake in the ocean. *Nature Review Microbiology*, 15, 573.

Zeher, J.P. and Capone, D.G. (2020). Changing perspectives in marine nitrogen fixation. *Science*, 368, eaay9514.

Zhang, P., Allahverdiyeva, Y., Eisenhut, M., Aro, E.M. (2009). Flavodiiron proteins in oxygenic photosynthetic organisms: Photoprotection of photosystem II by Flv2 and Flv4 in *Synechocystis* sp. PCC 6803. *PLOS One*, 4(4), e5331.

Zhu, S.-H. and Green, B.R. (2010). Photoprotection in the diatom *Thalassiosira pseudonana*: Role of LI818-like proteins in response to high light stress. *Biochimica et Biophysica Acta*, 1797, 1449–1457.

9

Future Responses of Marine Primary Producers to Environmental Changes

Kunshan GAO[1], Wenyan ZHAO[1] and John BEARDALL[1,2]

[1] *State Key Laboratory of Marine Environmental Science & College of Ocean and Earth Sciences, Xiamen University, China*

[2] *School of Biological Sciences, Monash University, Clayton, Victoria, Australia*

9.1. Introduction

Increasing atmospheric CO_2 concentration due to human activities is responsible for global warming and ocean warming, the latter enhancing stratification and shoaling of the upper mixed layer (UML). This hinders the upward transport of nutrients from deeper layers, and also causes organisms dwelling in the UML to be exposed to increased doses of PAR and UV radiation (see the review by Gao *et al.* (2020) and literature therein). Reduced O_2 solubility in seawater along with ocean warming leads to ocean deoxygenation and rapid expansion of hypoxic zones in nearshore waters, where human nutrient over-enrichment can deplete dissolved oxygen by promoting excessive levels of primary production and consequent elevated bacterial respiration as that organic matter is broken down (Schmidtko *et al.* 2017; Breitburg *et al.* 2018).

Marine primary producers account for about half of global photosynthetic carbon fixation. Their responses to ocean climate changes directly link to the efficiency of the marine biological CO_2 pump. Across most of the oceans, the dominant primary

Blue Planet, Red and Green Photosynthesis,
coordinated by Stephen Christopher MABERLY and Brigitte GONTERO.

producers are mainly phytoplankton, with macroalgae and seagrasses accounting for a higher proportional contribution in coastal waters. These organisms are exposed to both increasing pCO_2 and to declining pH, the latter effect being referred to as ocean acidification (OA) (Hurd *et al.* 2020). While increased CO_2 availability may potentially benefit photosynthesis, the associated acidic stress may be harmful. This is especially a problem during the nighttime, with respiratory CO_2 release leading to more pronounced impacts of OA during the dark period (Flynn *et al.* 2012).

While the responses of primary producers to OA have been extensively investigated (see recent reviews by Gao *et al.* (2012a), Hutchins and Fu (2017), Jin *et al.* (2020), and Ji and Gao (2020), Hurd *et al.* (2020) and references therein), contradictory results have often been reported regarding phytoplankton species or communities (Gao and Campbell 2014; Hutchins *et al.* 2015; Hong *et al.* 2017). In macroalgae, the tolerance of diel pH fluctuations and benefits from rising pCO_2 are evident (Gao *et al.* 1991, 1993a; Cornwall *et al.* 2012, 2015); in calcifying micro- and macro-algae, it is broadly accepted that OA decreases their calcification (Gao *et al.* 1993b, 2009; Riebesell *et al.* 2000; Feng *et al.* 2008; Gao and Zheng 2010; Sinutok *et al.* 2011). Notwithstanding, changes in levels of temperature, light, UV, and nutrients can modulate the responses of photosynthetic organisms to OA (see Gao *et al.* (2019) and literature therein).

Global change affects both marine and freshwater ecosystems. However, because of their greater quantitative importance in global primary productivity and biogeochemistry, in this chapter, we mainly focus on the responses of marine macroalgae and phytoplankton to major global climate drivers (OA, warming, UV, etc.), with reference to recent advances in our understanding of the combined effects of multiple stressors, and analyze the underlying processes and/or mechanisms involved.

9.2. Contemporary and future environmental changes

9.2.1. *Ocean acidification*

Human activities are responsible for the continuously rising CO_2 concentration in the atmosphere and in the oceans. The increased dissolution of CO_2 into seawater leads to a drop in pH and altered carbonate chemistry, known as ocean acidification. While the atmospheric CO_2 concentration has been rising by about 2–3 ppmv per year, the average pH drop of the global oceans is about 0.002 unit per year (Gattuso *et al.* 2015; Moreira and Pires 2016).

When CO_2 dissolves in water, it combines with the water to form carbonic acid, which dissociates to bicarbonate and hydrogen ions (see Chapter 4). Since bicarbonate ions in seawater account for more than 90% of the total dissolved inorganic carbon, the protons released during the first disassociation combine with carbonate ions to establish a new equilibrium, thereby leading to decreased levels of carbonate ions (Hurd *et al.* 2020). In short, OA results in an increase of bicarbonate and hydrogen ions and a decrease of carbonate ions and pH along with increased CO_2 concentration. Under Scenario RCP8.5 of the IPCC (business as usual with the current use of fossil fuels and associated CO_2 emissions), atmospheric CO_2 will increase to about 1,000 ppmv and, concurrently, the pH of surface ocean will drop by 0.4 unit (an increase of H^+ ions of 150%) by the end of this century. As the marine carbonate system not only provides inorganic carbon sources for primary producers, but also sustains a chemically stable environment for marine organisms, the altered chemical environment caused by OA is likely to influence the physiology of many species and ecosystem services (see Gao *et al.* (2020) and Hurd *et al.* (2020) and literature therein).

9.2.2. *Ocean warming*

With the accumulation of CO_2 and other greenhouse gases in the atmosphere, global warming is becoming progressively exacerbated. Such warming has been predicted to increase the frequency and magnitude of events associated with extreme climate changes, such as unexpected heatwaves and increased frequency of typhoons (Bigg and Hanna 2016).

The oceans have taken up more than 70% of Earth's total heat increase since the 1970s (Gattuso *et al.* 2015), leading to ocean warming. The SST (sea surface temperature) has shown a rise of 1°C over the past 100 years and been projected to rise by a further 2–4°C by 2100 (Gattuso *et al.* 2015). The SSTs of the Pacific, Atlantic and Indian Oceans have already risen, respectively, by 0.65°C, 0.41°C and 0.31°C, during the period of 1950–2009 (IPCC 2014), reflecting regionally differential responses.

Elongated periods of extreme hot weather can cause marine heatwaves (MHWs) that may extend for thousands of kilometers. With progressive global warming, MHWs are predicted to occur in a more frequent, extensive, intensified and long-lasting fashion (Frölicher *et al.* 2018), exposing marine organisms, including primary producers, to thermal shock. MHWs may result in more harm when combined with other stressors (Gao *et al.* 2019).

9.2.3. *Ultraviolet radiation*

Ultraviolet (UV) radiation (UVR) is the component of solar radiation that is usually classified into UV-A (315–400 nm), UV-B (280–315 nm) and UV-C (200–280 nm). The oxygen and the ozone layer in the atmosphere totally screen off UV-C and significantly reduce the level of UV-B that reaches the Earth's surface (Madronich 1992). Ozone-destructive compounds such as chlorofluorocarbons (CFCs) and other pollutants released by human activities damage the ozone layer, leading to more UV-B radiation reaching the Earth's surface. Although the emission of CFCs has been banned since the establishment of the Montreal Protocol in 1987 and partial restoration of the ozone layer has been observed since 2010 (Solomon *et al.* 2016), full recovery of this protective layer is still unclear, as ozone-destructive compounds may remain in the atmosphere for a long time (Weatherhead and Andersen 2006). On the other hand, global warming exacerbates cooling in the stratosphere, which can accelerate ozone depletion (McKenzie *et al.* 2003; Häder *et al.* 2007).

In addition to direct UV exposure, organisms living in shallow waters or within the upper mixing layers receive more UVR due to shoaling (shallowed depth) of the UML. Most phytoplankton dwelling within this layer will be subject to intensified UV exposure as the UML becomes shallower. Although this is less relevant to intertidal macroalgae, species distributed in bays with less tidal mixing can still be influenced by increased exposure to UVR. Considering regional differences, the Arctic and the North Sea have been suggested, on the basis of predictions for elevated UVR regimes in these regions, to suffer relatively more from UVR in future (Müller *et al.* 2008). Therefore, primary producers might be more susceptible to UV radiation in these areas.

9.2.4. *Ocean deoxygenation*

Some regions of the oceans are known to be showing progressively lower respiratory activity due to warming-induced deoxygenation, with rapid expansion of the extent of hypoxic (dissolved oxygen (DO)$\leqq$2 mg/L) waters (Breitburg *et al.* 2018). DO in the oceans has declined over the past 50 years (Schmidtko *et al.* 2017, Breitburg *et al.* 2018) and is predicted to worsen. Consequently, the respiration index ($RI=\log_{10}(pO_2/pCO_2)$) will become smaller with progressive ocean warming and acidification (Brewer and Peltzer 2009).

In addition to globally scaled deoxygenation, coastal eutrophication often triggers harmful algal blooms. The subsequent degradation of this organic matter by heterotrophic bacteria will stimulate oxygen consumption and cause hypoxia (Wang *et al.* 2017). As a consequence, the respiration index would be reduced and acidification aggravated by CO_2 released in respiration (Cai *et al.* 2011, Miyamoto *et al.* 2019).

9.3. Effects of CO_2 rise and ocean acidification

CO_2 is the only direct carbon source for carboxylation involving ribulose-1,5-bisphosphate carboxylase-oxygenase (Rubisco) in all primary producers, with only a few examples of algae capable of fixing HCO_3^- via C_4 photosynthesis (Beardall and Raven 2020a, 2020b). Phytoplankton and macroalgae are capable of utilizing both dissolved CO_2 and HCO_3^- for photosynthesis (see Chapters 4, 6 and 7). Utilization of HCO_3^- is either by the direct uptake of HCO_3^- ions or following its extracellular conversion to CO_2 facilitated by catalysis via an external periplasmic carbonic anhydrase which is often involved in CO_2 concentrating mechanisms (CCMs) (Beardall and Raven 2020a, 2020b). Therefore, both macroalgae (see Ji and Gao 2020 and literature therein) and phytoplankton (Beardall and Raven 2020b) are likely to be affected by the changes in carbonate chemistry due to OA.

While most macroalgae tested so far can utilize both CO_2 and HCO_3^-, a few other species (mainly green or red macroalgae) can only take up CO_2 (Raven *et al.* 1995; Hepburn *et al.* 2011; Koch *et al.* 2013). For these species and others without CCMs or only a low-level operation of CCMs, photosynthesis can potentially be limited by the CO_2 availability when light supply is sufficient, so increased CO_2 availability could stimulate photosynthesis and growth – as primarily demonstrated in macroalgae (Zou and Gao 2005; Zou *et al.* 2011; Suárez-Álvarez *et al.* 2012; Johnson *et al.* 2014; Celis-Plá *et al.* 2015; Kram *et al.* 2016; Celis-Plá *et al.* 2017; Roth-Schulze *et al.* 2018; Zweng *et al.* 2018). Elevated CO_2 has been shown to stimulate the growth rates of red (Gao *et al.* 1991; 1993a; Zou and Gao 2009), green (Gordillo *et al.* 2001; Xu and Gao 2012) and brown (Kübler *et al.* 1999; Wu *et al.* 2019) macroalgae. Even under fluctuating pH regimes, elevated CO_2 concentrations enhance the growth and net photosynthesis of the red alga *Gracilariopsis lemaneiformis* (Qu *et al.* 2017), suggesting that seaweeds adapted to diel pH changes are resilient to OA. For macroalgal species distributed in upper intertidal zones, photosynthetic rates become stimulated by elevated CO_2 concentrations in the air during emersion (low tide) periods when the thalli suffer dehydration (Gao *et al.* 1999; Zou and Gao 2002), with photoinhibition being repressed (Henley *et al.* 1992). Such enhancement by elevated CO_2 concentration in air could be attributed to the faster diffusion of CO_2 into the cells and decreased availability of bicarbonate during the dehydration at low tides. On the other hand, some macroalgae with CCMs show a preference for CO_2 instead of bicarbonate under elevated CO_2 levels, as HCO_3^- usage is energetically costly due to its complicated transport across membranes (Cornwall *et al.* 2012). Increased CO_2 availability could therefore benefit these species by reducing the energy expenditure that CCMs demand (Israel and Hophy 2002; Cornwall *et al.* 2012) provided they are resilient to the associated acidic stress. While non-CCM macrophytes are relatively abundant in higher CO_2 freshwater environments (Maberly *et al.* 2015), OA associated with increased pCO_2 can alter the physiology and community structure of marine macroalgae, since different taxa or species exhibit different capabilities for tolerating acidic stress (Johnson *et al.* 2014; Nunes *et al.* 2016; Celis-Plá *et al.* 2017; Hurd *et al.* 2020). Increasing CO_2 and declining pH drive macroalgal community shifts towards non-calcifying species (Hall-Spencer *et al.* 2008; Kuffner *et al.* 2008; Connell and Russell 2010, Schoenrock *et al.* 2016; Harvey *et al.* 2021). While the stimulation of photosynthesis in three red macroalgae lacking CCMs has been shown recently (Cornwall and Hurd 2020), the examination of 55 macroalgae possessing CCMs showed that 21 exhibited no response to OA, 15 increased growth and 5 decreased growth relative to that in ambient seawater (van der Loos *et al.* 2019). For the non-CCM species tested, some showed unaffected growth, and growth of others was enhanced (Kübler *et al.* 1999; Kram *et al.* 2016; Britton *et al.* 2019; van der Loos *et al.* 2019; Hurd *et al.* 2020). This suggests that OA can alter the interspecific competition among macroalgae, giving rise to genetically different communities compared with contemporary ones (Wu *et al.* 2008).

Marine phytoplankton contribute over 90% of the total oceanic primary productivity, playing an important role in the marine biological CO_2 pump. Diatom-dominated phytoplankton assemblages, for example, settle from the euphotic layer to the deep sea at a rate of 124–732 m per day, based on observations at 58 stations (Agusti *et al.* 2015). It is known that OA reduces the content of biogenic silica in diatoms (Milligan *et al.* 2004; Xu *et al.* 2015), which may affect the predation rate of diatoms by zooplankton as well as the density of fecal pellets (Liu *et al.* 2016). Consequently, OA can indirectly modulate the settling of particulate organic carbon through effects on diatom silicification. In the oligotrophic South China Sea, OA was shown to decrease photosynthetic carbon fixation, expressed either per unit water volume or per unit chlorophyll, with the abundance of diatoms being decreased (Gao *et al.* 2012b). Under limiting sunlight levels, OA promoted the growth of several diatom species, while under light-saturated conditions, it either had no effect or inhibited their growth (Boelen *et al.* 2011; Gao *et al.* 2012b, Qu *et al.* 2021). This has been attributed to changes in cellular energetics rather than CO_2 fertilization or acidic stress alone. The extra energy freed up due to the down-regulated operation of CCMs can aid in the growth of diatoms under low levels of light supply, though under excessive supply of light energy, growth can be decreased due to enhanced photoinhibition and photorespiration (Figure 9.1). On the other hand, coastal and pelagic diatoms, even those within the same genus (such as *Thalassiosira*), may display contrasting responses to decreasing pH, with the coastal species being more resilient than the pelagic species in terms of net photosynthesis and growth (Li *et al.* 2016). Such differences between coastal and pelagic species can be attributed to local adaptation to present natural variability (Vargas *et al.* 2017) as well as to macro-nutrient availability, higher levels of which in nearshore waters enable the supply of sufficient energy to cope with the acidic stress (Li *et al.* 2018; Gao *et al.* 2019). In addition to these plastic responses, adaptation to OA over 1,800 generations resulted in smaller cells and slower growth rates with suppressed net photosynthesis as well as decreased mitochondrial respiration (Li *et al.* 2017). In terms of biochemical responses, OA increased the level of toxic phenolic compounds in *Emiliania huxleyi*, diatoms and natural phytoplankton assemblages with enhanced expression of enzymes related to the metabolic pathways of ß-oxidation and Krebs cycle, as demonstrated in differently scaled experiments, including mesocosms (Jin *et al.* 2015).

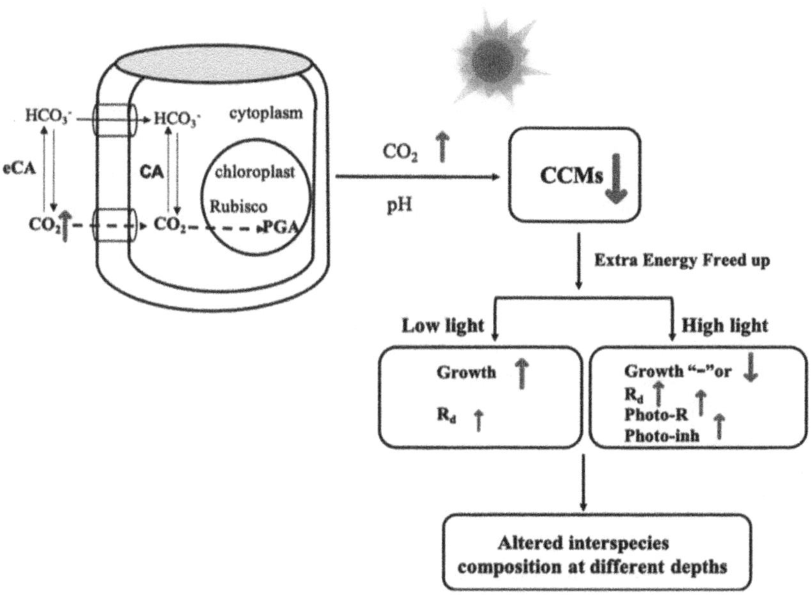

Figure 9.1. *Down-regulation of CO_2-concentrating mechanisms (CCMs) under the influence of ocean acidification (OA) saves energy, which acts to stimulate the growth of diatoms under growth-limiting light levels but inhibits it under excessive light intensity due to promoted photorespiration. Therefore, differential responses to OA can be expected under different weather conditions or at different water depths (redrawn, based on Gao* et al. *(2012b) and Qu* et al. *(2021))*

OA is reported to alter the community structure of marine primary producers. It appears to promote larger versus smaller phytoplankton species, at least in terms of growth rates (Wu *et al.* 2014; Bach and Taucher 2019), and OA-promoted shifts towards larger diatoms were observed in natural diatom communities (Tortell *et al.* 2008; Feng *et al.* 2010; Eggers *et al.* 2014; Bach *et al.* 2017; Taucher *et al.* 2017). Such phytoplankton community shifts may raise the trophic transfer of energy to animals (Sommer *et al.* 2002) and enhance production in higher trophic levels. At a natural CO_2 vent in the North Pacific Ocean, the large chain-forming diatom *Biddulphia biddulphiana* greatly increases in abundance as pH declines with increased pCO_2 toward the seep center (Harvey *et al.* 2019). Along this same gradient, the abundance of calcifying animals such as gastropods and sea urchins declined (Harvey *et al.* 2019), suggesting that OA can alter the food-web structure by altering the phytoplankton community composition.

9.3.1. *Effects of ocean warming and its combination with OA*

While the growth of most algal species tested exhibits positive, neutral or negative responses to moderate warming for short periods, their photosynthesis and respiration usually show immediate responses to temperature changes, as biochemical reactions are rapidly affected by a change in temperature (Ji and Gao 2020 and literature therein). The optimal temperature values for photosynthesis are usually lower in colder than in species growing in warmer waters or seasons (Gao 1990; Müller *et al.* 2008), implying a seasonal or adaptive trait. Respiration usually increases with a rise of temperature, following a linear pattern that differs from photosynthetic carbon fixation, which can be attributed to the fact that the enzymatic reactions related to respiration can be more influenced by temperature than are the photochemical reactions in photosynthesis (Davison 1991). The optimal temperatures of photosynthesis for most subtropical or temperate marine macroalgae can be found between 25°C and 30°C (Ji and Gao 2020). Acclimation to warming leads to a decreased ratio of respiration to gross photosynthesis (Zou and Gao 2013, 2014). This substantiates the hypothesis that macroalgae may benefit from moderate ocean warming when the cells are not limited by nutrients (Ji and Gao 2020). However, such apparently positive performance may disappear during the microscopic stages of macroalgal development, since most macroalgae such as *Pyropia* (*Porphyra)* and *Laminaria* plants live through a microscopic phase in late spring. In addition, warming negatively impacts most of the life cycle stages in some brown algae (Ladah and Zertuche 2007; Leal *et al.* 2017; Figure 9.2), as the release of spores, germination and early sporophyte growth require low temperature (Buschmann *et al.* 2004). Consequently, the microscopic stages of macroalgae can be more sensitive to ocean warming (Ji and Gao 2020). Nevertheless, little has been documented on the effects of warming on different life stages of macroalgae, and how macroalgae respond to warming through their life cycles remains to be examined. However, it is clear that higher temperatures, above current levels of tolerance, can lead to reduced performance and growth of macroalgae and changes in seaweed community composition (Harley *et al.* 2012; Krumhansl and Scheibling 2012; Koch *et al.* 2013; Brodie *et al.* 2014). This can lead to the retraction/disappearance of kelp forests, particularly at their low latitude limits (Tuya *et al.* 2012). In some circumstances, such warming can be sudden and acute rather than gradual (i.e. take the form of "marine heatwaves"). These can have very significant effects on macroalgal populations such that kelp forests become reduced or disappear, with major consequences for community structure, including invertebrates and fish (Smale and Wernberg 2013; Wernberg *et al.* 2013; Arafeh-Dalmau *et al.* 2019).

OA and warming are coincident major marine environmental problems that are gaining increasing attention. How these are likely to interactively influence primary producers is a key to understand future changes in marine ecosystem services and functions in the context of climate change (Gao *et al.* 2020). Different species are reported to show diverse responses to OA and warming in differently designed or scaled experiments. For example, OA and warming enhanced the growth of the picophytoplankton *Synechococcus* but did not show any impact on *Prochlorococcus* (Fu *et al.* 2007). For diatoms, OA and a 4°C increase in temperature stimulated the growth of *Skeletonema marinoi* (Kremp *et al.* 2012) but did not affect that of *Thalassiosira* or *Chaetoceros* (Hyun *et al.* 2014). While there have been a handful of studies that have examined the long-term evolutionary responses of phytoplankton to warming alone (Padfield *et al.* 2016; Chakravarti *et al.* 2017; Schaum *et al.* 2017, 2018; O'Donnell *et al.* 2018, 2019; Jin and Agustí 2018; Jin *et al.* 2020), little has been documented on the long-term evolutionary responses of phytoplankton to the combination of OA and warming. In a long-term experimental study (~400 days) on a coccolithophore *Emiliania huxleyi*, particulate inorganic (PIC) and organic (POC) carbon contents were higher by up to 100% and 50% under a combined OA and warming treatment, respectively, compared to non-adapted controls (Schlüter *et al.* 2014). In addition, the maximum persistence temperature (T_{max}) of *Emiliania huxleyi* significantly decreased after long-term selection under OA (Listmann *et al.* 2016), reflecting a narrowed thermal norm under the influence of elevated pCO_2. A quantitative meta-analysis focusing on multispecies studies at the community level indicated that primary productivity by temperate noncalcifying autotrophs increases with OA and warming, whereas tropical phytoplankton shows decreased productivity because of acidification (Nagelkerken and Connell 2015).

OA associated with CO_2 rise and warming together act synergistically to enhance the growth of the fleshy red alga *Plocamium cartilagineum*, while OA alone decreased its growth rate (Kram *et al.* 2016). In contrast, synergistic effects of these two drivers reduced biomass production in the brown alga *Sargassum muticum* and *Cystoseira tamariscifolia* (Olabarria *et al.* 2013) and suppressed early-stage development of the kelp *Macrocystis pyrifera*, increasing mortality of its spores (Gaitán-Espitia *et al.* 2014). On the other hand, no interactions of these two drivers were found during the meiospore phase in the brown macroalgae *Macrocystis, Pyrifera* and *Undaria pinnatifida* (Leal *et al.* 2017). In the calcifying macroalgae *Corallina officinalis* and *Amphiroa anceps*, OA and warming interact to additively reduce their calcification (Graba-Landry *et al.* 2018). These two drivers can also act together to decrease the growth (Bender-Champ *et al.* 2017) and calcification (Sinutok *et al.* 2011; Graba-Landry *et al.* 2018) of coral reef macroalgae. In contrast, a warming treatment was shown to counteract the negative impacts of OA on a crustose coralline alga (Kim *et al.* 2020), and was credited with stimulating growth and photochemical efficiency. In the calcifying green macroalga *Halimeda cuneata*, photosynthetic and enzymatic activities were insensitive to OA and warming (Scherner *et al.* 2016). Nonetheless, such positive, neutral and negative effects of OA combined with warming could be location-dependent and species-specific, and might also be attributed to different environmental conditions such as light levels, the presence of UVR and/or fluctuating light regimes (Jin *et al.* 2013; Hoppe *et al.* 2015; Gao *et al.* 2019). Detailed comparisons of the responses of different macroalgal species to climate change drivers are available in tables presented in a recent review (Ji and Gao 2021).

For all macro- and microalgae, including cyanobacteria, the temporary responses to the combination of OA and warming only reflect plastic acclimation. However, evolutionary responses to the combination of these two drivers over a long time can also differ under different light levels or regimes and reflect their adaptations to minimize energetic demands to cope with the environmental stresses. Therefore, field observations are believed to provide more reliable information on the combined effects of OA and warming. Mechanistically, both OA and warming can result in more carbon loss during nighttime due to enhanced respiration, which may balance the energy and carbon gains during daytime. Subsequently, the modulation of metabolic pathways during acclimation would be the key for adaptation to these two key drivers.

9.3.2. *Effects of UV radiation and its combination with OA and warming*

Solar UV radiation (UVR, 280–400 nm) can damage DNA (Karentz *et al.* 1991, Häder and Gao 2015) and repress its repair in phytoplankton (Gao *et al.* 2008;

Rastogi *et al.* 2014), along with negative effects on physiology and metabolism, including shifts in fatty acid composition and particularly the ratio of polyunsaturated to saturated fatty acids (Liang *et al.* 2006). While UV-B irradiance accounts for less than 1% of the total solar energy, it is usually more harmful than UV-A, as UV-B photons are more energetic than UV-A, which is about 6–8% of the total solar energy. However, when there is active mixing or solar radiation is fluctuating, photosynthetic carbon fixation could be inhibited by UV to a lesser extent (Helbling *et al.* 2003; Jin *et al.* 2013; Li *et al.* 2013). More UV-B-induced damage to DNA was detected in smaller diatom cells (Karentz *et al.* 1991). Moreover, UVR can also generate active free oxygen radicals that lead to oxidative stress (Häder *et al.* 2015), which may result in lower photosynthetic rates (Wu *et al.* 2005). However, UVR, especially UV-A, can play a positive role in stimulating photosynthesis of diatoms (Wu and Gao 2009), coastal phytoplankton assemblages (Gao *et al.* 2007) and macroalgae (Xu and Gao 2010). The presence of UVR can also change the morphology of filamentous cyanobacteria (Wu *et al.* 2005) and morphological development of the red alga *Pyropia* sp. (Jiang *et al.* 2007).

Through evolutionary adaptation, phytoplankton and macroalgae have acquired mechanisms to cope with UV radiation. For instance, they synthesize UV-shielding pigments such as MAAs (mycosporine-like amino acids) and possess mechanisms to eliminate active oxygen free radicals and repair damaged proteins and DNA (Rastogi *et al.* 2014; Häder *et al.* 2015). Nevertheless, in the surface layer of the water column, phytoplankton cells are harmed by UV, decreasing carbon fixation by 20–30% (Helbling *et al.* 2003; Häder and Gao 2015). Usually, phytoplankton species with larger cell sizes or the macroscopic stage of macroalgae are more resilient to solar UV radiation than small single-celled species or microscopic phases of macroalgae (Jiang and Gao 2008, Jiang *et al.* 2009; Wu *et al.* 2015; Ji and Gao 2020).

Ocean warming along with progressive OA exposes phytoplankton cells not only to higher temperatures and lower pH but also to higher levels of solar UV radiation associated with a shallower mixed depth (Beardall *et al.* 2009, 2014; Gao *et al.* 2019). Photosynthetic processes of marine primary producers respond to changes in UV radiation within minutes. Around noon, when solar radiation is the highest, photochemical performance detected by chlorophyll fluorescence techniques decreases faster in the presence of UVR. For instance, the diatom *Cylindrotheca closterium f. minutissima* decreased its PSII photochemical efficiency faster when the cells were exposed to OA conditions under incident sunlight with UVR, even after the diatom had previously been acclimated to the combined treatment of OA and UVR (Wu *et al.* 2012). On the other hand, the diatom *Skeletonema costatum* showed an increased activity of periplasmic carbonic anhydrase (CAe) when exposed to moderate UVR levels, with less enhancement under elevated CO_2 concentration, implying that UVR, especially UVA, can modulate CO_2 acquisition efficiency (Wu and Gao 2009). For the red tide alga *Phaeocystis globosa* grown under OA at 1,000 ppmv CO_2 in air (about 1,000 μatm in water) and solar radiation with UVR, photochemical efficiency showed the lowest values at noon (Chen and Gao 2011). Regardless of sunny or cloudy days, the combined effects of OA and UV-B lowered the photochemical yield in this red tide alga but raised its NPQ (non-photochemical quenching) only on sunny days (Chen and Gao 2011). Under simulated mixing conditions (fluctuating solar radiation), OA antagonistically acted to the enhancement of photosynthetic carbon fixation by UV-A, while lessening the photochemical inhibition caused by UV-B (Jin *et al.* 2013), suggesting that the combined impacts of OA and UVR differ for different UV wavelengths.

The enhancement of stratification, as well as increasing the average PAR light and UVR exposure of phytoplankton, also results in a reduction of nutrient supply to the upper mixing layer due to upward export from deeper layers. Nutrient limitation in turn exacerbates the damaging effects of UVR through reducing the capacity of cells to make MAAs (enhancing damage rates) or decreasing the capacity of cells to carry out repair (Beardall *et al.* 2009, 2014; Li *et al.* 2021).

The effects of OA and UV synergistically impact on the calcification of both calcifying macro- and microalgae (Gao and Häder 2017). OA suppresses calcification in algal calcifiers and consequently makes the calcified layers thinner, allowing more transmission of harmful UVR (Gao *et al.* 2009; Gao and Zheng 2010). Reduced calcification is thus likely to give rise to more damage to the cells, which in turn further decreases the calcification and photosynthetic carbon fixation (Figure 9.2). By examining the combined effects of OA and warming, PAR light intensity fluctuation and UVR on a coccolithophorid *Gephyrocapsa oceanica*, Jin *et al.* (2019) found that the combination of OA and warming synergistically increased the photochemical efficiency and photosynthetic carbon fixation rate under a lower frequency of light fluctuations, with insignificant impacts of UVR being observed. However, in another coccolithophorid *Emiliania huxleyi*, while a combination of OA and warming reduced the ratio of calcification to photosynthesis under PAR, the addition of UVR increased this ratio, reflecting an antagonistic interaction between UVR and greenhouse treatments (Tong *et al.* 2019). The increased ratio of calcification to photosynthesis is suggested to be linked to higher proportions of coccoliths, which play a protective role against UVR (Gao *et al.* 2009) (Figure 9.2). The presence of UVR is known to stimulate the production of particulate inorganic carbon (calcification) in *Emiliania huxleyi* at the sacrifice of its growth rate (Guan and Gao 2010).

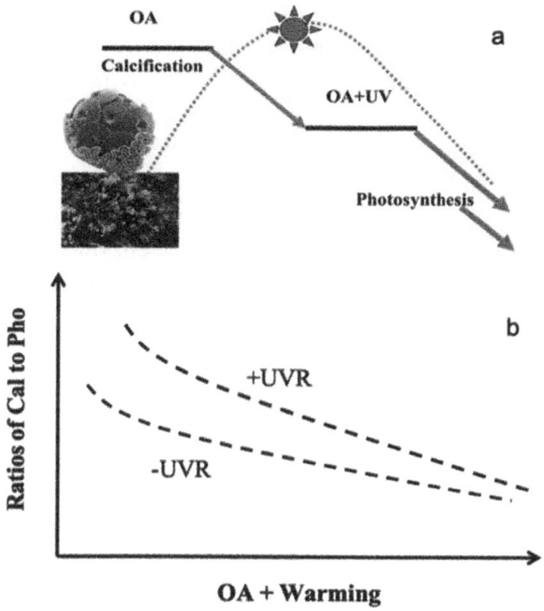

Figure 9.2. *Illustration of the responses of algal calcifiers to UV and ocean acidification (OA) (a) and to OA + warming with or without UVR (b). Both micro- and macro-calcifying algae decrease their calcification under ocean acidification (OA), and the addition of solar UV radiation further reduces their calcification and photosynthesis, reflecting a synergy between the impacts of OA and UV (a). While a combination of OA and warming decreases the ratios of photosynthesis (Pho) to calcification (Cal), the addition of UVR antagonistically increases the ratio, showing a stimulating effect of UVR on calcification (see Gao* et al. *2009; Guan and Gao 2010)*

Warming or moderate increase of temperature may alleviate UV-related harm due to increased activities of enzymes involved in repair processes (Beardall *et al.* 2014; Wong *et al.* 2015). By acclimating the diatom *Phaeodactylum tricornutum* to two CO_2 concentrations (390 and 1,000 μatm) for more than 20 generations, Li *et al.* (2012) found that growth under OA increased the NPQ of cells and partially counteracted the harm to PSII (photosystem II) caused by UV-A and UV-B, which, however, was moderated by increased temperature levels. Increased NPQ was inversely related to the ratio of repair to UV-B-induced damage, reflecting the induction of NPQ when repair capacity was less than the damage rate, and OA treatment partially counteracted UV-B-induced harm (Li *et al.* 2012). As for photosynthetic carbon fixation, the elevated levels of CO_2 and temperature synergistically reduced the inhibition caused by UV-B and thus increased the carbon fixation (Li *et al.* 2012). In contrast, OA at 700 μatm pCO2 interacted with both incident UV-A and UV-B levels to reduce the photosynthetic carbon fixation of a coastal phytoplankton assemblages (Wu and Gao 2010). In addition, a seasonal comparison showed that UV-induced inhibition of primary productivity of coastal phytoplankton assemblages was higher during winter (temperature low) than during summer (temperature high) periods, which is counterintuitive since lower levels of UVR occur during the winter period. A plausible explanation is that higher temperatures during the summer period can enhance enzyme-driven repair processes although community composition differences during different seasons cannot be ruled out.

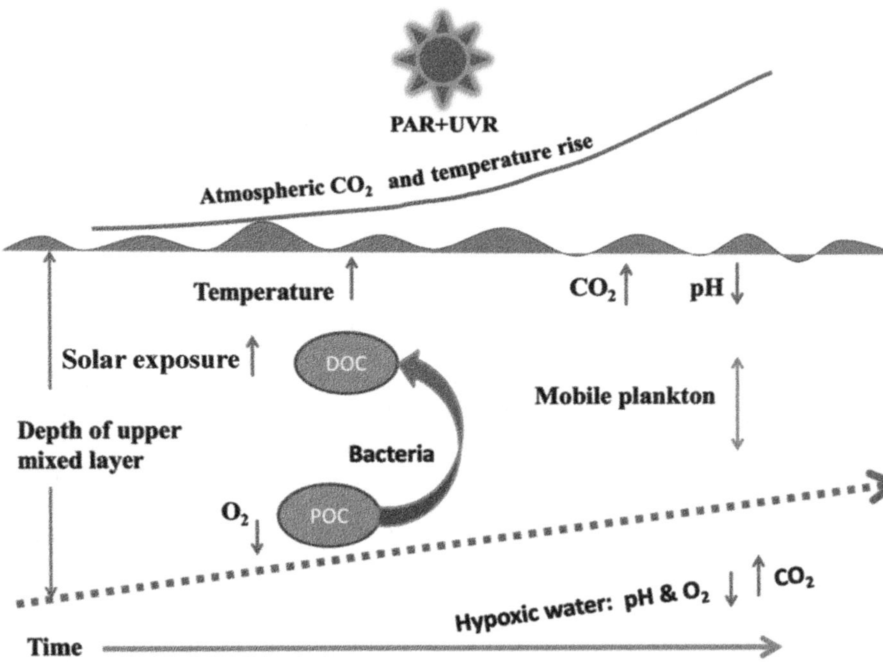

Figure 9.3. *Illustration of the top-down and bottom-up opposing stress hypotheses under the influences of ocean acidification, warming, deoxygenation and solar UV radiation. Warming enhances stratification (leading to a shallower upper mixing layer, blue arrows) that results in more integrated exposure to solar UV radiation, cascading a top-down pressure, and bacterial degradation of organic matter enhanced by warming results in less* O_2*, expanding hypoxic waters where pH and* O_2 *are lowered, forming a bottom-up pressure. When mobile organisms migrate downward to escape heat and UV stresses, they encounter low levels of pH and* O_2 *(redrawn after Gao 2018 and Gao* et al. *2020)*

Many microalgae and spores of many macroalgae actively migrate vertically and horizontally by flagellar movement in addition to being passively mixed in the water column. They may swim towards light by positive phototaxis under low light but away from the surface to avoid high light and UV-related harm in a similar way as occurs in zooplankton (see Gao *et al.* 2020 and literature therein). OA was shown to impair the swimming ability of flagellated microalgae, suggesting that their capacity to survive will be threatened in a high CO_2 world (Wang *et al.* 2020), especially under multiple stressors when they are exposed to top-down pressures (associated with warming and UV radiation) and bottom-up stresses (due to low levels of pH and O_2 in subsurface hypoxic waters) (Figure 9.3).

9.4. Other interactions

While, as discussed in the sections above, marine primary productivity can be influenced by climate change factors in different directions and to differing extents, interactions of primary producers with their grazers or viruses have been scarcely documented. Algal (including cyanobacterial) viruses are known to play an important role in algal ecology and evolution (Short *et al.* 2020). Marine viruses are able to tolerate a wide range of pH changes but are reported to be sensitive to warming and UV radiation (see the review by Gao *et al.* 2020 and literature therein). OA has been shown to exacerbate viral attack on a red tide microalga (Chen *et al.* 2015) and to modulate the interaction of a virus and its host picophytoplankton (Traving *et al.* 2014). On the other hand, algae grazers, such as zooplankton and molluscs, can be directly affected by OA (see the reviews by Gao *et al.* 2020 and Hurd *et al.* 2020 and literature therein) and are indirectly affected by food quality altered by climate change factors. For instance, Rossoll *et al.* (2012) showed that elevated CO_2 affected the fatty acid concentration and composition of the diatom *Thalassiosira pseudonana*. Furthermore, changes in fatty acid composition, especially the ratio of polyunsaturated fatty acids to saturated fatty acids, were translated to copepods (*Acartia tonsa*) fed the diatom grown at high CO_2 and these copepods showed reduced growth and development and decreased fecundity (Rossoll *et al.* 2012). In another study, cultures of diatoms, coccolithophores and natural phytoplankton assemblages increased their content of potentially toxic phenolic compounds when grown under OA conditions as a consequence of alterations in β-oxidation metabolic pathways (Jin *et al.* 2015). Copepods fed with OA-grown phytoplankton had a higher phenolics content compared to the control (Jin *et al.* 2015), reflecting a potential food chain effect of OA from primary to secondary producers with a plausible explanation for degraded taste of shrimps grown under OA (Dupont *et al.* 2014). While warming might stimulate the degradation of toxic phenolics due to the enhanced activity of related enzymes, biogeochemical cycling of these compounds in the oceans may be impacted. In the brown macroalga *Saccharina japonica,* OA treatment raised the accumulation of iodide and its content in abalone fed on the kelp (Xu *et al.* 2019). To cope with environmental stressors, the grazers tend to eat more and respire more (Li and Gao 2012); therefore, the enhanced accumulation of these secondary metabolites is expected. In short, OA can affect food quality in both micro- and macroalgae by influencing their secondary metabolites and their transfer to higher trophic levels, consequently driving significant consequences for food webs, though impacts of multiple drivers on marine food chains remain to be investigated (Jin *et al.* 2020).

9.5. Summary

The future changes in marine environments associated with anthropogenic activities include ocean warming, acidification associated with increased dissolved CO_2 concentrations and expansion of anoxic waters, in addition to coastal eutrophication and pollution. The effects of OA associated with pCO_2 rise can affect primary producers in different ways and to different extents, depending on strain- or taxon-specific physiology and/or other factors. Warming-induced enhancement of stratification exacerbates solar UVR impacts on photosynthetic organisms dwelling in the UML and simultaneously exposes them to a decreased nutrient availability due to diminished upward transport of the biogenic elements. All of these environmental changes interact to affect marine primary production and consequent air–sea CO_2 fluxes. For both macro- and micro-calcifying algae, the combined impacts of acidification and/or elevated temperature reduce their calcification and exacerbate UV-induced harm. In marine diatoms, OA and light interact to enhance their growth at low levels of sunlight but inhibit it at high levels, which is often attributed to changes in cellular energetics related to down-regulation of CCMs.

In most photosynthetic nitrogen fixers (diazotrophs), acidification associated with elevated CO_2 may enhance their N_2 fixation activity, but interactions with other environmental factors such as trace metal availability or pollutants and/or UV may neutralize or even reverse these effects. Most macroalgae, on the other hand, appear to benefit from elevated CO_2 with enhanced growth rates and tolerance to lowered pH, which is likely to speed up their life cycles, though temperature rises can have adverse effects, particularly on their microscopic stages. As to effects of reduced availability of O_2, due to both warming-enhanced deoxygenation and increased bacterial respiration associated with both warming and enhanced coastal eutrophication, though poorly documented to date, both macro- and microalgae are expected to benefit from lower oxygen levels due to the decreased ratio of O_2 to CO_2 availability surrounding Rubisco, an enzyme that catalyzes both carboxylation and oxygenation. Overall, primary producers in coastal waters and the open ocean can exhibit contrastingly different responses to future climate changes. In the upper mixing layers of the open oceans, especially that of oligotrophic waters, carbon fixation by phytoplankton will decline under influences of the multiple drivers; while in coastal and/or upwelling regions, a sustained sufficient supply of nutrients is expected to meet the extra energy demand to cope with the environmental stressors. Subsequently, it is predicted that primary productivity in surface oceans will decrease from nearshore waters to open oceans due to enhanced respiratory carbon loss, reduced photosynthesis and enhanced UV damage, along with decreased nutrient availability, under the influences of the multiple drivers.

9.6. Perspectives

Both micro- and macroalgae are expected to adapt and evolve to ocean climate changes, though there might be "losers", such as the calcifying species (Gao *et al.* 2019). In microalgae, evolutionary responses to elevated CO_2 over hundreds to thousands of generations have been documented (Collins 2006; Brennan and Collins 2015; Li *et al.* 2017; Tong *et al.* 2018), showing irreversible down-regulation of physiological performance. In contrast, there has been little documentation of the evolutionary responses to OA combined with other ocean climate changes in macroalgae, mainly due to their complex and long life cycles (Ji and Gao 2020). With regard to N_2 fixing cyanobacteria, OA treatment over several years brought about irreversible enhancement of N_2 fixation and growth, which appears to be associated with DNA methylation (Hutchins *et al.* 2015; Walworth *et al.* 2016).

OA and warming can interactively perturb the biological performance of many species and biogeochemical processes, and thereby threaten ecosystem services (Gattuso *et al.* 2015). Coastal ecosystems, where macroalgal productivity accounts for most of the primary productivity, are under more pressure from human activities and global ocean changes, since they are adjacent to highly populated areas. OA and warming together could additively increase respiratory carbon loss in both nearshore and offshore waters, though macroalgae are known to be more resistant to OA. This is because respiration increases with warming and coping with OA-induced pH stress requires extra energy and enhances respiration (Gao *et al.* 2019). As a result, OA and warming may interact to affect the marine biological pump (MBP) and microbial carbon pump (MCP), both of which are considered to be sensitive to warming (Jiao and Zheng 2011). Nevertheless, the combined effects of OA and warming along with other stressors on MBP/MCP are yet to be investigated, which is very important since ocean climate changes regulate ocean carbon uptake and the magnitude of ocean carbon sinks.

9.7. Acknowledgments

This work was funded by the National Natural Science Foundation (41720104005, 41806141, 41721005). The authors are grateful to Xianglan Zeng for her help for assistance in organizing the literature.

9.8. References

Agusti, S., González-Gordillo, J.I., Vaqué, D., Estrada, M., Cerezo, M.I., Salazar, G., Gasol, J.M., Duarte, C.M. (2015). Ubiquitous healthy diatoms in the deep sea confirm deep carbon injection by the biological pump. *Nature Communications*, 6(1), 7608.

Arafeh-Dalmau, N., Montaño-Moctezuma, G., Martínez, J.A., Beas-Luna, R., Schoeman, D.S., Torres-Moye, G. (2019) Extreme marine heatwaves alter kelp forest community near its equatorward distribution limit. *Frontiers in Marine Science*, 6, 499.

Bach, L.T. and Taucher, J. (2019). CO_2 effects on diatoms: A synthesis of more than a decade of ocean acidification experiments with natural communities. *Ocean Science*, 15(4), 1159–1175.

Bach, L.T., Alvarez-Fernandez, S., Hornick, T., Stuhr, A., Riebesell, U. (2017). Simulated ocean acidification reveals winners and losers in coastal phytoplankton. *PLOS ONE*, 12(11), e0188198.

Beardall, J. and Raven, J.A. (2020a). Structural and biochemical features of carbon acquisition in algae. In *Photosynthesis in Algae: Biochemical and Physiological Mechanisms. Advances in Photosynthesis and Respiration (Including Bioenergy and Related Processes)*, Larkum, A., Grossman, A., Raven, J. (eds). Springer, Cham.

Beardall, J. and Raven, J.A. (2020b). Acquisition of inorganic carbon by microalgae and cyanobacteria. In *Microbial Photosynthesis*, Wang, Q. (ed.). Springer, Singapore.

Beardall, J., Sobrino, C., Stojkovic, S. (2009). Interactions between the impacts of ultraviolet radiation, elevated CO_2, and nutrient limitation on marine primary producers. *Photochemical & Photobiological Sciences*, 8, 1257–1265.

Beardall, J., Stojkovic, S., Gao, K. (2014). Interactive effects of nutrient supply and other environmental factors on the sensitivity of marine primary producers to ultraviolet radiation: Implications for the impacts of global change. *Aquatic Biology*, 22, 5–23.

Bender-Champ, D., Diaz-Pulido, G., Dove, S. (2017). Effects of elevated nutrients and CO_2 emission scenarios on three coral reef macroalgae. *Harmful Algae*, 65, 40–51.

Bigg, G. and Hanna, E. (2016). Impacts and effects of ocean warming on the weather. In *Explaining Ocean Warming: Causes, Scale, Effects and Consequences*, Laffoley, D. and Baxter, J.M. (eds). IUCN (International Union for Conservation of Nature), Gland.

Boelen, P., van de Poll, W.H., van der Strata, H.J., Neven, I.A., Beardall, J., Buma, A.G.J. (2011). Neither elevated nor reduced CO_2 affects the photophysiological performance of the marine Antarctic diatom *Chaetoceros brevis*. *Journal of Experimental Marine Biology and Ecology*, 406, 38–45.

Breitburg, D., Levin, L.A., Oschlies, A., Grégoire, M., Chavez, F.P., Conley, D.J., Garçon, V., Gilbert, D., Gutiérrez, D., Isensee, K., Jacinto, G.S., Limburg, K.E., Montes, I., Naqvi, S.W.A., Pitcher, G.C., Rabalais, N.N., Roman, M.R., Rose, K.A., Seibel, B.A., Telszewski, M., Yasuhara, M., Zhang, J. (2018). Declining oxygen in the global ocean and coastal waters. *Science*, 359(6371), eaam7240.

Brennan, G. and Collins, S. (2015). Growth responses of a green alga to multiple environmental drivers. *Nature Climate Change*, 5(9), 892–897.

Brewer, P.G. and Peltzer, E.T. (2009). Limits to marine life. *Science*, 324(5925), 347–348.

Britton, D., Mundy, C.N., Mcgraw, C.M., Revill, A.T., Hurd, C.L. (2019). Responses of seaweeds that use CO_2 as their sole inorganic carbon source to ocean acidification: Differential effects of fluctuating pH but little benefit of CO_2 enrichment. *ICES Journal of Marine Science*, 76(6), 1860–1870.

Brodie, J., Williamson, C.J., Smale, D.A., Kamenos, N.A., Mieszkowska, N., Santos, R., Cunliffe, M., Steinke, M., Yesson, C., Anderson, K.M., Asnaghi, V., Brownlee, C., Burdett, H.L., Burrows, M.T., Collins, S., Donohue, P.J.C., Harvey, B., Foggo, A., Noisette, F., Nunes, J., Ragazzola, F., Raven, J.A., Schmidt, D.N., Suggett, D., Teichberg, M., Hall-Spencer, J.M. (2019). The future of the northeast Atlantic benthic flora in a high CO_2 world. *Ecology and Evolution*, 4, 2787–2798.

Buschmann, A.H., Vásquez, J.A., Osorio, P., Reyes, E., Filún, L., Hernández-González, M.C., Vega, A. (2004). The effect of water movement, temperature and salinity on abundance and reproductive patterns of *Macrocystis* spp. (Phaeophyta) at different latitudes in Chile. *Marine Biology*, 145(5), 849–862.

Cai, W.-J., Hu, X., Huang, W.-J., Murrell, M.C., Lehrter, J.C., Lohrenz, S.E., Chou, W.-C., Zhai, W., Hollibaugh, J.T., Wang, Y., Zhao, P., Guo, X., Gundersen, K., Dai, M., Gong, G.-C. (2011). Acidification of subsurface coastal waters enhanced by eutrophication. *Nature Geoscience*, 4(11), 766–770.

Celis-Plá, P.S.M., Hall-Spencer, J.M., Horta, P.A., Milazzo, M., Korbee, N., Cornwall, C.E., Figueroa, F.L. (2015). Macroalgal responses to ocean acidification depend on nutrient and light levels. *Frontiers in Marine Science*, 2, 26.

Celis-Plá, P.S.M., Martínez, B., Korbee, N., Hall-Spencer, J.M., Figueroa, F.L. (2017). Ecophysiological responses to elevated CO_2 and temperature in *Cystoseira tamariscifolia* (Phaeophyceae). *Climatic Change*, 142(1), 67–81.

Chakravarti, L.J., Beltran, V.H., Van Oppen, M.J.H. (2017). Rapid thermal adaptation in photosymbionts of reef-building corals. *Global Change Biology*, 23(11), 4675–4688.

Chen, S.W. and Gao, K. (2011). Solar ultraviolet radiation and CO_2-induced ocean acidification interacts to influence the photosynthetic performance of the red tide alga *Phaeocystis globosa* (Prymnesiophyceae). *Hydrobiologia*, 675(1), 105–117.

Chen, S.W., Gao, K.S., Beardall, J. (2015). Viral attack exacerbates the susceptibility of a bloom-forming alga to ocean acidification. *Global Change Biology*, 21(2), 629–636.

Collins, S., Sültemeyer, D., Bell, G. (2006). Changes in C uptake in populations of *Chlamydomonas reinhardtii* selected at high CO_2. *Plant, Cell & Environment*, 29(9), 1812–1819.

Connell, S.D. and Russell, B.D. (2010). The direct effects of increasing CO_2 and temperature on non-calcifying organisms: Increasing the potential for phase shifts in kelp forests. *Proceedings of the Royal Society B: Biological Sciences*, 277(1686), 1409–1415.

Cornwall, C.E. and Hurd, C.L. (2020). Variability in the benefits of ocean acidification to photosynthetic rates of macroalgae without CO_2-concentrating mechanisms. *Marine and Freshwater Research*, 71(3), 275–280.

Cornwall, C.E., Hepburn, C.D., Pritchard, D., Currie, K.I., McGraw, C.M., Hunter, K.A., Hurd, C.L. (2012). Carbon-use strategies in macroalgae: Differential responses to lowered pH and implications for ocean acidification. *Journal of Phycology*, 48(1), 137–144.

Cornwall, C.E., Revill, A.T., Hurd, C.L. (2015). High prevalence of diffusive uptake of CO_2 by macroalgae in a temperate subtidal ecosystem. *Photosynthesis Research*, 124(2), 181–190.

Davison, I.R. (1991). Environmental-effects on algal photosynthesis – Temperature. *Journal of Phycology*, 27(1), 2–8.

Dupont, S., Hall, E., Calosi, P., Lundve, B. (2014). First evidence of altered sensory quality in a shellfish exposed to decreased pH relevant to ocean acidification. *Journal of Shellfish Research*, 33(3), 857–861.

Eggers, S.L., Lewandowska, A.M., Barcelos e Ramos, J., Blanco-Ameijeiras, S., Gallo, F., Matthiessen, B. (2014). Community composition has greater impact on the functioning of marine phytoplankton communities than ocean acidification. *Global Change Biology*, 20(3), 713–723.

Feng, Y., Warner, M.E., Zhang, Y., Sun, J., Fu, F.-X., Rose, J.M., Hutchins, D.A. (2008). Interactive effects of increased pCO_2, temperature and irradiance on the marine coccolithophore *Emiliania huxleyi* (Prymnesiophyceae). *European Journal of Phycology*, 43(1), 87–98.

Feng, Y., Hare, C.E., Rose, J.M., Handy, S.M., Ditullio, G.R., Lee, P.A., Smith Jr., W.O., Peloquin, J., Tozzi, S., Sun, J., Zhang, Y., Dunbar, R.B., Long, M.C., Sohst, B., Lohan, M., Hutchins, D.A. (2010). Interactive effects of iron, irradiance and CO_2 on Ross Sea phytoplankton. *Deep-Sea Research Part I – Oceanographic Research Papers*, 57(3), 368–383.

Flynn, K.J., Blackford, J.C., Baird, M.E., Raven, J.A., Clark, D.R., Beardall, J., Brownlee, C., Fabian, H., Wheeler, G.L. (2012). Changes in pH at the exterior surface of plankton with ocean acidification. *Nature Climate Change*, 2(7), 510–513.

Frolicher, T.L., Fischer, E.M., Gruber, N. (2018). Marine heatwaves under global warming. *Nature*, 560(7718), 360–364.

Fu, F.X., Warner, M.E., Zhang, Y.H., Feng, Y.Y., Hutchins, D.A. (2007). Effects of increased temperature and CO_2 on photosynthesis, growth, and elemental ratios in marine *Synechococcus* and *Prochlorococcus* (Cyanobacteria). *Journal of Phycology*, 43(3), 485–496.

Gaitan-Espitia, D.J., Hancock, J.R., Padilla-Gamino, J.L., Rivest, E.B., Blanchette, C.A., Reed, D.C., Hofmann, G.E. (2014). Interactive effects of elevated temperature and pCO_2 on early-life-history stages of the giant kelp *Macrocystis pyrifera*. *Journal of Experimental Marine Biology and Ecology*, 457, 51–58.

Gao, K.S. (1991). Effects of seawater current speed on the photosynthetic oxygen evolution of *Sargassum thunbergii* (Phaeophyta). *Japanese Journal of Phycology*, 39, 291–293 [In Japanese with English summary].

Gao, K.S. (2018). Ecological and physiological effects of ocean acidification and their correlations with warming, UV radiation and deoxygenation. *Journal of Xiamen University (Natural Science)*, 57, 800–810 [In Chinese with English summary].

Gao, K.S. and Campbell, D.A. (2014). Photophysiological responses of marine diatoms to elevated CO_2 and decreased pH: A review. *Functional Plant Biology*, 41(5), 449–459.

Gao, K.S. and Häder, D.-P. (2017). Effects of ocean acidification and UV radiation on marine photosynthetic carbon fixation. In *Systems Biology of Marine Ecosystems*, Kumar, M. and Ralph, P. (eds). Springer International Publishing, Cham.

Gao, K.S. and Zheng, Y. (2010). Combined effects of ocean acidification and solar UV radiation on photosynthesis, growth, pigmentation and calcification of the coralline alga *Corallina sessilis* (Rhodophyta). *Global Change Biology*, 16(8), 2388–2398.

Gao, K.S., Aruga, Y., Asada, K., Ishihara, T., Akano, T., Kiyohara, M. (1991). Enhanced growth of the red alga *Porphyra yezoensis* Ueda in high CO_2 concentrations. *Journal of Applied Phycology*, 3(4), 355–362.

Gao, K.S., Aruga, Y., Asada, K., Kiyohara, M. (1993a). Influence of enhanced CO_2 on growth and photosynthesis of the red algae *Gracilaria* sp. and *G. chilensis*. *Journal of Applied Phycology*, 5(6), 563–571.

Gao, K.S., Aruga, Y., Asada, K., Ishihara, T., Akano, T., Kiyohara, M. (1993b). Calcification in the articulated coralline alga *Corallina pilulifera*, with special reference to the effect of elevated CO_2 concentration. *Marine Biology*, 117(1), 129–132.

Gao, K.S., Ji, Y., Aruga, Y. (1999). Relationship of CO_2 concentrations to photosynthesis of intertidal macroalgae during emersion. *Hydrobiologia*, 398–399, 355–359.

Gao, K.S., Wu, Y.P., Li, G., Wu, H.Y., Villafane, V.E., Helbling, E.W. (2007). Solar UV radiation drives CO_2 fixation in marine phytoplankton: A double-edged sword. *Plant Physiology*, 144(1), 54–59.

Gao, K.S., Li, P., Watanabe, T., Helbling, E.W. (2008). Combined effects of ultraviolet radiation and temperature on morphology, photosynthesis, and DNA of *Arthrospira* (Spirulina) *platensis* (Cyanophyta). *Journal of Phycology*, 44(3), 777–786.

Gao, K.S., Ruan, Z.X., Villafane, V.E., Gattuso, J.-P., Helbling, E.W. (2009). Ocean acidification exacerbates the effect of UV radiation on the calcifying phytoplankter *Emiliania huxleyi*. *Limnology and Oceanography*, 54(6), 1855–1862.

Gao, K.S., Helbling, E.W., Haeder, D.-P., Hutchins, D.A. (2012a). Responses of marine primary producers to interactions between ocean acidification, solar radiation, and warming. *Marine Ecology Progress Series*, 470, 167–189.

Gao, K.S., Xu, J.T., Gao, G., Li, Y.H., Hutchins, D.A., Huang, B.Q., Wang, L., Zheng, Y., Jin, P., Cai, X.N., Haeder, D.-P., Li, W., Xu, K., Liu, N.N., Riebesell, U. (2012b). Rising CO_2 and increased light exposure synergistically reduce marine primary productivity. *Nature Climate Change*, 2(7), 519–523.

Gao, K.S., Beardall, J., Haeder, D.-P., Hall-Spencer, J.M., Gao, G., Hutchins, D.A. (2019). Effects of ocean acidification on marine photosynthetic organisms under the concurrent influences of warming, UV radiation, and deoxygenation. *Frontiers in Marine Science*, 6, 322.

Gao, K.S., Gao, G., Wang, Y.J., Dupont, S. (2020). Impacts of ocean acidification under multiple stressors on typical organisms and ecological processes. *Marine Life Science & Technology*, 2(3), 279–291.

Gattuso, J.P., Magnan, A., Bille, R., Cheung, W.W.L., Howes, E.L., Joos, F., Allemand, D., Bopp, L., Cooley, S.R., Eakin, C.M., Hoegh-Guldberg, O., Kelly, R.P., Poertner, H.O., Rogers, A.D., Baxter, J.M., Laffoley, D., Osborn, D., Rankovic, A., Rochette, J., Sumaila, U.R., Treyer, S., Turley, C. (2015). Contrasting futures for ocean and society from different anthropogenic CO_2 emissions scenarios. *Science*, 349(6243), aac4722.

Gordillo, F.J.L., Niell, F.X., Figueroa, F.L. (2001). Non-photosynthetic enhancement of growth by high CO_2 level irm the nitrophilic seaweed *Ulva rigida* C. Agardh (Chlorophyta). *Planta*, 213(1), 64–70.

Graba-Landry, A., Hoey, A.S., Matley, J.K., Sheppard-Brennand, H., Poore, A.G.B., Byrne, M., Dworjanyn, S.A. (2018). Ocean warming has greater and more consistent negative effects than ocean acidification on the growth and health of subtropical macroalgae. *Marine Ecology Progress Series*, 595, 55–69.

Guan, W.C. and Gao, K.S (2010). Impacts of UV radiation on photosynthesis and growth of the coccolithophore *Emiliania huxleyi* (Haptophyceae). *Environmental and Experimental Botany*, 67(3), 502–508.

Häder, D.-P. and Gao, K.S. (2015). Interactions of anthropogenic stress factors on marine phytoplankton. *Frontiers in Environmental Science*, 3, 14.

Häder, D.P., Kumar, H.D., Smith, R.C., Worrest, R.C. (2007). Effects of solar UV radiation on aquatic ecosystems and interactions with climate change. *Photochemical & Photobiological Sciences*, 6(3), 267–285.

Häder, D.-P., Williamson, C.E., Wängberg, S.-Å., Rautio, M., Rose, K.C., Gao, K., Helbling, E.W., Sinha, R.P., Worrest, R. (2015). Effects of UV radiation on aquatic ecosystems and interactions with other environmental factors. *Photochemical & Photobiological Sciences*, 14(1), 108–126.

Hall-Spencer, J.M., Rodolfo-Metalpa, R., Martin, S., Ransome, E., Fine, M., Turner, S.M., Rowley, S.J., Tedesco, D., Buia, M.-C. (2008). Volcanic carbon dioxide vents show ecosystem effects of ocean acidification. *Nature*, 454(7200), 96–99.

Harley, C.D.G., Anderson, K.M., Demes, K.W., Jorve, J.P., Kordas, R.L., Coyle, T.A., Graham, M.H. (2012). Effects of climate change on global seaweed communities. *Journal of Phycology*. 48, 1064–1078.

Harvey, B.P., Agostini, S., Kon, K., Wada, S., Hall-Spencer, J.M. (2019). Diatoms dominate and alter marine food-webs when CO_2 rises. *Diversity-Basel*, 11(12), 242.

Harvey, B.P., Kon, K., Agostini, S., Wada, S., Hall-Spencer, J.M. (2021). Ocean acidification locks algal communities in a species-poor early successional stage. *Global Change Biology*, 27(10), 2174–2187.

Helbling, E.W., Gao, K.S., Goncalves, R.J., Wu, H.Y., Villafane, V.E. (2003). Utilization of solar UV radiation by coastal phytoplankton assemblages off SE China when exposed to fast mixing. *Marine Ecology Progress Series*, 259, 59–66.

Henley, W.J., Lindley, S.T., Levavasseur, G., Osmond, C.B., Ramus, J. (1992). Photosynthetic response of *Ulva-rotundata* to light and temperature during emersion on an intertidal sand flat. *Oecologia*, 89(4), 516–523.

Hepburn, C.D., Pritchard, D.W., Cornwall, C.E., McLeod, R.J., Beardall, J., Raven, J.A., Hurd, C.L. (2011). Diversity of carbon use strategies in a kelp forest community: Implications for a high CO_2 ocean. *Global Change Biology*, 17(7), 2488–2497.

Hong, H.Z., Shen, R., Zhang, F.T., Wen, Z.Z., Chang, S.W., Lin, W.F., Kranz, S.A., Luo, Y.W., Kao, S.J., Morel, F.M.M., Shi, D.L. (2017). The complex effects of ocean acidification on the promenant N_2-fixing cyanobacterium *Trichodesmium*. *Science*, 356(6337), 527–530.

Hoppe, C.J.M., Holtz, L.-M., Trimborn, S., Rost, B. (2015). Ocean acidification decreases the light-use efficiency in an Antarctic diatom under dynamic but not constant light. *New Phytologist*, 207(1), 159–171.

Hurd, C.L., Beardall, J., Comeau, S., Cornwall, C.E., Havenhand, J.N., Munday, P.L., Parker, L.M., Raven, J.A., McGraw, C.M. (2020). Ocean acidification as a multiple driver: How interactions between changing seawater carbonate parameters affect marine life. *Marine and Freshwater Research*, 71(3), 263–274.

Hutchins, D.A. and Fu, F. (2017). Microorganisms and ocean global change. *Nature Microbiology*, 2(6), 17058.

Hutchins, D.A., Walworth, N.G., Webb, E.A., Saito, M.A., Moran, D., Mcilvin, M.R., Gale, J., Fu, F.-X. (2015). Irreversibly increased nitrogen fixation in *Trichodesmium* experimentally adapted to elevated carbon dioxide. *Nature Communications*, 6, 8155.

Hyun, B., Choi, K.-H., Jang, P.-G., Jang, M.-C., Lee, W.-J., Moon, C.-H., Shin, K. (2014). Effects of increased CO_2 and temperature on the growth of four diatom species (*Chaetoceros debilis, Chaetoceros didymus, Skeletonema costatum* and *Thalassiosira nordenskioeldii*) in laboratory experiments. *Journal of Environmental Science International*, 23(6), 1003–1012.

IPCC (2014). Contribution of working groups I, II and III to the fifth assessment report of the intergovernmental panel on climate change. Synthesis report, IPCC, Geneva, 151.

Israel, A. and Hophy, M. (2002). Growth, photosynthetic properties and Rubisco activities and amounts of marine macroalgae grown under current and elevated seawater CO_2 concentrations. *Global Change Biology*, 8(9), 831–840.

Ji, Y. and Gao, K.S. (2020). Effects of climate change factors on marine macroalgae: A review. *Advances in Marine Biology* [Online]. Available at: https://doi.org/10.1016/bs.amb.2020.1011.1001.

Jiang, H.X. and Gao, K.S. (2008). Effects of UV radiation on the photosynthesis of conchocelis of *Porphyra haitanensis* (Bangiales, Rhodophyta). *Phycologia*, 47(3), 241–248.

Jiang, H.X., Gao, K.S., Helbling, E.W. (2007). Effects of solar UV radiation on germination of conchospores and morphogenesis of sporelings in *Porphyra haitanensis* (Rhodophyta). *Marine Biology*, 151(5), 1751–1759.

Jiang, H.X., Gao, K.S., Helbling, E.W. (2009). The conchocelis of *Porphyra haitanensis* (Rhodophyta) is protected from harmful UV radiation by the covering calcareous matrix. *Journal of Phycology*, 45(6), 1270–1277.

Jiao, N.Z. and Zheng, Q. (2011). The microbial carbon pump: From genes to ecosystems. *Applied and Environmental Microbiology*, 77(21), 7439–7444.

Jin, P. and Agusti, S. (2018). Fast adaptation of tropical diatoms to increased warming with trade-offs. *Scientific Reports*, 8, 17771.

Jin, P., Gao, K., Villafane, V.E., Campbell, D.A., Helbling, E.W. (2013). Ocean acidification alters the photosynthetic responses of a coccolithophorid to fluctuating ultraviolet and visible radiation. *Plant Physiology*, 162(4), 2084–2094.

Jin, P., Wang, T., Liu, N., Dupont, S., Beardall, J., Boyd, P.W., Riebesell, U., Gao, K. (2015). Ocean acidification increases the accumulation of toxic phenolic compounds across trophic levels. *Nature Communications*, 6, 8714.

Jin, P., Liu, N., Gao, K. (2019). Physiological responses of a coccolithophore to multiple environmental drivers. *Marine Pollution Bulletin*, 146, 225–235.

Jin, P., Hutchins, D.A., Gao, K. (2020). The impacts of ocean acidification on marine food quality and its potential food chain consequences. *Frontiers in Marine Science*, 7, 543979.

Johnson, M.D., Price, N.N., Smith, J.E. (2014). Contrasting effects of ocean acidification on tropical fleshy and calcareous algae. *Peer J*, 2, e411.

Karentz, D., Cleaver, J.E., Mitchell, D.L. (1991). Cell survival characteristics and molecular responses of Antarctic phytoplankton to ultraviolet-B radiation. *Journal of Phycology*, 27, 326–341 [Online]. Available at: https://doi.org/10.1111/j.0022-3646.1991.00326.x.

Kim, J.-H., Kim, N., Moon, H., Lee, S., Jeong, S.Y., Diaz-Pulido, G., Edwards, M.S., Kang, J.-H., Kang, E.J., Oh, H.-J., Hwang, J.-D., Kim, I.-N. (2020). Global warming offsets the ecophysiological stress of ocean acidification on temperate crustose coralline algae. *Marine Pollution Bulletin*, 157, 111324.

Koch, M., Bowes, G., Ross, C., Zhang, X.-H. (2013). Climate change and ocean acidification effects on seagrasses and marine macroalgae. *Global Change Biology*, 19(1), 103–132.

Kram, S.L., Price, N.N., Donham, E.M., Johnson, M.D., Kelly, E.L.A., Hamilton, S.L., Smith, J.E. (2016). Variable responses of temperate calcified and fleshy macroalgae to elevated pCO_2 and warming. *ICES Journal of Marine Science*, 73(3), 693–703.

Kremp, A., Godhe, A., Egardt, J., Dupont, S., Suikkanen, S., Casabianca, S., Penna, A. (2012). Intraspecific variability in the response of bloom-forming marine microalgae to changed climate conditions. *Ecology and Evolution*, 2(6), 1195–1207.

Kübler, J.E., Johnston, A.M., Raven, J.A. (1999). The effects of reduced and elevated CO_2 and O_2 on the seaweed *Lomentaria articulata. Plant, Cell & Environment*, 22(10), 1303–1310.

Kuffner, I.B., Andersson, A.J., Jokiel, P.L., Rodgers, K.U.S., Mackenzie, F.T. (2008). Decreased abundance of crustose coralline algae due to ocean acidification. *Nature Geoscience*, 1(2), 114–117.

Krumhansl, K.A. and Scheibling, R.E. (2012). Production and fate of kelp detritus. *Marine Ecology Progress Series*, 467, 281–302.

Ladah, L.B. and Zertuche-Gonzalez, J.A. (2007). Survival of microscopic stages of a perennial kelp (*Macrocystis pyrifera*) from the center and the southern extreme of its range in the Northern Hemisphere after exposure to simulated El Niño stress. *Marine Biology*, 152(3), 677–686.

Leal, P.P., Hurd, C.L., Fernndez, P.A., Roleda, M.Y. (2017). Meiospore development of the kelps *Macrocystis pyrifera* and *Undaria pinnatifida* under ocean acidification and ocean warming: Independent effects are more important than their interaction. *Marine Biology*, 164(1), 7.

Li, W. and Gao, K.S. (2012). A marine secondary producer respires and feeds more in a high CO_2 ocean. *Marine Pollution Bulletin*, 64(4), 699–703.

Li, Y.H., Gao, K.S., Villafane, V.E., Helbling, E.W. (2012). Ocean acidification mediates photosynthetic response to UV radiation and temperature increase in the diatom *Phaeodactylum tricornutum*. *Biogeosciences*, 9(10), 3931–3942.

Li, G., Che, Z.W., Gao, K.S. (2013). Photosynthetic carbon fixation by tropical coral reef phytoplankton assemblages: A UVR perspective. *Algae*, 28(3), 281–288.

Li, F.T., Wu, Y.P., Hutchins, D.A., Fu, F.X., Gao, K.S. (2016). Physiological responses of coastal and oceanic diatoms to diurnal fluctuations in seawater carbonate chemistry under two CO_2 concentrations. *Biogeosciences*, 13(22), 6247–6259.

Li, F.T., Beardall, J., Collins, S., Gao, K.S. (2017). Decreased photosynthesis and growth with reduced respiration in the model diatom *Phaeodactylum tricornutum* grown under elevated CO_2 over 1800 generations. *Global Change Biology*, 23(1), 127–137.

Li, F.T., Beardall, J., Gao, K.S. (2018). Diatom performance in a future ocean: Interactions between nitrogen limitation, temperature, and CO_2-induced seawater acidification. *ICES Journal of Marine Science*, 75(4), 1451–1464.

Li, Z.Z., Lan, T., Zhang, J.J., Gao, K.S., Beardall, J., Wu Y.P. (2021). Nitrogen limitation decreases the repair capacity and enhances photoinhibition of photosystem II in a diatom. *Photochemistry and Photobiology*, 97, 745–752.

Liang, Y., Beardall, J., Heraud, P. (2006). Effects of nitrogen source and UV radiation on the growth, chlorophyll fluorescence and fatty acid composition of *Phaeodactylum tricornutum* and *Chaetoceros muelleri* (Bacillariophyceae). *Journal of Photochemistry and Photobiology B*, 82, 161–172.

Listmann, L., Leroch, M., Schlueter, L., Thomas, M.K., Reusch, T.B.H. (2016). Swift thermal reaction norm evolution in a key marine phytoplankton species. *Evolutionary Applications*, 9(9), 1156–1164.

Liu, H.B., Chen, M., Zhu, F., Harrison, P.J. (2016). Effect of diatom silica content on copepod grazing, growth and reproduction. *Frontiers in Marine Science*, 3, 89.

Maberly, S.C., Berthelot, S.A., Stott, A.W., Gontero, B. (2015). Adaptation by macrophytes to inorganic carbon down a river with naturally variable concentrations of CO_2. *Journal of Plant Physiology*, 172, 120–127.

Madronich, S. (1992). Implications of recent total atmospheric ozone measurements for biologically-active ultraviolet-radiation reaching the earth's surface. *Geophysical Research Letters*, 19(1), 37–40.

McKenzie, R.L., Bjorn, L.O., Bais, A., Ilyasd, M. (2003). Changes in biologically active ultraviolet radiation reaching the Earth's surface. *Photochemical & Photobiological Sciences*, 2(1), 5–15.

Milligan, A.J., Varela, D.E., Brzezinski, M.A., Morel, F. (2004). Dynamics of silicon metabolism and silicon isotopic discrimination in a marine diatom as a function of pCO_2. *Limnology and Oceanography*, 49(2), 322–329.

Miyamoto, Y., Nakano, T., Yamada, K., Hatakeyama, K., Hamaguchi, M. (2019). Combined effects of drift macroalgal bloom and warming on occurrence and intensity of diel-cycling hypoxia in a eutrophic coastal lagoon. *Estuaries and Coasts*, 42(2), 494–503.

Moreira, D. and Pires, J.C.M. (2016). Atmospheric CO_2 capture by algae: Negative carbon dioxide emission path. *Bioresource Technology*, 215, 371–379.

Müller, R., Wiencke, C., Bischof, K. (2008). Interactive effects of UV radiation and temperature on microstages of *Laminariales* (Phaeophyceae) from the Arctic and North Sea. *Climate Research*, 37(2–3), 203–213.

Nagelkerken, I. and Connell, S.D. (2015). Global alteration of ocean ecosystem functioning due to increasing human CO_2 emissions. *Proceedings of the National Academy of Sciences of the United States of America*, 112(43), 13272–13277.

Nunes, J., McCoy, S.J., Findlay, H.S., Hopkins, F.E., Kitidis, V., Queiros, A.M., Rayner, L., Widdicombe, S. (2016). Two intertidal, non-calcifying macroalgae (*Palmaria palmata* and *Saccharina latissima*) show complex and variable responses to short-term CO_2 acidification. *ICES Journal of Marine Science*, 73(3), 887–896.

O'Donnell, D.R., Hamman, C.R., Johnson, E.C., Kremer, C.T., Klausmeier, C.A., Litchman, E. (2018). Rapid thermal adaptation in a marine diatom reveals constraints and trade-offs. *Global Change Biology*, 24(10), 4554–4565.

O'Donnell, D.R., Du, Z.-Y., Litchman, E. (2019). Experimental evolution of phytoplankton fatty acid thermal reaction norms. *Evolutionary Applications*, 12(6), 1201–1211.

Olabarria, C., Arenas, F., Viejo, R.M., Gestoso, I., Vaz-Pinto, F., Incera, M., Rubal, M., Cacabelos, E., Veiga, P., Sobrino, C. (2013). Response of macroalgal assemblages from rockpools to climate change: Effects of persistent increase in temperature and CO_2. *Oikos*, 122(7), 1065–1079.

Padfield, D., Yvon-Durocher, G., Buckling, A., Jennings, S., Yvon-Durocher, G. (2016). Rapid evolution of metabolic traits explains thermal adaptation in phytoplankton. *Ecology Letters*, 19(2), 133–142.

Qu, L.M., Xu, J.T., Sun, J.Z., Li, X.S., Gao, K.S. (2017). Diurnal pH fluctuations of seawater influence the responses of an economic red macroalga *Gracilaria lemaneiformis* to future CO_2-induced seawater acidification. *Aquaculture*, 473, 383–388.

Qu, L.M., Campbell, D., Gao, K.S. (2021). Ocean acidification interacts with growth light to suppress CO_2 acquisition efficiency and enhance mitochondrial respiration in a coastal diatom. *Marine Pollution Bulletin*, 163, 11208.

Rastogi, R.P., Singh, S.P., Incharoensakdi, A., Haeder, D.P., Sinha, R.P. (2014). Ultraviolet radiation-induced generation of reactive oxygen species, DNA damage and induction of UV-absorbing compounds in the cyanobacterium *Rivularia* sp HKAR-4. *South African Journal of Botany*, 90, 163–169.

Raven, J.A., Walker, D.I., Johnston, A.M., Handley, L.L., Kubler, J.E. (1995). Implications of ^{13}C natural-abundance measurements for photosynthetic performance by marine macrophytes in their natural-environment. *Marine Ecology Progress Series*, 123(1–3), 193–205.

Riebesell, U., Zondervan, I., Rost, B., Tortell, P.D., Zeebe, R.E., Morel, F.M.M. (2000). Reduced calcification of marine plankton in response to increased atmospheric CO_2. *Nature*, 407(6802), 364–367.

Rossoll, D., Bermúdez, R., Hauss, H., Schulz, K.G., Riebesell, U., Sommer, U., Winder, M. (2012). Ocean acidification-induced food quality deterioration constrains trophic transfer. *PLOS ONE*, 7(4), e34737.

Roth-Schulze, A.J., Thomas, T., Steinberg, P., Deveney, M.R., Tanner, J.E., Wiltshire, K.H., Papantoniou, S., Runcie, J.W., Gurgel, C.F.D. (2018). The effects of warming and ocean acidification on growth, photosynthesis, and bacterial communities for the marine invasive macroalga *Caulerpa taxifolia*. *Limnology and Oceanography*, 63(1), 459–471.

Schaum, C.E., Barton, S., Bestion, E., Buckling, A., Garcia-Carreras, B., Lopez, P., Lowe, C., Pawar, S., Smirnoff, N., Trimmer, M., Yvon-Durocher, G. (2017). Adaptation of phytoplankton to a decade of experimental warming linked to increased photosynthesis. *Nature Ecology and Evolution*, 1(4), 0094.

Schaum, C.E., Buckling, A., Smirnoff, N., Studholme, D.J., Yvon-Durocher, G. (2018). Environmental fluctuations accelerate molecular evolution of thermal tolerance in a marine diatom. *Nature Communications*, 9, 1719.

Scherner, F., Pereira, C.M., Duarte, G., Horta, P.A., Barreira e Castro, C., Barufi, J.B., Barreto Pereira, S.M. (2016). Effects of ocean acidification and temperature increases on the photosynthesis of tropical reef calcified macroalgae. *PLOS ONE*, 11(5), e0154844.

Schlüter, L., Lohbeck, K.T., Gutowska, M.A., Groeger, J.P., Riebesell, U., Reusch, T.B.H. (2014). Adaptation of a globally important coccolithophore to ocean warming and acidification. *Nature Climate Change*, 4(11), 1024–1030.

Schmidtko, S., Stramma, L., Visbeck, M. (2017). Decline in global oceanic oxygen content during the past five decades. *Nature*, 542(7641), 335–339.

Schoenrock, K.M., Schram, J.B., Amsler, C.D., Mcclintock, J.B., Angus, R.A., Vohra, Y.K. (2016). Climate change confers a potential advantage to fleshy Antarctic crustose macroalgae over calcified species. *Journal of Experimental Marine Biology and Ecology*, 474, 58–66.

Short, S.M., Staniewski, M.A., Chaban, Y.V., Long, A.M., Wang, D. (2020). Diversity of viruses infecting eukaryotic algae. *Current Issues in Molecular Biology*, 39, 29–61.

Sinutok, S., Hill, R., Doblin, M.A., Wuhrer, R., Ralph, P.J. (2011). Warmer more acidic conditions cause decreased productivity and calcification in subtropical coral reef sediment-dwelling calcifiers. *Limnology and Oceanography*, 56(4), 1200–1212.

Smale, D.A. and Wernberg, T. (2013). Extreme climatic event drives range contraction of a habitat-forming species. *Proceedings of the Royal Society B: Biological Sciences*, 280, 20122829.

Solomon, S., Ivy, D.J., Kinnison, D., Mills, M.J., Neely III, R.R., Schmidt, A. (2016). Emergence of healing in the Antarctic ozone layer. *Science*, 353(6296), 269–274.

Sommer, U., Stibor, H., Katechakis, A., Sommer, F., Hansen, T. (2002). Pelagic food web configurations at different levels of nutrient richness and their implications for the ratio fish production: Primary production. *Hydrobiologia*, 484(1–3), 11–20.

Suarez-Alvarez, S., Luis Gomez-Pinchetti, J., Garcia-Reina, G. (2012). Effects of increased CO_2 levels on growth, photosynthesis, ammonium uptake and cell composition in the macroalga *Hypnea spinella* (Gigartinales, Rhodophyta). *Journal of Applied Phycology*, 24(4), 815–823.

Taucher, J., Bach, L.T., Boxhammer, T., Nauendorf, A., Achterberg, E.P., Alguero-Muniz, M., Aristegui, J., Czerny, J., Esposito, M., Guan, W., Haunost, M., Horn, H.G., Ludwig, A., Meyer, J., Spisla, C., Sswat, M., Stange, P., Riebesell, U., Aberle-Malzahn, N., Archer, S., Boersma, M., Broda, N., Budenbender, J., Clemmesen, C., Deckelnick, M., Dittmar, T., Dolores-Gelado, M., Dorner, I., Fernandez-Urruzola, I., Fiedler, M., Fischer, M., Fritsche, P., Gomez, M., Grossart, H.-P., Hattich, G., Hernandez-Brito, J., Hernandez-Hernandez, N., Hernandez-Leon, S., Hornick, T., Kolzenburg, R., Krebs, L., Kreuzburg, M., Lange, J.A.F., Lischka, S., Linsenbarth, S., Loscher, C., Martinez, I., Montoto, T., Nachtigall, K., Osma-Prado, N., Packard, T., Pansch, C., Posman, K., Ramirez-Bordon, B., Romero-Kutzner, V., Rummel, C., Salta, M., Martinez-Sanchez, I., Schroder, H., Sett, S., Singh, A., Suffrian, K., Tames-Espinosa, M., Voss, M., Walter, E., Wannicke, N., Xu, J., Zark, M., Gran Canaria, K.C. (2017). Influence of ocean acidification and deep water upwelling on oligotrophic plankton communities in the subtropical North Atlantic: Insights from an *in situ* mesocosm study. *Frontiers in Marine Science*, 4, 85.

Tong, S.Y., Gao, K.S., Hutchins, D.A. (2018). Adaptive evolution in the coccolithophore *Gephyrocapsa oceanica* following 1,000 generations of selection under elevated CO_2. *Global Change Biology*, 24(7), 3055–3064.

Tong, S.Y., Hutchins, D.A., Gao, K.S. (2019). Physiological and biochemical responses of *Emiliania huxleyi* to ocean acidification and warming are modulated by UV radiation. *Biogeosciences*, 16(2), 561–572.

Tortell, P.D., Payne, C.D., Li, Y., Trimborn, S., Rost, B., Smith, W.O., Riesselman, C., Dunbar, R.B., Sedwick, P., Ditullio, G.R. (2008). CO_2 sensitivity of southern ocean phytoplankton. *Geophysical Research Letters*, 35(4), L04605.

Traving, S.J., Clokie, M.R.J., Middelboe, M. (2014). Increased acidification has a profound effect on the interactions between the cyanobacterium *Synechococcus* sp WH7803 and its viruses. *FEMS Microbiology Ecology*, 87(1), 133–141.

Van Der Loos, L.M., Schmid, M., Leal, P.P., McGraw, C.M., Britton, D., Revill, A.T., Virtue, P., Nichols, P.D., Hurd, C.L. (2019). Responses of macroalgae to CO_2 enrichment cannot be inferred solely from their inorganic carbon uptake strategy. *Ecology and Evolution*, 9(1), 125–140.

Vargas, C.A., Lagos, N.A., Lardies, M.A., Duarte, C., Manriquez, P.H., Aguilera, V.M., Broitman, B., Widdicombe, S., Dupont, S. (2017). Species-specific responses to ocean acidification should account for local adaptation and adaptive plasticity. *Nature Ecology and Evolution*, 1(4), 0084.

Walworth, N.G., Fu, F.-X., Webb, E.A., Saito, M.A., Moran, D., Mcllvin, M.R., Lee, M.D., Hutchins, D.A. (2016). Mechanisms of increased *Trichodesmium* fitness under iron and phosphorus co-limitation in the present and future ocean. *Nature Communications*, 7, 12081.

Wang, B., Chen, J.F., Jin, H.Y., Li, H.L., Huang, D.J., Cai, W.J. (2017). Diatom bloom-derived bottom water hypoxia off the Changjiang estuary, with and without typhoon influence. *Limnology and Oceanography*, 62(4), 1552–1569.

Wang, Y.T., Fan, X., Gao, G., Beardall, J., Inaba, K., Hall-Spencer, J.M., Xu, D., Zhang, X.W., Han, W.T., Mcminn, A., Ye, N.H. (2020). Decreased motility of flagellated microalgae long-term acclimated to CO_2-induced acidified waters. *Nature Climate Change*, 10(6), 561–567.

Weatherhead, E.C. and Andersen, S.B. (2006). The search for signs of recovery of the ozone layer. *Nature*, 441(7089), 39–45.

Wernberg, T., Smale, D.A., Tuya, F., Thomsen, M.S., Langlois, T.J., de Bettignies, T., Bennett, S., Rousseaux, C.S. (2013). An extreme climatic event alters marine ecosystem structure in a global biodiversity hotspot. *Nature Climate Change*, 3, 78–82.

Wong, C.-Y., Teoh, M.-L., Phang, S.-M., Beardall, J. (2015). Interactive effects of temperature and UV radiation on photosynthesis of *Chlorella* spp. from polar, temperate and tropical environments: Differential impacts on repair and damage. *PLOS ONE*, 10(10), e0139469.

Wu, H.Y. and Gao, K.S. (2009). Ultraviolet radiation stimulated activity of extracellular carbonic anhydrase in the marine diatom *Skeletonema costatum*. *Functional Plant Biology*, 36(2), 137–143.

Wu, Y.P. and Gao, K.S. (2010). Combined effects of solar UV radiation and CO_2-induced seawater acidification on photosynthetic carbon fixation of phytoplankton assemblages in the South China Sea. *Chinese Science Bulletin*, 55(32), 3680–3686.

Wu, H.Y., Gao, K.S., Villafane, V.E., Watanabe, T., Helbling, E.W. (2005). Effects of solar UV radiation on morphology and photosynthesis of filamentous cyanobacterium *Arthrospira platensis*. *Applied and Environmental Microbiology*, 71(9), 5004–5013.

Wu, H.Y., Zou, D.H., Gao, K.S. (2008). Impacts of increased atmospheric CO_2 concentration on photosynthesis and growth of micro- and macro-algae. *Science In China Series C – Life Sciences*, 51(12), 1144–1150.

Wu, X.J., Gao, G., Giordano, M., Gao, K.S. (2012). Growth and photosynthesis of a diatom grown under elevated CO_2 in the presence of solar UV radiation. *Fundamental and Applied Limnology*, 180(4), 279–290.

Wu, Y.P., Campbell, D.A., Irwin, A.J., Suggett, D.J., Finkel, Z.V. (2014). Ocean acidification enhances the growth rate of larger diatoms. *Limnology and Oceanography*, 59(3), 1027–1034.

Wu, Y.P., Li, Z.Z., Du, W.J., Gao, K.S. (2015). Physiological response of marine centric diatoms to ultraviolet radiation, with special reference to cell size. *Journal of Photochemistry and Photobiology B*, 153, 1–6.

Wu, H.L., Feng, J.C., Li, X.S., Zhao, C.Y., Liu, Y.H., Yu, J.T., Xu, J.T. (2019). Effects of increased CO_2 and temperature on the physiological characteristics of the golden tide blooming macroalgae *Sargassum horneri* in the Yellow Sea, China. *Marine Pollution Bulletin*, 146, 639–644.

Xu, J.T. and Gao, K.S. (2010). Use of UV-A energy for photosynthesis in the red macroalga *Gracilaria lemaneiformis*. *Photochemistry and Photobiology*, 86(3), 580–585.

Xu, J.T. and Gao, K.S. (2012). Future CO_2-induced ocean acidification mediates the physiological performance of a green tide alga. *Plant Physiology*, 160(4), 1762–1769.

Xu, D., Wang, D.S., Li, B., Fan, X., Zhang, X.W., Ye, N.H., Wang, Y.Y., Mou, S.L., Zhuang, Z.M. (2015). Effects of CO_2 and seawater acidification on the early stages of *Saccharina japonica* development. *Environmental Science and Technology*, 49(6), 3548–3556.

Xu, D., Brennan, G., Xu, L., Zhang, X.W., Fan, X., Han, W.T., Mock, T., McMinn, A., Hutchins, D.A., Ye, N.H. (2019). Ocean acidification increases iodine accumulation in kelp-based coastal food webs. *Global Change Biology*, 25(2), 629–639.

Zou, D.H. and Gao, K.S. (2002). Effects of desiccation and CO_2 concentrations on emersed photosynthesis in *Porphyra haitanensis* (Bangiales, Rhodophyta), a species farmed in China. *European Journal of Phycology*, 37(4), 587–592.

Zou, D.H. and Gao, K.S. (2005). Ecophysiological characteristics of four intertidal marine macroalgae during emersion along Shantou coast of China, with a special reference to the relationship of photosynthesis and CO_2. *Acta Oceanologica Sinica*, 24(3), 105–113.

Zou, D.H. and Gao, K.S. (2009). Effects of elevated CO_2 on the red seaweed *Gracilaria lemaneiformis* (Gigartinales, Rhodophyta) grown at different irradiance levels. *Phycologia*, 48(6), 510–517.

Zou, D.H. and Gao, K.S. (2013). Thermal acclimation of respiration and photosynthesis in the marine macroalga *Gracilaria lemaneiformis* (Gracilariales, Rhodophyta). *Journal of Phycology*, 49(1), 61–68.

Zou, D.H. and Gao, K.S. (2014). The photosynthetic and respiratory responses to temperature and nitrogen supply in the marine green macroalga *Ulva conglobata* (Chlorophyta). *Phycologia*, 53(1), 86–94.

Zou, D.H., Gao, K.S., Luo, H.J. (2011). Short- and long-term effects of elevated CO_2 on photosynthesis and respiration in the marine macroalga *Hizikia fusiformis* (Sargassaceae, Phaeophyta) grown at low and high N supplies. *Journal of Phycology*, 47(1), 87–97.

Zweng, R.C., Koch, M.S., Bowes, G. (2018). The role of irradiance and C-use strategies in tropical macroalgae photosynthetic response to ocean acidification. *Scientific Reports*, 8, 9479.

List of Authors

John BEARDALL
State Key Laboratory of Marine
Environmental Science & College of
Ocean and Earth Sciences
Xiamen University
and
School of Biological Sciences
Monash University
Clayton
Victoria
Australia

Chris BOWLER
Institut de Biologie de l'École Normale
Supérieure (IBENS)
École Normale Supérieure
CNRS, INSERM
Université PSL
Paris
France

Richard G. DORRELL
Institut de Biologie de l'École Normale
Supérieure (IBENS)
École Normale Supérieure
CNRS, INSERM
Université PSL
Paris
France

Kunshan GAO
State Key Laboratory of Marine
Environmental Science & College of
Ocean and Earth Sciences
Xiamen University
China

Caterina GEROTTO
Laboratorio di Fisiologia delle Alghe
e delle Piante
Dipartimento di Scienze della Vita
e dell'Ambiente
Università Politecnica delle Marche
Ancona
Italy

Brigitte GONTERO
Aix-Marseille Université
CNRS, BIP UMR 7281
Marseille
France

Sven A. KRANZ
Florida State University
Department of Earth, Ocean and Atmospheric Science
Tallahassee
USA

Timothy M. LENTON
Global Systems Institute
University of Exeter
United Kingdom

Stephen Christopher MABERLY
Lake Ecosystems Group
UK Centre for Ecology & Hydrology
Lancaster
United Kingdom

Yusuke MATSUDA
Department of Bioscience
Kwansei Gakuin University
Sanda
Hyogo
Japan

Moritz T. MEYER
Department of Molecular Biology
Princeton University
USA

Hermanus NAWALY
Department of Bioscience
Kwansei Gakuin University
Sanda
Hyogo
Japan

Charlotte NEF
Institut de Biologie de l'École Normale Supérieure (IBENS)
École Normale Supérieure
CNRS, INSERM
Université PSL
Paris
France

Alessandra NORICI
Laboratorio di Fisiologia delle Alghe e delle Piante
Dipartimento di Scienze della Vita e dell'Ambiente
Università Politecnica delle Marche
Ancona
Italy

Juan José PIERELLA KARLUSICH
Institut de Biologie de l'École Normale Supérieure (IBENS)
École Normale Supérieure
CNRS, INSERM
Université PSL
Paris
France

John A. RAVEN
Division of Plant Science
University of Dundee at the
James Hutton Institute
Invergowrie
United Kingdom
and
University of Technology, Climate
Change Cluster, Faculty of Science
Sydney
and
School of Biological Sciences
University of Western Australia
Crawley
Australia

Sebastian D. ROKITTA
Alfred-Wegener-Institute – Helmholtz
Center for Polar and Marine Research
Bremerhaven
Germany

Björn ROST
Alfred-Wegener-Institute – Helmholtz
Center for Polar and Marine Research
Bremerhaven
and
University of Bremen,
FB Biology/Chemistry
Germany

Kohei YONEDA
Department of Bioscience
Kwansei Gakuin University
Sanda
Hyogo
Japan

Wenyan ZHAO
State Key Laboratory of Marine
Environmental Science & College of
Ocean and Earth Sciences
Xiamen University
China

Index

C, D

E, F

M, N

O, P

Q, R

S, T

U, V

W, X, Z

Printed and bound by CPI Group (UK) Ltd, Croydon, CR0 4YY

13/09/2026

14960995-0006